BIODIVERSITY RESERVE

A Goldmine for Aquaculture

BIODIVERSITY RESERVE

A Goldmine for Aquaculture

P. Ray

2010
DAYA PUBLISHING HOUSE
DELHI-110 035

ISBN 81-7035-639-3
ISBN 978-81-7035-639-4

Published by : **Daya Publishing House**
1123/74, Deva Ram Park
Tri Nagar, Delhi - 110 035
Phone: 27383999
Fax: (011) 23260116
e-mail : dayabooks@vsnl.com
website : www.dayabooks.com

Showroom : 4760-61/23, Ansari Road, Darya Ganj,
New Delhi - 110 002
Phone: 23245578, 23244987

Laser Typesetting : **Vaishnav Graphics & Systems**
Jaipur - 302004

Printed at : **Chawla Offset Printers**
Delhi - 110 052

PRINTED IN INDIA

The Curious World which we

inhabit is more

WONDERFUL

than it is

CONVENIENT

more beautiful than it is

USEFUL

it is more to be admired than

it is to be

USED

Preface

I feel happy to add more informations of the past and present of Sundarbans in this present edition as ecological changes has taken place so rapidly due to changes in the morpho-ecological conditions and human interferences. This has affected the biodiversity of Sundarbans resulting major changes of biomass of the delta.

There was once a time (6000-7000 years back) when (1) The entire Gangtic delta was under sea, (2) Heavy silt and clay particles were carried down by the Ganga and Brahmaputra river system of the Himalayas, and even from the Chhottanagpur hills which after settling formed the great Sundarban Delta, (3) The stones and soils of Chhottanagpur-hills were discovered from the depths of 300 m. at Sagar Island and in several Western portions of the Sundarbans Delta, (4) The flow of the great Ganges (in the 14th Century AD) was on the western part of the famous port of Tambralipta - now known as Tamluk of Midnapur District and (5) The present metropolis of Calcutta was under mangrove swamps of Sundarban delta around 5000 years back. Such interesting findings are the results of investigations carried out by various workers (Chanda, 1977; Maity, 1976; Gupta, 1957; Ghosh, 1941; Chanda and Mukherjee, 1969 ; Mukherjee, 1972; Vishnu Maitre and Gupta, 1972 and Sen and Banerjee, 1984).

This delta* has been formed by the action of three great rivers, the Ganges, the Brahmaputra and the Meghna with the assistance of their innumerable distributories and redistributories formed principally in course of oscillation of these streams.

The behaviour of different streams within the delta is certainly peculiar and are guided by their own whims coupled with the oscillation of the parent streams. Some of these streams Bhagirathi (and probably the Saraswaty), the Bhairab, the Ayral Khan (Plot No. 1) served the different periods as principal outfalls of the Ganges to the sea. The others like the Kobodak, the Chitra, the Kumar, the Nabaganga, the Jalangi, the Mathabhanga etc. are or were only subsidiary streams the partially helped to raises the delta face.

The present delta face chiefly consists of black clay and fine sand silts which are known to expand or to contract according to the forces of the current with remarkable rapidity almost like an elastic substance. Underneath this formation, lies a lower delta or older alluvium consisting of yellow sand and clay and sometimes compact red clay (rangamati) of ferruginious origin. The extent of this alluvium has been shown in the excellent plan

---

*Report of the Committee to enquire into "The Drainage condition of Calcutta and adjoining areas, Govt. of West Bengal, 1945."

prepared by Captain Sherwill in 1857 and partly reproduced in Plan No. 1. The formation of this older allunivium cannot perhaps be attributed to the activities of the present streams. Exposure of this red clay rising to a considerable height above the highest known flood levels of the present day is found near Burdwan on the West and Dacca on the east.

Natural laws guiding the movement of these rivers are still mysterious. It is true they do not and infact cannot flow into the same channel for any great length of time against the so far recognised functions of all perennial delta building rivers. Curiously enough these rivers have either altered their courses or been intersected by other rivers of recent origin before they fulfilled their natural functions sufficiently. The change of course of the parent stream, the Ganga from the south along the Bhagirathi-Hooghly to south east along the present Padma Course probably took place in or about the year 1521 AD.

At Tribeni (about 10 miles above the Jubilee Bridge) the Ganges had an important distributary, the Jamuna which after a tortuous course entered the Ichhamati near Tibbi about 10 miles east of Gobordanga, the combined stream Ichhamati-cum-Jamuna then flowed into the sea by the Jamuna estuaries, independent of the Raymangal and running parallel to it. Iswaripur, the Capital of Pratapaditya (1553 AD) was situated on this river about 14 miles south of the present Kaligunj.

From the Jamuna itself important radistributarius like (i) the Nowee or Labonyamoii, (ii) the Sunthi or Sandhyabati, (iii) the Western Bidyadhari or Dudbibi Khal, (iv) Central Bidyadhari or the Nonagang and (v) the eastern Bidyadhari or the Sealdaha Gong and the Rumarjola and the Tambuldaha in the lower reach of the Eastern and Central Bidyadhari were all active delta-building streams. All these channels empted into the Matla estuary.

Within Calcutta itself the existance of a Creek along the present Creek Row flowing from the Hooghly into the salt water lakes was also traced by historians. The huge depression known as salt water lakes were very deep and far more extensive in those days than they are at present and served as a huge spill reservoir of the Bidyadhari and the Matla series of rivers.

The position is totally different today. In the western delta when rivers have flowed in their present courses with few changes for many years (as should naturally be the case with this moribund portion of the delta) it was found that a complete change has taken place within the last 80 years and the system was threatened with extinction.

The disastrous tidal inundation in the Magrahat swamp in 1830 and in subsequent years also in the east of Calcutta over an extensive area of about 100 sq. miles in 1944, and 50 sq. miles in 1945 and also the disastrous floods of the Damodar in 1913, 1917, 1935 and 1943 are only fore-runners of nature's ravange.

The parent stream Ganges by changing her course has lowered the flood level by about 12 ft. in the Western delta. In the lower reach from Kalighat to the surface, there was a recognised and flourishing trade route along the old course of the Ganges dotted with prosperous towns and villages on both the banks. The Saptamukhi (or Satmukhi) estuary and the cannal creck or the Murag Ganga which poured bulk of upland water supply of the Gangas up to the 15th century are now showing sings of deterioration. The distributories of the Jamuna have now been cut off from the parent stream over virtually converted to tidal channel.

In the Matla esturay, the Bidyadhari and Peali have undergone unprecedented deterioration. The Atharabanka, once a powerful tidal channel issuing from Matla at Port canning and a recognised trade route even upto 1880 is now hardly traceable. The Karati another flourishing channel running North to South from Kultigong and Matla and navigable by stremers even a little over a decade ago is now high and dry in its central portion. Deterioration of the Karati started over 50 years ago, but it received new lease of life after a breach of Pinabad lot in 1913, when nature attempted to resume her delta-building work but was again forcibly prevented.

A new series of rivers have been developed in the east known as Kulti series due perhaps to predominance of tidal forces Northwards and might be for want of perennial supply above. This has altered the natural directions of drainage.

The Indian part of Sundarban delta covers an area 9630 sq. kms. (9630125.78 ha.) bounded by the estuary of Hooghly on the West to Ichhamati-Raimangal in the east, Bay of Bengal on the South and an imaginary Dampier - Hedges line on the North and is spread over two districts of 24-Parganas, North and South in the State of West Bengal.

The area consists of a group of islands which are about 54 in number interspersed by innumerable rivulets, tidal rivers, cracks and was originally covered with thick and impenetrable mangrove forests.

With the gradual increased population the forest was gradually cleared from 18th century onwards for habitation and settlement. It is interesting to see that areas of reclamation from forests increased alarmingly with time and population increase, and as a result, the forest area could increase only 6.66 percent as indicated from Table 1.

Characteristics of the Sundarban delta.

Year	*Area of Sundarban delta including forest (ha)*	*Reclaimed area (ha)*	*Actual forest area (ha)*
1947 Ray (1993)	720000	320000	400000
1987 Naskar and Guha Baksi (1987)	9680 Sq. Km. (963012.78 ha)	5363.4 Sq. km. (536347.12 ha)	4266.6 Sq. Km. (426665.66 ha)
Increase in 40 years	33.75 %	67.60%	6.66%
		Position of Population	
1951	-	Population of the Sundarbans	1159559
1991	-	-Do-	3205528
40 years.	Increase of population 176% in 40 years.		

Natural calamities like storm, Cyclone, flood etc. a common phenomenon associated with earthquake have brought considerable changes in the ecosystem. Appearance of inundated shallow areas (bils) of various sizes resulted from natural calamities was reported perhaps for the first time in 1870-73 by W.W. Hunter in his valuable document "Statistical Accounts of Bengal (Gangopadhaya 1983). The largest one being Boyra Bil whose area was 40 sq. miles followed by Datbhanga bil (22 sq. miles), Ballibil (10 sq. miles), Kulgachi, Boriti Bil, Bil Borti, Dholkhera, Ghajolmari, Nagarghata, Khalish Khali etc. None of them exist now and are converted into agricultural lands. Sir K.G. Gupta (1908) in his valuable document on "Results of enquiry into the fisheries and fishery matters of Bengal" remarked that Calcutta had a valuable fishery resource of immense potentiality. His prediction came to be true when in subsequent years a significant change in the ecosystem was brought through natural calamities i.e. high flood water of 22 feet of Bidyadhari associated with cyclone inundated the entire Sundarbans resulting formation of a number of shallow inundated water areas (bils) in the year 1922. (Thakur, 1986). Each of the water areas was a mile wide and 2-3 miles long and Goberia bhery (7.5 thousand bighas) alone used to earn 4 crores of rupees as foreign exchange by exporting prawn (*P. monodan*) alone. This shows importance of Biodiversity Reserve which was once a gold mine for aquaculture. This tempted people to shift their preference to fish culture than agriculture. In the subsequent years various categories of bheries both in fresh and brackishwater region came into existance as fishery industry, the principal of them are salt lake bheries of fresh water, Bidyadhari Spilt areas, Dettaic brackishwater impoundment, Parennial bheries, integrated fish farming and converted monoculture paddy field etc.

With the growing population it has become essential to construct bridges, chanals and lock gates, agricultural fields and swamps in the salt lake area of Sundarbans for human habitation and comfort. As a result the 100 sq. miles (25589.5 ha.) of salt lake area was gradually reduced to only 35.8 sq. miles (8951.2 ha) and the entire flooded spill water caused by Bidyadhari became completely cut off from the parent Bidyadhari (Sen Gupta, 1980), while in the case of Sundarbans, Banerjee (1964) stated that about 565 sq. km. (56500.75 ha.) of Sundarban forest in West Bengal as reclaimed cleared for agriculture and human habitation during the last 63 years and it was of great concern to many naturalist and foresters that this rapid deforestation rate of Sundarbans may cause serious natural disbalance for this tract in near future (Rao, 1959; Naskar, 1985). If this rapid trent of deforestation is allowed, it is evident that in 100 years 1280-1540 sq. km. (128001.7 - 154002.04 ha) of forest will be cleared in West Bengal Sundarbans. The famous mangrove forests will be disappear by the present rate of human intrusion and interference with the forest products (Naskar 1985). Ray has shown almost similar information that during 1947-1987 although the delta has increased by 33.7% due to silting, the area reclaimed was alarmingly high i.e. 67.6% which is more than that recorded by Banerjee (1964) when the forest area increased by only 6.66% Again Raha (1999) has given further alarming information recorded through satellite imagery which indicates further sign of deterioration of the forest by another 4000 sq. km. (400005 ha). Further, increase of Sundarban saline soil which is due to reduction of freshwater inflow during non-monsoon months is considered responsible for higher plant mortality. Thus the information on forest reported by various scientists indicate that the famous mangroove forests may disappear by the present rate of human intrusion and interference with the forest products. All these together have greatly affected the bio-diversity of Sundarbans which was being used as a natural habitat for esturaine and marine fauna since centuaries. It is worthwhile to mention that this forest was once a gold mine for commercial important Crustaceans like *P. monodan*. Population of some commercial important species are now appreciably affected.

The gear which was once used in deep water of the Hooghly-Matla estuarine system are no longer in use due to shallow water and has been replaced by simple gear normally used in shallow waters. The extent of silting in the estuarine systems has gone up to such an extent that occurrence of flood has become occasional these days. Even in mighty river Hooghly, hilsa, the most prized fish cannot ascend up streaming for breeding purposes.

Thus by putting the facts sequentially *viz.*, delta formation, creation of shallow swampy water used for fish culture, diversion of flow of major rivers, suspension of freshwater supply, high rate of silting in all the estuarine system by surging effect of tidal waters, population influx, degradation of forest area due to high saline soil and human activities which resulted suffering of marine fauna, fish and other threatened endemic fauna, loss of fish population, increase of fishing intensity with simple gear, low migration of prized hilsa for breeding purposes and occasional flood etc. Will now high light the document in four chapters. Extensive study made in the case of Hooghly flowing on the west of Sundarbans will now cover facts in more details.

Informations presented on the above facts will appraise the readers with complete picture of Culture and Capture fishery of Sundarban delta she is controlling since long indicating her destiny in future.

The entire document is divided into four major heads are:

I. Culture fishery (Fresh and Brackishwater).

II. Capture fishery

   1. The Kulti estuary

   2. The Bhagirathi-Hooghly esturay.

III. The Sundarban Delta.

IV. Past and Present of Sundarban delta.

A seperate pictorial document on the formation of the Sundarban Delta and its sequential changes with the passing of time, shown through a number of plates under the caption "Sundarbans at a glance" is also given.

Acknowledgements

Dr. P.R. Sen, Principal investigator CIFA (ICAR) (Retd.) my friend and colleague who not only welcomed my ideas in writing on the Biodiversity Reserve – a Gold mine for Aquaculture but also encouraged me to reveal valuable information which was lacking earlier. Dr. Sen has always been cooperative in periodic discussion and valuable suggestions. My greatest obligations has been incurred to him.

My sincere thanks are also due to Dr. K.R. Naskar, National fellow (ICAR) for allowing me to use data from his publication.

A ward of appreciation is also due to editors of different daily news papers of Kolkata like - Ananda Bazar Patrika, Bartaman, Aajkal and especially to *The Statesman* (Kolkata) whose published valuable photographs and other facts presented in this document will be of immense help to readers to understand the actual conditions of ecological imbalance of the Sundarbans.

My heart felt thanks are also due to my dear nephews, Dr. Sujit Roy for some valuable suggestions and Bidyut Roy for his constant inspiration throughout the preparation of this manuscript.

Contents

I.
Culture Fishery

1. Aquaculture in Freshwater Impoundment
2. Aquaculture in Brackishwater Impoundment

1
Aquaculture in Freshwater Impoundment

1. INTRODUCTION

Taking into consideration historic account. River Bhagirathi is said to receive major discharges from the Ganga prior to 12th century. Since then the Bidyadhari river was known to erode river banks and carry silt from the Hooghly (Bhagirathi) to the Bay of Bengal through one of its tributaries, the Jamuna and its offshoots (Fig. 1). The Damodar had one of its outfalls into the Hooghly near Kalna which existed at least till 1660 and the second near Niassarai till about the middle of 18th century (located above the offtake of the Jamuna at Tribeni (Fig. 2a). During this period the backwater of the Bidyadhari used to innundate about 100 sq. miles (256 sq. km.) area on the eastern part of Calcutta. It was during this period that the Salt lake was known for its abundance of fish fauna consisting mainly of brackishwater species like *Lates calcarifer, Mugil parsia, Mugil tade, Mystus gulio* etc. (Gupta, 1908). *Hilsa ilisha*, as stated by Sen (1941) "was also not uncommon when these Salt lakes used to be innundated by water from tidal zone of the Bidyadhari". Such abundance of fish is also reported in the Duncan Stewart's evidence given during the Municipal enquiry of 1836 which states as — "for 3 miles to the eastward the saltwater lake overflows in September and October and Prodigious numbers of fish resort there but in November and December when floods are dissipated, those fishes are left to die with their putrefaction affecting the air with thick stinking vapours which the North-east winds bring with them to Fort William. Thus that a great yearly mortality is caused by them".

It was during 1795-1905, due to sudden dangerous geographical annihilation in Rajmahal and its neighbouring areas, the Ganga changed its course of flow shifting the major floods to the Padma (Sen Gupta, 1980). Further, diversion of the Damodar down its present channel with its outfall below Uluberia finally sealed the fate of the Jamuna which together with the spill channels began to deteriorate rapidly (Fig. 2b). Deprived of the upland water supply, the Bidyadhari was then thrown entirely on her other resources, *viz.*, local drainage and spill areas to prolong her life (Fig. 2b). At this stage another characteristic peculiar to tidal waters carrying a large proportion of silt hastened her decay. This process of silting was further supplemented by local rainfall of Nadia (average 843 mm) and North 24 Parganas district (average 1043 mm), floods etc., causing further silting up of the river beds of Bidyadhari-Kulti complex. Simultaneously, the deltaic district was studded with a large number of marshes and swamps either in the bed of disconnected rivers or channels or spill areas of varying sizes which in the later years were converted into fishery ponds. Thus, a major change in the ecosystem took place which in the later years could be termed as 'historical transformation.' The vast spilled water areas of the Bidyadhari and the Kulti, geographically located between 88°30' – 89°N and 22°15 – 23°15'5 E, in subsequent years under natural process, emerged in disguise as highly profitable area for fish and paddy culture — an important industry of the deltaic region of the Ganga in two divisions — the Northern delta of old formation and the South Sundarbans towards the sea. The metamorphosis of the Bidyadhari-Kulti complex is discussed under the heads viz.:

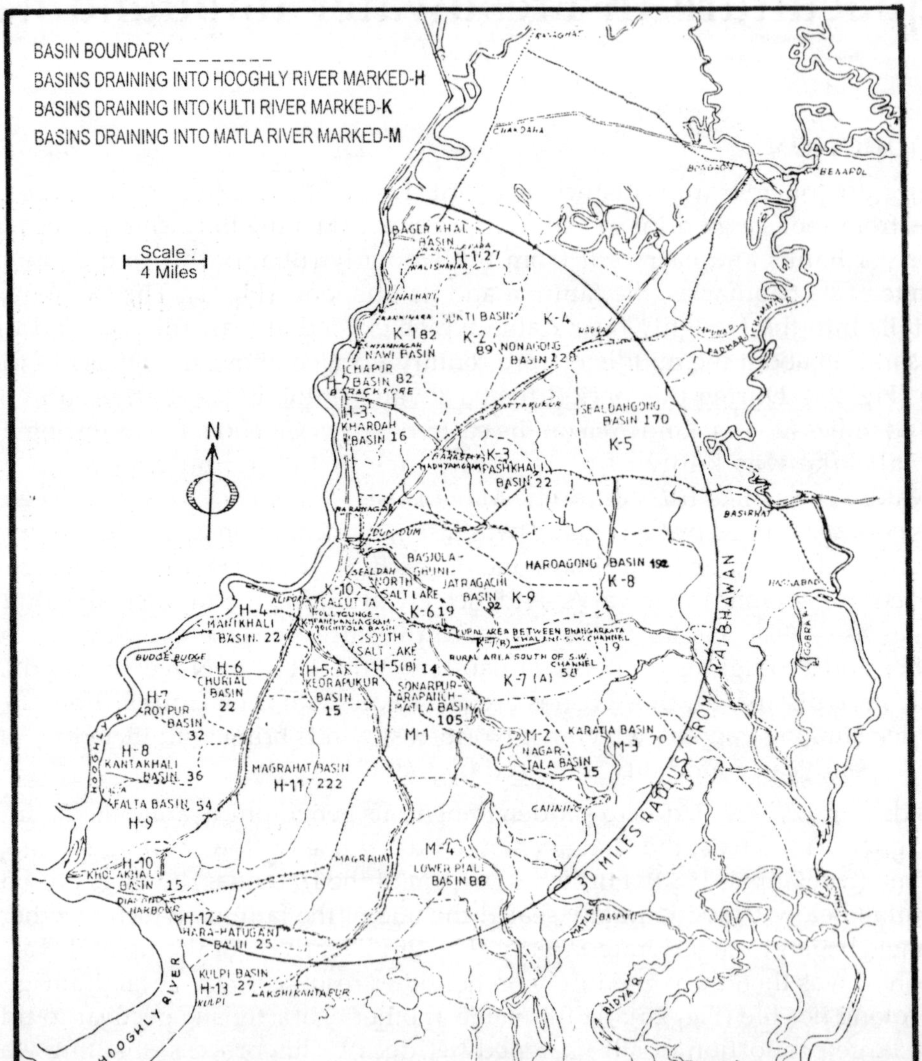

Fig. 1 : Tributories of the Hooghly and its offshoots with zonal and basinwise distribution of area. Basin area represented by numerical numbers are in sq. miles.

(Source : Mukherjee, 1959)

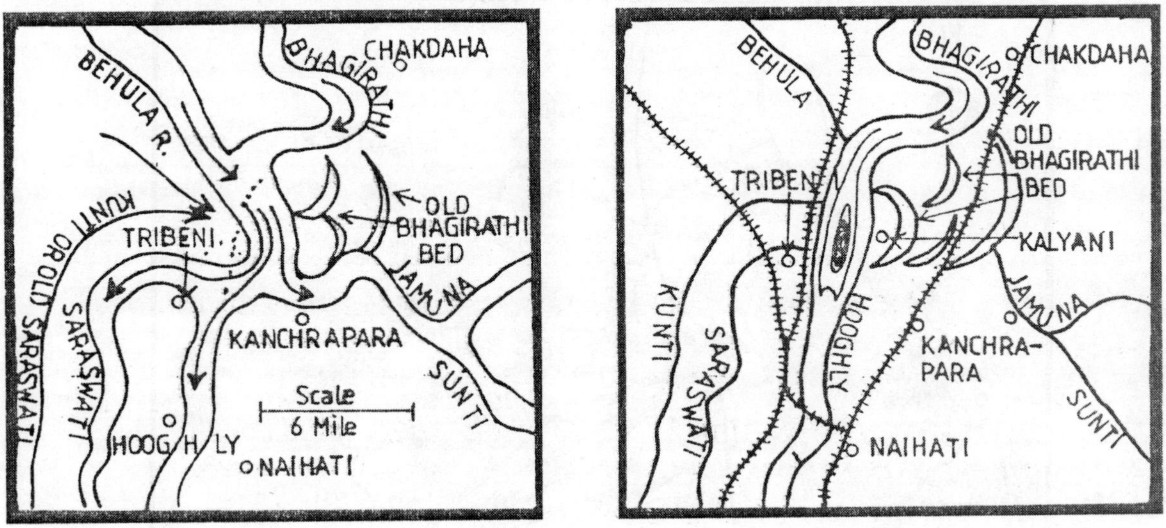

(a) (b)

Fig. 2 : (a) Active Ganga, Jamuna and Saraswati during seventeenth century. (b) Conditions of these rivers after the Ganga and the Damodar (Behula) shifted their courses.

1. The Salt lake ; The Bidyadhari spill area and its progressive changes. (Cultury fishery A).
2. Metropolitan waste and its impact on aquaculture.

2. GENESIS OF AQUACULTURE IMPOUNDMENTS

The Salt lake, being situated near Calcutta lies between 22°27′ and 22°96′ N and 88°24′ and 88°30′E latitudes (Fig. 3) occupying a large portion of the alluvial plains of lower Bengal, the whole of which is comparatively of recent geological formation. The Salt lake is the bed of some large distributory of the present Hooghly which is blocked by sand banks deposited somewhere in the south after having its overflow discharged into the Matla and the Bidyadhari rivers. This bed together with the 'beels' along the bank of the dead river forms what we call the Salt lake which was being fed by tidal water coming from the Bidyadhari and the Matla rivers until about 1930. The southern saltlake was to a large extent open to a free tidal spill coming from the Matla, the Piali, and the Bidyadhari series of rivers which finally entered into the Central lake channel and other channels connected with them. The principal Khals by which tide enters this lake are the Nowbhaga khal, the Paranchaprasi khal and Hatgachia Khal. These khals vary from 100 to 250 ft in width at their mouths and the depths of water have been found to be 3.5 to 5 fathoms at low water. From these main khals there are endless ramifications of small crecks by which the water spreads over the face of the swamps (Fig. 1).

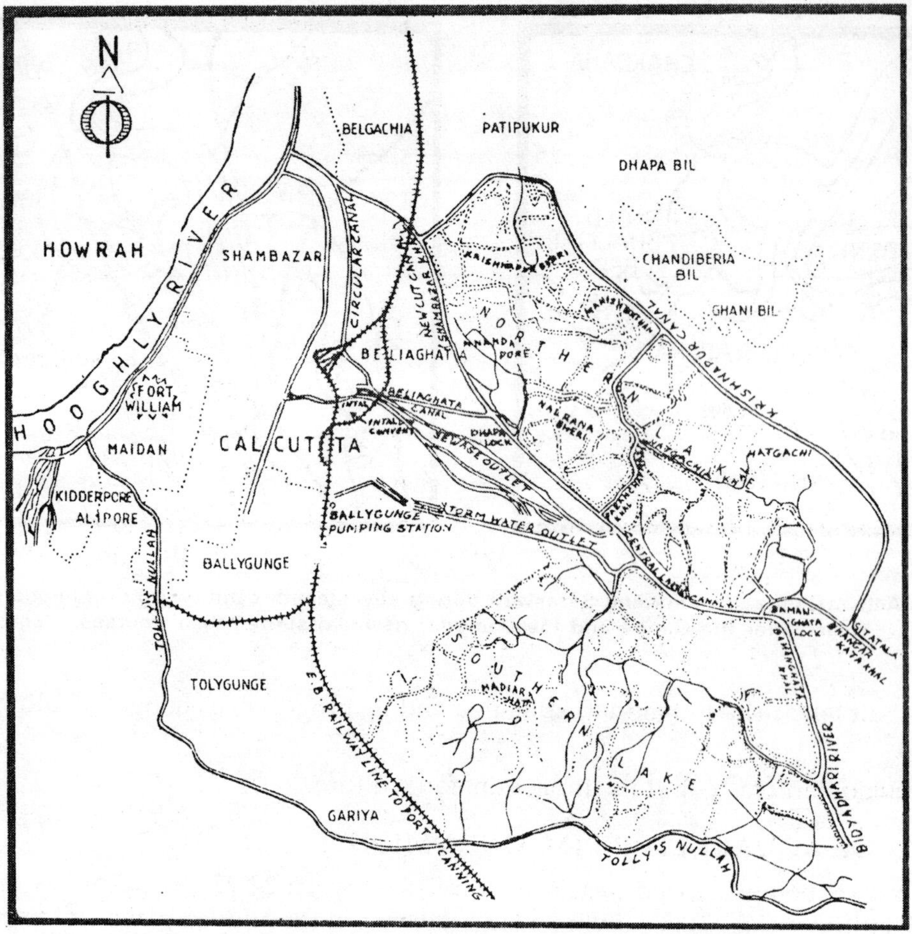

Fig. 3 : North and South Salt lakes of 24-Parganas.

Biswas (1927) while giving an account of vegetation of Salt Lake reported that these swamps and the bottom of the lake were gradually raised by constant deposition of silt and were overgrown by *Sudema maritinna* which is interspersed at some places by *Cynodon ductylon* and *Fimbristylis farruginea.* These plants act as sand binders in arresting the seeds of many littoral plants, chiefly of *Suada marilima Avicennia officinalis, Acanthus ilicifolius* and others which are carried away by the tide along with the silt and humus. The seeds after being deposited on muddy soil, germinate and give rise to mangrove formation. These littoral plants soon send out a network of suckers from which develops a regular assemblage of pneumatophores which help further in arresting the silt and thus raise the level of the land and give origin to the formation of regular swamps forests. These forests in course of time are cut down by fishermen and the land soon becomes dry and quite habitable.

The manner in which the Bidyadhari and other rivers of the region are silting up preventing the salt water from extending upwards into the salt lake system is interesting (Sewell, 1934). Data covering several years for only a single place are given below in Table 1.

Table 1. Showing Silting Up of Salt Lake Area

Location	Years			
	1926 May	1928 Feb	1932-33 Dec-Feb	1933 Mar
Canal off Lans down Jute Mill and Dakshindari village	18.48	9.60	2.20	3.60

From the above table it is seen that the water is gradually getting fresh. This fact is also supported by the rapid fall in the salinity of the salt lake during 1928-1940 as indicated by Sen (1941) in the following Table 2.

Table 2. Salinity status of the Salt Lake during 1928-1940

Year	(Chlorine content per 100,000 parts)
1928	1,499
1936	950
1939	640
1940	130

In the year 1930 a zamindar of the locality finding his vast area of land not yielding revenue made an attempt to culture carps by letting in sewage in small quantity into the enclosed spilled water. The results obtained were very encouraging and within a short period practically the whole area was converted into ponds for culturing fresh water fishes when salinity dropped to 500-600 ppm.

With the initiation of fish culture this system gradually intensified and thus a large number of bheries came into existence both in North and South Salt Lake (Appendices I and II). The interest in fish culture by allowing small quantity of sewage in the bhery and the results obtained being more encouraging, people gradually took more interest for fish culture by using sewage water. It is since then the importance of knowing characteristics of sewage waste and using it properly or scientifically was gaining importance.

Such an attempt increased mosquito menace, Dr. S.L. Hora in 1944 came out with suggestion of controlling mosquito larvae through fish culture. Subsequently, in 1940 construction of two canals, 28 km each, solved the problems to drain out the sewage and rain water into the Kulti estuary. Sewage was drained through a dry weather flow canal, while accumulated rain water of the city was allowed to drain through the storm water flow canal which was wide. Both were connected near Bantala through an escape regulator gate. These arrangements were so efficiently made that at the time of high tide in the Kulti estuary the water used to rush in the storm water flow canal as back water and flushed the deposits and was allowed to enter the dry weather flow canal through the escape regulator gate with high velocity and disposed back into the Kulti estuary during low tide. The Kulti water was thus used both for sewage dilution in bheries around Bantala and for flushing out the deposits from both the canals. Such arrangements, however, lasted for a short period only. It was between 1940 and 1956 shallow impounded waters, locally known as bheries, gradually came into existence due entirely by the efforts of private sector. An appraisal of survey of bheries made by secretary, fish Producers Association in 1959 is given in Appendices I & II. According to the "Master Plan for Fisheries Development Part I (1974), the total number of bheries in the North and South Salt lake areas, were 58 (Total area 2447 ha) and 153 (3529.73 ha) respectively which got further reduced to 167 (4442 ha), as 44 bheries (1534 ha) were acquired by the Government for city extension and other purposes. A classified statement of different size groups of bheries of North and South Salt lake is given in Table 3.

Table 3. Classified list of Bheries under Different Size Groups

Size group of bheries (bigha)	North Salt lake		South Salt lake		Total	
	No.	(area in bigha)	No.	(area in bigha)	No.	(area in bigha)
Upto 50	16	385	30	1,025	46	1,408
50-100	5	406	30	2,063	35	2,469
100-500	25	6,494	49	11,882	74	18,376
500-1000	8	5,465	—	—	8	5,465
1000-1500	4	5,600	—	—	4	5,600
1500-2000	—	—	—	—	—	—
Above 2000	—	—	—	—	—	—

Banerjee (1931) has given an interesting account of tidal amplitude at Dhapa around a few points commencing as far back as 1830, which leads us to draw certain definite conclusions (cf. Table 4).

Table 4. Tidal Amplitude at Dhapa for Different Years

Years	Range of tide	High waters	Low waters
1830	2.53 ft	1.50 (average)	- 1.25 (Min.)
1865	4.67 ft	4.90 (average)	- 4.30 (Min.)
1881*	11.0 ft	7.40 (max.)	- 3.60 (Min.)
1901	12.58 ft	8.75 (max.)	- 3.83 (Min.)
1920	8.34 ft	9.17 (max.)	- 0.83 (Min.)

*Dhapa Lock was constructed.

The above table clearly shows that there was a steady increase in the amplitude of the tide up to 1901, but shortly after there was a decrease. In recent years this tidal impulse had steadily receded off the upper reaches of the Bidyadhari. During neap tides the tidal impulse does not even reach up to Bamanghata now a days. The progressive deterioration of the Bidyadhari at Bamanghata is also evident from the Table 5 (Mukherjee, 1959).

Thus the river totally lost its characteristics by the year 1942.

Observations on the salinity in various parts of the lake carried out from time to time clearly show the degree of variation that has occurred in the past years. An indication of salinity at 4 different points of the salt lake during dry and wet months of 1928 is mentioned in Table 6.

Table 5. Progressive Deterioration of the Bidyadhari River

Years	Cross Sectional area at Bamanghata at high water (Sq. ft)	Lowest Bed O.M.S.L.
1883	13,674	- 46.2
1904	9,702	- 17.2
1913	6,690	0.6
1926	2,560	5.9
1934	250	10.7
1942	Dry	Bed 4-6 ft above ground level.

Table 6. Salinity Level of the Salt Lake at Different Points During 1928*

Area	Dry month (April)	Wet month (July)
Parancha prasi's Khal	15.48 ppt	5.97 ppt
Salt lake proper	14.99 ppt	4.87 ppt
Bidyadhari river	13.73 ppt	5.17 ppt
Dhapa Lock	13.53 ppt	4.67 ppt
*Biswas (1932)		

*Biswas (1932).

From Table 6 it is evident that the highest salinity (ranging 13.53-15.48 ppt) occurs during the dry months of April and early May. The lowest salinity occurs during monsoon.

The periphery of the salt lake and the associated canal systems, as discussed by Sewell (1934), are bound on the west by the Hooghly river and are in direct communication with Chitpur canal. This canal after a course of about 1.6 km bifurcates, one branch remaining towards the south for about 3.2 km before it turns east past Chingrighatta runs to Dhapa lock; and the other runs first east and then south east for about a mile and then again bifurcates — the main line of the canal continues towards the south-east to meet Chingrighatta Canal just north of Dhapa lock and the second branch is turned into the Kistopore canal and runs at first towards the north-east, but after a course of about 3.2 km turns round and runs in the south eastern direction for about 9.6 km and finally turns southwards to enter the Bamanghatta Canal at right angle about 1.6 km away on east of Bamanghatta Lock. Dhapa and Bamanghatta Locks are connected by the Central lake canal into which sewage and stream water flow from Calcutta pours and this canal continues off Bamanghatta lock to join the Bidyadhari river. The other canal–Tolly's Nallah takes off from the Hooghly river and runs eastward across the Bidyadhari river in the vicinity of Samookpota. The construction of sluice gate near Samookpota to regulate the flow of tides from far end of the nallah causing thereby siltation resulted in further injury to the Bidyadhari river. In addition, the construction of various engineering works, such as bridges, canals etc., that have interfered with the free flow of the water and their silt which is carried up the rivers and streams during flood tide and is deposited at stack waters (9 m in 21 years recorded in 1904 and subsequently additional 5.5 m recorded in 1912 in Bidyadhari river) hastened the process. Thus the sewage instead of flowing into the river joined the water-logged swampy areas. In the absence of proper outflow and dilution, the concentrated raw sewage used to accumulate heavily during dry months and flood vast shallow areas during wet season. The water become heavily polluted. This made the cultivators to reclaim the silted river bed of the Bidyadhari for agricultural uses. This is how the existence of once mighty Bidyadhari river ceased. Contrary to this, the entire swampy salt lake area receiving city sewage became a permanent source of malaria outbreak and threatened the city life. The Salt Lake basin.

In order to protect the health of citizens of Calcutta city the construction of bunding a number of canals and lock gates under different projects were completed up to 1945. These were :

a. 1777 — Excavation or Tolly's Nullah.

b. 1790 — Bunding up of the Dud Bibi Khal or the western Bidyadhari near Tardaha thus preventing spilling north ward.

c. 1829 — Excavation of Circular Khal.

d. 1830-34 — Excavation of several east and west cuts known as Bhangar Khal thereby disturbing the tidal equilibrium of the spill channels of the Bidyadhari and the Kulti series of rivers.

e. 1833 — Construction of old Chitpur lock.

f. 1843 — Premature reclamation started in the Kultigong area by throwing embankments.

g. 1859 — Excavation of New Cut Canal.

h. 1880-90 — Gradual premature reclamation of Salt water lakes for fish culture.

i. 1981-82 — Construction of Dhapa and Chitpur Locks.

j. 1896 — Cross-damming the Kantatola gang : one of the most powerful tidal spill channels of Bidyadhari.

k. 1897-98 — Canalisation of Bhangar Khal.

l. 1902 — Construction of corporation drainage canals across the natural line of flow.

m. 1910 — Excavation of Kistopur canal.

n. 1914 — Construction of Uttarbhag lock and sluice.

o. 1920 — Construction of Samukpota sluice.

p. 1922 — Forced diversion of the Arapanch basin from Tolly's Nullah to the Piali, at the cost of Tolly's Nullah.

q. 1934-45 — The Kulti outfall scheme.

r. 1962 — Filling up of the Salt lake area for construction of a township.

In doing so the salt lake got reduced from 100 sq. miles (256.9 sq km) to only 35 sq. miles (90.615 sq km) but the problem still remained acute and complicated for proper disposal of sewage. It was around the year 1900 some villagers put embankments in some areas of the spilled water for culture of Brackish water fishes like *Lates calcarifer, Mugil parsia, Mugil tade* and prawns in these embankments by trapping their seeds during high tide. But this procedure could not last long due to (i) restricted ingress and egress of river water which resulted in heavy silting of the Bidyadhari from raw sewage and storm water suspensoids and (ii) Vast shallow area which used to get flooded during wet season.

3. METROPOLITAN WASTES

3.1 Disposal

Calcutta produced 0.0682 mm^3/d of waste water in 1975. With increase in population its volume rose considerably viz., 0.109 mm^3 in 1891, 0.3046 mm^3 in 1914, 0.5910 mm^3 in 1943 and 0.73 mm^3 in 1973 and 0.75 mm^3/d in 1986 with the rise in population from 6 m to 9.1 m during 1970-81 of the 0.75 mm^3 waste, 0.23 mm^3 is of industrial origin. In addition, 0.41 mm^3/d of unfiltered water, which was being used for fire fighting, street washing, flushing gallies, etc., increased the dilution as well as the volume of the effluent. To dispose off such enormous volume of water the city has about 1412 km length of drainage system. In this system the average annual deposition of rubbish or silt is 14 m cubic ft prior to 1973, on an average, 6-7 m cubic ft of sludge which subsequently increased to 9 m cu ft used to be removed. The average yearly rainfall 16" (1600 mm) being high further helped in increasing the silt accumulation in the drainage. The sedimentation tanks constructed in 1942 remained in operation for a short period. About 70-80% of the sludge (6000 cu ft/day) collected in 7 lagoon tanks used to be removed and sold as manure at a concessional rate to different districts of West Bengal. The supernatant sewage water together with 20-30% of sludge was being used for pisciculture and the rest drained into the Storm Water Flow (SWF) canal for their disposal into the Kulti estuary.

The capacity of storm water flow (SWF) canal is 3000 cu ft/sec against 250 cu ft/sec for dry weather flow (DWF) canal. Nair (1944) observed that owing to the narrowed cross-section of silted Bidyadhari the crude sewage used to be discharged into the storm water canal. The velocity of the sewage water dropped considerably and practically there was no appreciable flow in the canal itself. At the point of discharge, little flow was noticed for a length of about 1.6 km or 3.2 km and then at the outfall end for adult the same length when the sluices of the Kulti were opened. The whole canal, especially at Bantala and Kulti ends functioned as an open sedimentation tank. The sludge of the canal bed became septic and filled with gases. Masses of gas filled sludge used to float at the surface and fouled the whole canal with septic sewage and was toxic to fish life. No other fish except air breathers like *Clarias, Anabas, Channa etc.*, were found to survive. The water of the central length of the canal between 8th and 22.4th km mile, where the sewage became clear to some extent and further drainage from the adjacent areas helped in improving the condition of foul sewage. The entire quantity of which was approximately 130 m gallen day (during forties), took about 34 hrs to reach the Kulti. Moreover, at certain places in the canal, small pools and puddels were formed where settleable solid rots decomposed producing disagreeable gases. The bed of the canal was covered with black residual matter. The surplus water during the monsoon drained into the sewage canal and diluted the sewage to a considerable extent.

The large list of industries presented by Mitra (1954) in his valuable document *Census 1951, West Bengal* clearly indicated that the metropolitan wastes were mainly organic in character and there was hardly any source that can cause metallic pollution. Lately new industrial complexes had come up rapidly. These contribute wastes like acids, various chemicals, paints, varnish, battery, cosmetics and perfumenry, drugs and pharmaceuticals, electroplating, fertilizer, match manufacturing, metallurgical ferrous and non-ferrous, galvanizing, pulp and paper, soap detergents, textiles and good number of tanneries located in the eastern Calcutta. 160 large size tanneries existed in China town alone and more than 2000 smaller ones were scattered here and there in the city. These industries came into existence sometime in 1963 and in a period of 20 years, their numbers increased considerably. This means that the sewage effluent prior to 1963 was almost free or had negligible concentration of metallic contamination. Added to it the supply of additional unfiltered water used for washing is almost suspended since 1969. As a result of defunct sedimentation tank, the raw effluent, however, in course of flow in an open canal for about 8 km gets most of the suspensoids settled or adsorbed by thick rooted *Eichhornia crassipes* growing in abundance at the mouth of the regulation gate. It helps partly to convert raw-sewage as light-black coloured odour-free waste or enter into the bhery. A sizeable volume of such waste periodically or almost continuously (depending upon sewage availability) is being used through a separate drain from the main canal. Raw sewage is also drawn by pumps at a few points for use in the neighbouring agriculture fields. The remaining waste is drained into the DWF Canal for its final disposal into the Kulti estuary, 27 km away from Bantala. The SWF canal is used for disposal of rain water for 3-4 months during wet season. About 6000 ha of encroached land lying in between DWF and SWF canal are under regular cultivation and have become a source of constant erosion during monsoon. Added to it, the untreated raw waste released through Bantala lock gate carries enough of suspensoids since after the tank has become non-functional. All these have caused about 8 ft deep black bottom ooze in the DWF canal and arrested the flow of the effluent. Further, all the waste water coming out of the bheries and joining the SWF canal is also drawn for use in the adjoining agricultural fields with the help of pumps at more than 15 points on the SWF canal. This has resulted in considerable decrease in waste water flow. During summer months at several points, the flow gets restricted to a small channel only on the bed of the SWF canal. The exposed areas are regularly being cultivated by the local villagers. After the Government of West Bengal had taken over the responsibility of the canal from Calcutta Corporation in 1968, the canal was dredged only once during 1974-75. Since then no care was taken.

3.2 Characteristics

Major factors so far controlling the characteristics of waste flowing during different times are : (1) Until around 1963, the waste which was dominated by organic complexes, used to receive additional unfiltered water (0.41 mm^3/d). This water was being used for fire fighting, washing of streets, gullies etc. This additional water diluted the sewage considerably; (2) during the period the sedimentation tanks remained in operation, the bheries used to receive sewage free from major suspension; (3) with the growing industrial complexes in the matropolis since 1963 considerable volume of the waste contained various toxic chemicals; (4) Suspension of unfiltered water deprived the waste from dilution and finally (5) the role of water hyacinth (*Eichhornia crassipes*) in cleansing the waste was also affected.

However, based on observation of actual flow at six centres spread over the 27 km long DWF canal, NEERI (1978) calculated the average pollution load for the year in terms of total solids, suspended solids, dissolved solids, bio-chemical oxygen demand, chemical oxygen demand, total nitrogen, and total phosphorus. Seasonwise studies indicated that average flow per 12 hour was highest during monsoon (92448000 l) followed by summer (72576000 l) and winter (68256000 l). There was hardly any appreciable improvement in pH. Total pollution load calculated in terms of total solids and suspended solids were in the order 74.99 t, 149.92 t and 60.63 t; and 8.56 t, 12,19 t and 6.28 t during monsoon, summer and winter respectively. The corresponding organic (BOD_5) and inorganic (COD) loads were 5.77 t, 3.28 t and 4.20 t; and 15.35 tm 9.87 t and 11.33 t/12 hrs respectively. Seasonal characteristics of waste water at different sampling points spread over the 27 m DWF canal are indicated in Table 7.

Sludge collected from six points spread over the DWF canal between Bantala and Ghusighata were analysed for metals like lead, copper, zinc, Cadmium, Nickel, Cobalt, Chromium and manganese. The results expressed in mg/kg of soil are given in Table 8.

These values clearly indicate that suspended solids chilate metals and settle. There is thus a considerable control of toxic metals of the influent sewage which otherwise could have caused hazard to the bheries and Kulti estuary.

However, characteristics of raw sewage (Table 9), introduced into the bheries during different years during 1932-84 as studied by different workers, are discussed below.

Physical : The waste was always light black in colour with offensive smell. In early thirties suspended solids were lowest (37.7 ppm). Except with a slight depression (64-196 ppm) during 1974-75, values during subsequent years were on the increase. The values were 160-420 ppm in 1958, 968 ppm in 1974 and 38-172 ppm in 1979. Values for total solids recorded during 1974 and 1975 were fairly high (5388 and 792-2260 ppm).

Chemical : pH more or less maintained a level of 7.0-7.5 with nil dissolved oxygen. The corresponding BOD_5 values were moderately high during the years 1974, 1974-75 and 1984 ranging between 60 and 80 ppm, while during 1979 and 1983 the values were fairly high (400-530 ppm and 200-650 ppm). Ghosh (1983) observed a range 200-650 ppm. Carbondioxide estimated during 1958 (20-96 ppm), 1974 (25.3 ppm) and 1979 (25-45 ppm) were fairly high. With a slight depression during 1979, total alkalinity indicated moderately high values (170-490 ppm in 1958; 255 ppm in 1974; 206-440 ppm in 1974-75) but in 1984 the values ranged between 136 and 316 ppm. Organic matter represented as OC was lowest during 1932, rose as high as 24-88.8 ppm, in 1958; 73.5 ppm in 1974 and 30-75.5 ppm in 1979. Chlorinity had almost similar fluctuations (120-130 ppm) with a fairly higher level (832 ppm) during 1974-75.

Table 7. Seasonal characteristics of metropolitan wastes at different sampling points of DWF canal

Parameters	*Topsta 0 km.*		*Batchtola wooden Bridge 2.2 km*		*Before sedi-mentation tank 6.2 km.*		*After sedi mentation tank 6.3 km*			*Bamanghata 9.4 km 25.0 km*				*Ghatakupur*		
Flow, m^3/sec	2.38 (W)	2.49 (S)	4.59 (M)	1.13 (M)	1.42 (W)	1.76 (S)	-	-	-	-	1.67 (M.W)		2.35 (S)	0.14 (S)	1.19 (M)	0.85 (W)
pH	7.2 (W.S)		7.4 (M)	7.3 (S)	7.4 (M)	7.6 (W)	7.4 (M)	7.5 (W)	7.3 (M)	7.5 (W)	7.5 (M)	7.7 (S)	7.8 (W)	8.0 (W)	8.6 (S)	8.1 (M)
Total solids	752 (W)	876 (M)	2331 (S)	874 (M)	908 (W)	2260 (S)	792 (M)	792 (W)	682 (M)	976 (W)	588 (M)	1829 (S)	890 (W)	832 (W)	1818 (S)	1044 (M)
Suspended solids	128 (W)	168 (S)	244 (M)	80 (M)	196 (S)	208 (W)	64 (W)	68 (M)	48 (M)	52 (W)	48 (M)		68 (WS)	32 (W)	240 (S)	68 (M)
Chlorides (cl)	103 (W)	138 (M)	895 (S)	118 (W)	175 (M)	832 (S)	92 (M)	216 (W)	88 (M)	189 (W)	88 (M)	197 (W)	665 (S)	146 (M)	690 (S)	172 (W)
T.alkalinity (CaCo$_3$)	248 (M)	316 (S)	328 (W)	266 (M)	388 (W)	354 (S)	206 (M)	44 (W)	224 (M)	440 (W)	200 (M)	386 (S)	408 (W)	252 (M)	268 (W)	322 (S)
Sulphate (SO$_4$)	8 (W)	24 (M)	88 (S)	12 (W)	36 (M)	78 (S)	28 (W)	31 (M)	28 (M)	31 (W)	18 (W)	44 (M)	62 (S)	18 (W)	36 (M)	46 (S)
T.phosphorus (p)	0.063 (M)	4.6 (S)	2.6 (W)	1.24 (M)	6.3 (W)	4 (S)	0.33 (M)	25.1 (W)	2.51 (M)	23.0 (W)	2.12 (M)	5.3 (S)	19.8 (W)	0.63 (M)	13.8 (W)	3.9 (S)
Ammoniacal Nitrogen (N)	7.20 (M)	7.8 (W)	11.5 (S)	11.6 (M)	14 (S)	14.2 (W)	8.48 (W)	11.4 (M)	6.8 (W)	13.6 (M)	5.01 (W)	0.4 (M)	18.1 (S)	— —		8.8 (W)
T. Nitrogen-N	14.6 (W)	15.7 (M)	17.6 (S)	18.2 (M)	21.6 (S)	26.0 (W)	16.5 (M)	35.0 (W)	18.8 (M)	30.8 (W)	15.7 (M)	24.6 (S)	27.2 (W)	3.1 (M)		17.4 (W)
D.O.	nil (M)		0.5 (MW)	nil		nil	nil	nil (M)	nil (W)	0.8	nil		nil	—		—
BOD	40 (M)	42 (S)	52 (W)	55 (S)	79 (M)	88 (W)	84 (M)	96 (W)	78 (W)	81 (M)	39 (W)	68 (S)	75 (M)	15 (M)		16 (W)
COD	117 (S)	152 (W)	189 (M)	140 (M)	222 (S)	236 (W)	151 (M)	208 (W)	159 (M)	196 (W)	136 (W)	147 (M)	228 (S)	30 (M)	68 (W)	114 (S)

Except for pH and flow results are exprssed to mg/t.

Seasons : Summer(S), Monsoon(M) and Winter(W).

Modified from the report of NEERI (1978).

Table 8. Metal Concentration in the Sludge of DWF Canal

Metals	Range of value mg/kg soil	Mean value
Lead	12.78 — 181.23	98.03
Copper	36.78 — 246.72	113.92
Zinc	101.90 — 760.1	366.1
Cadmium	Nil — Nil	
Nickel	23.19 — 50.84	39.22
Cobalt	10.95 — 16.32	13.86
Chromium	25.35 — 378.84	197.05
Manganese	33.69 — 444.01	333.24

Table 9. Characteristics of Waste Water during different periods

Parameters	*1932*[a]	*1958*[b]	*1974*[c]	*1974-75*[d]	*1979*[e]	*1983*[f]	*1984*[g]
Physical Character							
Temperature	—	24-35°C	31.8°C	—	—	—	—
Turbidity (units)	—	—	110	—	—	—	100
Colour	—	—	Blackish	—	—	—	Light brackish
Odour	—	—	Offensive	—	—	—	Organic
Total Solids	—	—	5388	792-2260			
Suspended-S	37.7	160-420	968	64-196	380-972	—	—
Dissolved-S	—	—	4420	—	—	—	—
Chemical Parameters							
pH	—	6.9-7.3	7.0	7.3-7.5	7.0-7.4	—	7.1-7.5
Dissolved Oxygen		Nil	Nil	Nil	Nil		Nil
Carbon dioxide	—	20-96	25.3	—	25-45	—	—
Alkalinity ($CaCO_3$)	—	170-490	254.8	206-440	78-292	—	136-316
BOD_5	—	—	55.0	55-96	400-530	200-650	60-80
COD	—	—	—	151-222	—	—	—
OC (by $KMnO_4$, 30 mts.)	4.9	24-88.8	73.5[1];	10.5[2]; 662[3]	30-75.5		
H_2S	—	2.4-48	Nil	—	3.5-5.8		Nil
Sulphate	—	—	—	28-78	—		—
Chloride	—	—	120	92-832	120-130		—
Sp. Conductivity (micro mhos/cm)	—	—	—		1260-1543		719-1028
Nitrogen							
Ammonia-N	—	12-63.6	9.0	8.48-14.0	9-25	—	—
Albuminoid-N	0.71	1.1-16.0	Trace	—	—	—	—
Nitrite-N	—	Nil-0.08	Nil	—	Nil-1.5	—	—
Nitrate-N	—	0.01-0.33	0.15	—	9-18.14	—	9-17.2
Total-N	—	—	—	16.5-35.0	—	—	—
Phosphorus							
Inorganic-P	—	0.12-14.5 (P_2O_5)	5.0 (PO_4)	—	3-5.0 (PO_4)	—	1.8-4.0 (PO_4)
Total-P	—	—	—	0.33-25.1	—	—	—
Phenol	—	—	—	—	0.2	—	—
Metals							
Zinc	—	—	—	—	1.6,3.8*	1.8946	1.5-2.4
Copper	—	—	—	—	7.5,6.0*	0.410	0.3-0.5
Manganese	—	—	—	—	90,65*	—	13-60
Arsenic	—	—	—	—	—	<0.06	—
Iron	—	—	—	—	115,40*	1.5139	65-150

contd...

Table 9. – *contd...*

Parameters	*1932*[a]	*1958*[b]	*1974*[c]	*1974-75*[d]	*1979*[e]	*1983*[f]	*1984*[g]
Cadmium	—	—	—	—	—	<0.06	—
Chromium	—	—	—	—	—	0.0137	—
Nickel	—	—	—	—	—	0.018	—
Lead	—	—	—	—	—	<0.06	—
Mercury	—	—	—	—	—	<0.04	—
(Results are expressed in ppm. except pH and sp. conductivity)							
Microbiol load							
TPC	—	—	—	—	—	—	3×10^4-4×10^6
Coliform Bacteria	—	—	—	—	—	—	1.8-2×10^6**
E. Coli	—	—	—	—	—	—	
Staph auries	—	—	—	—	—	—	4-7×10^4**
Pseudomonas	—	—	—	—	—	—	6×10^4**

(a) P.C. Bose (1944); (Average of analysis throughout the year 1932)
(b) Saha *et al.*, (1958)
(c) B.B. Ghosh (Personal Communication);
(d) NEERI (1974-75);
(e) Ray and Pakrasi (1979);
(f) Dhrubajyoti Ghosh (1983);
(g) Ray and Bandopadhyaya (1984).
1. O.C. 30 mts by $KMnO_4$
2. .... 3 mts by $KMnO_4$
3. .... 4 hrs by $KMnO_4$

* From soil - data not available, ** From sludge (1981)

Nutrient Status : Nitrate nitrogen indicated fairly high values in 1979 (9.0-18.14 ppm) and 1984 (9.0-17.2 ppm). During 1958 (0.01-0.33 ppm) and 1974 (0.15 ppm) the values were much lower. Phosphorus concentration was fairly high in 1974 (5.0 ppm), 1979 (3.0-5.0 ppm) and 1984 (1.8-4.0 ppm).

Metals : Zinc, copper, manganese and iron estimated for the first time during 1979 were 1.6, 7.5, 90 and 115 ppm respectively. The estimated values for Zinc, Copper, Iron, Cadmium, Chromium, Nickel Lead, Arsenic and Mercury as indicated by Ghosh (1983) were 1.8946, 0.410, 1.5139, <0.06, 0.0137, 0.018, <0.06, <0.06 and <0.04 ppm respectively. In the year 1984 the estimated values for zinc, copper, manganese and iron were in the order of 1.5-2.4 ppm, 0.3-0.5 ppm, 13-60 ppm and 65-150 ppm respectively. Mercury content ranged between 0.4 and 0.48 ppm from Tapsia to Bantala which got reduced to 0.37 ppb further down*.

Microbiol load : Microbial load estimated during 1984 ranged between 3×10^4 and 4×10^6 as TPC, while in the sludge the values were 1.8-2×10^6.

---

* Personal communication with Dr. H.C. Josi, CIFRI (1986).

Further, there are 10 types of biomedical wastes, such as human anatomical wastes, animal wastes, microbiology and biotechnology waste, waste sharps, discarded medicines and cytotoxic drugs, solid waste, liquid waste, incineration ash and chemical waste. KMC is collecting the untreated waste and dumping it in Dhapa. Only Eastern command Hospital has an operational incenerator in a city that produces 10 tonnes of bio-medical waste every day. The hospitals in the city continue to function without such treatment facility (The Statesman Bengal Plus 29.7.2002.)

4. BACKGROUND INFORMATION

Albeit recycling of organic wastes in aquaculutre in the topic of the day; their conservation rather than destruction in essential. Calcutta Metropolitan wastes which were once a real problem or menace to people of the greater Calcutta, for their unscientific disposal, are now being used indiscriminately in bheries (shallow impoundments) and estuarine areas. Scientific information on the effect of such wastes on aquatic biomass of the Bidyadhari complex in the context of pisciculture operation is scanty. Geographical conditions and hydraulics of the river system were discussed by Banerjee (1931), while Biswas (1927) studied the flora and vegetation of Salt lake, Sewell (1934) studied the fauna of the lake during the transitional period when the Bidyadhrai river was heading rapidly towards deterioration due to silting and consequent changes of brackishwater into freshwater conditions. Studies on piscicultural aspects in the salt lake area were initiated by Sen (1941) followed by Bose and Nair separately in the year 1944. Basu (1950) made observations on ecology and bionomics of Indian major carps while Saha *et al.* (1958) covered the effect of Calcutta sewage on fish life. Ray *et al.* (1981) reported on pisciculture practices both under fresh and brackishwater conditions. Ray and Bandyopadhyaya (1984) made observations on different aspects of hydrobiology, bacteriology and presence of heavy metal in raw sewage, water, soil and fish flesh in randomly selected five bheries in the north salt lake. The secretary, Fish Producers' Association, made a valuable survey of bheries both in the north and south salt lake area in the year 1959 with regard to their number, area, production/ha and annual returns.

5. BHERY MANAGEMENT

The existing bhery ecosystem is the creation of impoundments over the raised bed of the original Bidyadhari river and its vast spill areas by putting dikes of various sizes. It is interesting to note the gradual changes that occurred in the fishery management since 1900. Sen (1941), perhaps, for the first time reported about scientific management of bheri and the fish culture aspects. Usually towards the end of March, the ponds designed for carp culture were completely dried by draining the river into the adjacent lakes while several of the shallow ponds dried up naturally during the month. The ponds were then cleared of debris and vegetation and the edges were trimed. The mud exposed to the sun soon got completely dried and converted into dust. After these preliminary measures the ponds were filled with water from the adjoining bheries towards the end of May or in June through small cuts. Sometimes water was also introduced either with the help of a *donga,* which is a dugout trunk of a palm tree, or with a 'seoni', an improvised basket made of palmyra leaves or of tin sheets. Rain water further increased the volume of water. This was followed by the introduction of fry by the end of May or early June. Sometime during September, stocking was done with fingerlings. It was believed that heavy showers within two days of introduction of tny fry (spawn) were detrimentl to them. Carp normally grew to 1/4 seer (234.4 g) in one year although in some cases they attained up to 3/4 seer (703.125 g). Bhetki also grew to 1/4 seer (234.4) in course of 1 year. When the carps attained a suitable size they were occasionally disturbed by beating the water with bamboo poles as this was supposed to result in the growth of fish to a size of one seer (937.5 g) or more in the course of two years. Rotten water-hyacinth, sullage-water and algae were supplied to the fry as food. The dried mud from the bed of sullage canals and also the algae and other micro-organisms growing around grasses on the edges of ponds were used as food. Some preferred Lemna which was largely eaten by fry (Chatterji, 1934). The food of carps constituted mostly of vegetable matter (Gupta, 1908), while the animal food consumed was composed of insect larvae, smaller crustacean, molluscs and other similar organisms (Mookerji, 1938). It was observed that tiny fry used to grow to a size of one to three inches within three months. Gupta (1908) noted carps growth as 16 or 17 cm in length in the first year. The favourite food of tiny fry consisted of microplankton such as Infusoria (*Paramacium* according to Chatterji, 1934), diatoms and minute unicellular algae, while fingerlings consumed minute crustaceans such as *Daphnia* and *Cyclops* in addition to algae; juveniles and adults consumed earthworms, Chironomid larvae and higher crustaceans such as shrimps in addition to microscopic forms like *Daphnia, Cyclops* and algae.

Fry mortality was commonly caused by congestion and/or lack or proper food. The usual practice followed was to distributed the fry to other ponds whch ensured better results. Besides, a regular practice of dragging nets helped in removal of slimy matter, often of algal origin, external parasites from the body of fish and also partial removal of gaseous formation at the substratum of ponds. De (1910) recommended planting of bamboo sticks and posts in the middle of ponds to enable fish to rub their bodies for removal of their external parasites. Usually precautions were taken for the safety of the spawn from frogs carnivorous animals like *Babus, Ambasis, Notopterus* and *Wallago* which found their way along with the spawn when released into the bheri. Aquatic vegetation like *Hydrilla* and *Ceratophyllum* were prevented from growing into the bheri.

Nair (1944) later studied the management more scientifically when more modifications were made in the bhery ecosystem. For the first time he mentioned about the introduction of sewage into the bhery. According to him larger bheries were dewatered by the end of March or beginning of April and salt water was drawn from Kistopore canal to a depth of 6″ (15 cm). This was followed by intake of raw, foul smelling and dark coloured sewage to an average depth of 3 ft (0.9 m). Within 2-3 weeks' time the sediments settled, the water became clear and free from any odour. Rich growth of algae appeared along with the clarity and greenish colour of water. After initial introduction of sewage, inflow of sewage in the bhery was allowed once a month or so at 1 : 4 sewage and water ratio. Such diluted water was generally taken for 4 to 5 days or say 100 hrs to maintain a depth of 3 ft. A common belief existed among the pisciculturists that the suspensoids of sewage was lost before being drawn to the fishery during its flow for about 5 km distance (from Topse to Bantala) before it was drawn for fishery purposes. Besides, a further dilution of the sewage to a certain extent in the bhery by adding freshwater ensured complete stabilisation. Proper attention was paid for aeration, purification and fish movement by exercising daily fishing operations from 9 a.m. to 2 p.m. and by beating the surface water with bamboo poles. Kistopore canal-water was used due to its salinity (2%) as it helped to control the parasite infection such as *Argulus, Myxosporidium* etc. This otherwise could cause a great loss to the fishery by spreading in an epidemic. Thus treated fishery pond was stocked with major carp fingerlings (3″-6″ size) 7.5-15 cm. In case of mixed stocking the sizes chosen were of the order of about 20 in number per seer and the total number stocked was 3 lakhs. Such fingerlings in 5-6 months' time attained marketable size (catla - 1.5 sr or 1406.25g; rohu - 1 sr or 937.5 g and mrigal - 1.5 sr or 1406.25 g) and were sold. The undersized were put back and the entire sale was completed by February. The bhery was again prepared for stocking or allowed to be fallow until March-April. Nair (1944) further stated that *Mystus gulio, Mugil parsia* and *P. sophore* also formed important by-product of the catch at no extra cost. A mortality of about 6% by argulus infection was common.

Further modification was noticed around fifties. Basu (1950) reported that the fishery (bhery) ponds were as usual shallow and 2′-3′ of water was maintained. Sewage was fed once a month during summer and winter while during monsoon months, the fisheries were drained off to check over flooding. Sewage was allowed to enter against the direction of wind and about 30-40 million gallons of sewage was mixed with 100-200 million gallons of fishery water slowly in the course of 5-10 days thereby reducing the chance of sudden pollution and fish mortality during sewage feeding. Fisheries were drained off for 1 or 2 months for a periodic cleaning of vegetation and drying the bottom. Sometimes, paddy was cultivated as an alternative crop to prevent soil getting sewage sick. Basu (1950) further stated that different concentrations of sewage in a number of experimental ponds, in which dosage and other conditions could be controlled were tried at Hanakhali (600 acres) and Charchania (250 acres).

In the later years around sixty further modification became essential for the pisciculturists to achieve better yields. The need for keeping separate nursery ponds where fry and fingerlings could be stocked, according to their availability, 4 or 5 times and fingerlings transferred to the adjoining larger rearing ponds was keenly felt by fish farmers and was put to practice. This system is still continuing.

Due to insufficient supply of Kistopore canal water, around forties, the dilution was arranged from Kulti estuary through the D.W.F. canal. This system also worked for some time but was abandoned due to silting of canal bed. With the suspension of canal water, rainwater became the only source of dilution during the wet months and the idea of using saline water for controlling parasitic infection completely disappeared.

Raw sewage is now drawn into the nursery pond and allowed to settle. In 2-3 weeks' time transparency in the water gradully appears. A thick growth of plankton is observed. This produces sufficient amount of oxygen which significantly helps in stabilisation of sewage. Spawn is heavily stocked and the stocking density per acre (0.4 ha) is @ 10,000 fry of 5-6.35 cm size of fingerlings 6.35-12.70 (Saha, 1970). An account of size and stocking density of spawn, fry and fingerling with growth attained in different sizes of bheries and followed by local pisciculturists is given in Table 10.

Table 10. Stocking Pattern of Fry and Fingerlings and their Growth

Size of bheri (bigha)	*Spawn size of fry stocked (no./kg.)*	*Stocking density*	*Size attained No./kg*	*No. of days*	*Mortality %*
2-5	2 cups (1-1.2 lakh)	1-1.2 lakh	10,000-40,000	3-5	25-30
5-10/12	10,000-40,000	—	700-1000	12-15	—
13-25	700-100	—	130-150	50-60	—
30-50	130-150	1000/bigha	30-40	120	—

(*Courtesy* : Secretary, Fish Producers Association)

In case 30-40 nos/kg of fry is reared in a 30-50 bigha (4-6.7 ha) bhery from end of November to 15th of April the size attained is 7-10 nos/kg. (100-143 g/kg). Thus in about 4 months time they grow to a size 100-143 g/kg and they are then transferred to the rearing pond. Sewage is allowed round the clock, dusk to dawn or occasionally depending upon its availability. Besides, supplementary feeding is also done by hanging gunny bag full of rice bran under water from the poles fixed at several points in a bhery. Irrespective of season, the water level of about 1 metre in a rearing bhery is maintained. Depth of bottom ooze normally varies between 6" and 10" (15-25 cm). In order to protect the bundhs from erosion due to wind action and also as a protective measure against any theft by cast netting operation, about 8-10 ft wide matting of water hyacinth in the marginal areas (properly fenced) is allowed to grow on all the sides of the bhery. Once in 3 or 4 years large size bheries (above 100 bigha or 13.3 ha) and every year in case of smaller bheries (< 100 bigha or 13.3 ha) are dewatered dried for 1-3 weeks and limed by calcium oxide, 2 mds/bigha (562.5 kg/ha). In cases where drying is not done, liming of the pond by broadcasting the powder directly in the water @ 40 kg/bigha (13.3 ha) is done with the help of a boat. In the past paddy was also cultivated in the same bheri. But this practice is now completely abandoned as pisciculture has proved to be more paying.

5.1 Hydrobiological Investigations

There is hardly any informations on the systematic study on qualitative aspects of physicochemical parameters, plankton and other biological features of the Salt lake ecosystem. Investigation carried out from time to time since 19th century until forties the time of inception of regular pisciculture, only on specific aspects like geographical conditions and hydraulics, are discussed by Banerjee (1931); flora and fauna of the Salt lake by Biswas (1927) and Sewell (1934). During the transitional period Sen (1941) prepared an exhaustive list of fish fauna and common crustaceans occuring during 1937-40. It is only from fifty onwards that an attempt has been made to study the abiotic and biotic parameters. Basu (1950), probably for the first time, made some efforts to study the ecology and bionomics of Indian major carps in bheries. This was followed by Saha *et al.* (1958) whose study was restricted to nature of dissipation of waste water in the bhery identifying the extended zone of pollution from sewage inlet point. Datta *et al.* (1983) added some information on abiotic factors and plankton community in two ponds and also on primary productivity by Datta and Chaudhury (1984). Ghosh (1983) attempted to reveal valuable information on concentration of metals in water (both at inlet and outlet points). It is only more recently the author has made a detailed investigation for a period of two years (1982-84) in randomly selected large and small 5 bheries in the North Salt lake area where fish culture is in vogue. Although the bheries are existing in the same basin due to irregular feeding of sewage it is difficult to estimate the total organic and inorganic load that are being introduced into the bhery. The results of investigation on qualitative and quantitative aspects of physicochemical characters of soil, water, sewage, plankton biomass, benthos, bioaccumulation of metals, bacteriological load and bioassay tests etc. for the bheries are discussed below. Hydrological parameters of water encountered are shown in Table 11A and for Soil Characteristics in Table 11B.

5.1.1 Physicochemical Characteristics

A. Nursery Pond

The water of the nursery pond after drawing the affluent and allowing sufficient time to attain stability indicate that the water become very rich in micro-organisms, considered as important food for initially stocked fry or spawn. The general abiotic conditions of water observed during August and September were in the ranges as follows.

Temperature 33 and 34.5°C, pH 8.7-9.3, dissolved oxygen 9.24 and 25.6 ppm with its corresponding saturation between 129.2 and 355%. All these are indicative of high phytoplankton abundance. This can control the unhealthy ecological conditions arising from high rate of biodegradation, by supplementing the oxidisable organic material in water and soil medium with enough of oxygen that becomes available through photosynthesis. But sometimes such supersaturation of oxygen may cause mortality of delicate fry. BOD_5 is thus kept at a lower level (3.2-21.7 ppm). Alkalinity, chloride and sp. conductivity ranges were 100-136; 32-100 ppm and 799-889 millimhos respectively. The nutrient status is also favourable and maintains an optimum level (Nitrate : 0.16-7.19 ppm and phosphate; 0-2-1.0 ppm).

Table 11A. Hydrological Conditions of water encountered during January-December 1983

Parameters	Bheri-A	Bheri-B	Bheri-C	Bheri-D	Bheri-E
Water temperature (°C)	22.5-32°C	20-30°C	21-30°C	22-31°C	22-33°C
pH	8.1-8.7	8.1-9.3	8.3-9.1	8.3-9.5	8.7-8.8
DO (ppm)	7.2-9.9	9.2-15	6.4-12.96	7.36-15.26	7.9-10.8
BOD_5 (ppm)	13.12-19.1	9.1-21	12-19	20-51.8	7.5-31
Alkalinity (ppm)	140-272	180-212	150-268	170-220	108-220
Hardness (ppm)	100-216	160-192	90-212	110-260	120-204
Chlorinity (ppm)	80-130	90-130	100-150	90-120	90-140
Sp. Conductivity (micro mohs/Cm)	576-1028	659-899	719-1028	719-1028	599-1028
Nitrate (ppm)	0.3-0.4	0.3-0.52	0.5-0.8	0.67-0.81	0.23-0.54
Phosphate (ppm)	0.1-0.6	0.5-1.4	0.6-1.8	0.8-1.3	0.6-1.8
Primary productivity (mg C/M³/hr)	143.8-450	250-350	250-781	150-912	250-472.5

Table 11B: Soil Characteristics (in ranges) of five bheries during Januray-February 1983

Parameters	*Bheri-A*	*Bheri-B*	*Bheri-C*	*Bheri-D*	*Bheri-E*
Soil temperature °C	21.31°C	18-34°C	19.5-31°C	19.5-31°C	23-32.5°C
pH	7.8-8.1	7.5-7.6	7.7-8.0	7.7-8.0	7.7-7.9
IOD (MgO_2/100 gm soil/5 mints)	10.6-11.2	10.6-21.9	1.2-9.3	2.6-9.9	10.2-14.1
Sp. conductivity (micro mohs/cm)	276-449	553-719	514-553	378-659	379-1945
T. nitrogen (%)	0.06-0.12	0.04-0.2	0.11-0.15	0.02-0.3	0.06-0.45
Organic carbon (%)	0.63-1.71	1.81-3.1	21.30-2.61	1.6-3.78	1.57-2.55
Av. nitrogen (mg/100 gm soil)	4.36-8.36	8.7-10.5	7.1-8.56	6.12-10.6	7.9-11.7
Av. phosphorus (mg/100 gm soil)	4.4-8.2	5.4-8.1	2.28-6.3	8.1-9.9	4.4-8.1
Free $CaCO_3$ (%)	19.5-34.2	19-35	20-37.2	16.5-33.7	19.3-35

B. Rearing Pond

Water temperature : Since the depth of water in a bhery is always maintained at a low level (about 1 metre) and is constantly exposed to sun, the water and soil temperature remain at a high level all through the year with a little fall during winter. Saha *et al.* (1958) recorded water temperature ranging between 24.0 and 32.3°C, the upper level being 28.0-32.3°C in eight months with the lower level varying as 24.0-26.5°C during winter (November-February). Dutta and Chaudhury (1982) also recorded a similar pattern of fluctuation, the upper level being 28.4-35.5°C followed by 21.6-27.8°C. The present author in his extensive, ranging between 20.0 and 22.5°C during winter while during summer the range was 30.0-33.0°C Soil temperature ranges were 18.0-23.0°C (winter) and 31.0-32.5°C (summer). Such high temperature is evidently ideal for high biodegradation of organics present both in water and soil media.

Turbidity, colour and odour : The bhery in general maintains almost clear water throughout the season (turbidity around 100) except during monsoon when turbidity increases slightly. A slight brown colour with organic smell always persists. The mechanical effect of wind action cannot cause appreciable turbidity by disturbing the neritic waters due to the presence of water hyacinth carpetted along the littoral zones. Further, turbulance in the water of such shallow depth also promotes low turbidity.

pH : In randomly selected five bheries of different sizes the author has observed the pH range of 8.1-9.3. The diurnal variation of pH studied indicates that in bheri A (1312 ha) pH ranged between 7.9 and 8.7, 7.2 and 8.8 and 7.5 and 8.1 respectively during summer, monsoon and winter. In the bhery B (337.5 ha) the range was 7.7 and 9.3, 8.2 and 9.1 and 7.9 and 8.4 during the said seasons. In the bhery C (2137 ha) the values were 7.7 to 9.9 during summer and 7.6 to 7.7 during monsoon, while in bhery D (957 ha) the ranges were 7.6 to 9.1 and 7.5 to 9.5 during the three seasons. In bhery E (900 ha) the variation during summer was 7.9-9.6. For other seasons data were not available.

Dissolved Oxygen : The lower level of dissolved oxygen concentration estimated in the five bheries ranged between 6.4 and 9.2 ppm rising as high as 9.2-15.26 ppm. The saturation of dissolved oxygen concentration calculated from the diurnal variation in each bhery (cf Tables 12 A-E and Fig. 4 A-E) indicated 5.1-171.2% during summer, 30-193% during monsoon and 18.8-164% during winter in bhery A; 13.3-217.4%, 46-180% and 33.3-167.4% in bhery B; 10.5-218.3% and 6.3-105% during summer and monsoon in bhery C; In bhery D the ranges were 7.7-188.7% and 74.3-142% in the above seasons and finally in bhery E such values were 15.4-285% during summer. Datta *et al.* (1983) while studying the diurnal variation of pH and dissolved summer. Datta *et al.* (1983) while studying the diurnal variation of pH and dissolved oxygen in another bhery (20°30′ N and 88°19′ E) observed a very poor saturation 39.4-73.6% when pH varied between 7.7 and 7.8. For a healthy growth of fish a desirable level of dissolved oxygen saturation between 125-130% in necessary, although coarse fish can live in water with dissolved oxygen content as low as 30% saturation or even low. Figs. 4, A-E show that out of 24 hours, saturation of dissolved oxygen remained below the optimum level for a longer period. Observations made by Datta *et al.* (1983) showed that the condition with regard to oxygen saturation in the pond they studied was still worse. This shows that the ecological conditions of bheries studied for dissolved oxygen saturation was not favouring the healthy growth of fish although they may not show any sign of distress for their survival.

From the analysis of variance in dissolved oxygen saturation in bheries during different seasons it is observed that there are no significant differences at 5% and 1% except the difference between the periods. Interaction between seasons and periods is however, significant at 5% and 1% levels which reveals that there is a relation between season and period. To find out the maximum dissolved oxygen saturation period, mean difference between maximum periods has been tested by 't' - test which shows that there is no difference between the two maxima of dissolved oxygen saturation periods (12.00 and 15.00 hrs) at 5% and 1% levels. It is hence concluded that maximum dissolved oxygen saturation period lies between 12.00 and 15.00 hours.

Biochemical Oxygen Demand : (BCD_5) : In all the five bheries BOD_5 never showed high values. It ranged between 13.12-19.1 and A, 9.1-21 and in B, 12-19 in C, 20-51.8 an D and 7.5-31 ppm in bheri E against 60 and 80 ppm in the raw sewage. Ghosh (1983) observed a range 25-65 ppm.

Alkalinity : Round the year study indicated alkalinity to range between 140-272 ppm in A, 180-212 in B, 150-268 ppm. in C; 170-220 ppm in D and 180-220 ppm in bhery E. The upper level, however, was estimated during summer. Alkalinity of the raw waste, however, ranged between 136 and 316 ppm.

Hardness : Hardness indicated slightly lower values than alkalinity. The ranges were 100-216 ppm in bhery A; 160-192 ppm in bhery B; 90-212 ppm in bhery C; 100-260 ppm in bhery D and lastly 120-204 ppm in bhery E.

Chlorinity : Chlorinity indicated still low concentration than alkalinity and hardness. The ranges were 80-130 ppm, 90-130 ppm, 100-150 ppm, 90-120 ppm and 90-140 ppm respectively in bheries A, B, C, D and in E against 120 and 130 ppm in the raw sewage.

Specific Conductivity : Concentration of soluble salts measured as specific conductivity in terms of micromhos/cm at 25°C indicated fairly values in all the bheries and their waters as productive. The values were 576-1028 in A; 659-899 in B; 719-1028 in C; 719-1028 in D and 599-1028 in Bhery E. In the raw sewage the concentration ranged between 719 and 1028 micromhos.

Nutrient Status: Domestic sewage, apart from being organically rich also receives effluents from a number of industries and contains appreciable concentration of nitrate - N (9.0-17.2 ppm) and phosphorus (1.8-4.0 ppm). Such wastes when allowed in the bhery enrich the nutrient status. Nitrate content estimated individually ranged 0.3-0.4 ppm in bhery A; 0.3-0.52 ppm in B; 0.5-0.8 ppm in C; 0.67-0.81 ppm in D and 0.23-0.51 ppm in bhery E. The corresponding phosphate contents were 0.1-0.6 ppm, 0.5-1.4 ppm, 0.6-1.8 ppm, 0.8-1.3 ppm and 0.6-1.8 ppm respectively in bheries A, B, C, D and E. Concentraion of such high values, present round the year in bheries provide an ideal environment for the growth of aquatic biomass.

5.1.2 Soil Characteristics

Soil temperature registered almost similar trends like water with appreciable seasonal changes. The lower temperature level was recorded during winter and higher mostly during summer. Since the bheries under investigation are located in the same periphery, the temperature variations between them were very little. However, individually the variations recorded were 21.0-31.0°C in bhery A; 18.5-34.0°C in B; 19.5-31.0°C in C; 19.5-31.0°C in D and 23.0-32.5°C in E. Other important parameters studied were pH, immediate Oxygen demand (IOD), sp. Conductivity, total nitrogen, organic carbon, available nitrogen and phosphorus and free calcium carbonate (Table 11.B)

Vatriation in pH observed round the year was 7.8-8.1 in bhery A; 7.5-7.6 in bhery B; 7.7-8.0 in C; 7.7-8.0 in D and 7.7-7.98 in bhery E. Likewise, immediate oxygen demand (IOD-mg 0°2°/100g soil/5 mts) by soil sediments ranged from 10.6-11.2 in bhery A; 10.6-21.9 in B; 1.2-9.3 in C; 2.6-9.9 in D and 10.2-14.1 in bhery E. The specific conductivity value ranged 276-449, 553-719, 514-553, 378-659 and 379-1945 micromhos in bheries A, B, C, D and E. respectively. The lower level was observed in winter while the upper value during summer. Organic carbon and nitrogen respectively ranged between 1.2-1.71% and 0.1-0.12%; 1.81-3.12% and 0.04-0.2%; 1.30-2.61% and 0.11-0.15%; 1.6-3.78 and 0.02-0.3, and 1.57-2.55 and 0.048-0.08% in the five bheries in alphabetical order. Similar observations on available nitrogen and phosphorus gave ranges respectively as follows, 4.36-8.36 and 4.4-8.2; 8.7-10.5 and 5.4-8.1; 7.1-8.56 and 2.28-6.3; 6.12-10.2 and 8.1-9.9 and lastly 7.9-11.7 and 4.4-8.1 mg/ 100 g soil. Free calcium carbonate was present in moderately high concentration viz., 19.5-34.2%; 19-35%; 20-33.2%, 16.5-33.7% and 19.3-35% in bheries A, B, C, D and E respectively.

5.1.3 Metal Contamination

Since the metropolitan wastes contain substantial percentage of effluents from various industrial complexes, contamination from metals such as zinc: 1.5-2.5 ppm; copper: 0.3-0.5 ppm, manganese: 13-60 ppm and iron 65-150 ppm were estimated. On introduction of such wastes into the bheries the metals partly get adsorbed or chilated with organic suspensoids and settle or get chemically bound and precipitated at appropriate pH. Since most of the bheries receive the effluent for a good length of time (almost continuously) presence of such metals in small concentration in the bhery water is not unlikely. Analysis of metals is water and soil (Table 13) of the bheries indicate that in bhery A concentration of copper ranged trace-1.5 ppm; zinc 0.07-0.12 ppm; iron trace-0.1 ppm and manganese 0.1-0.8 ppm. In the soil sediments such parameters were 15-18 ppm; 29.4-56 ppm; 90-350 ppm and 350-435 ppm in

Figure 4 (A)

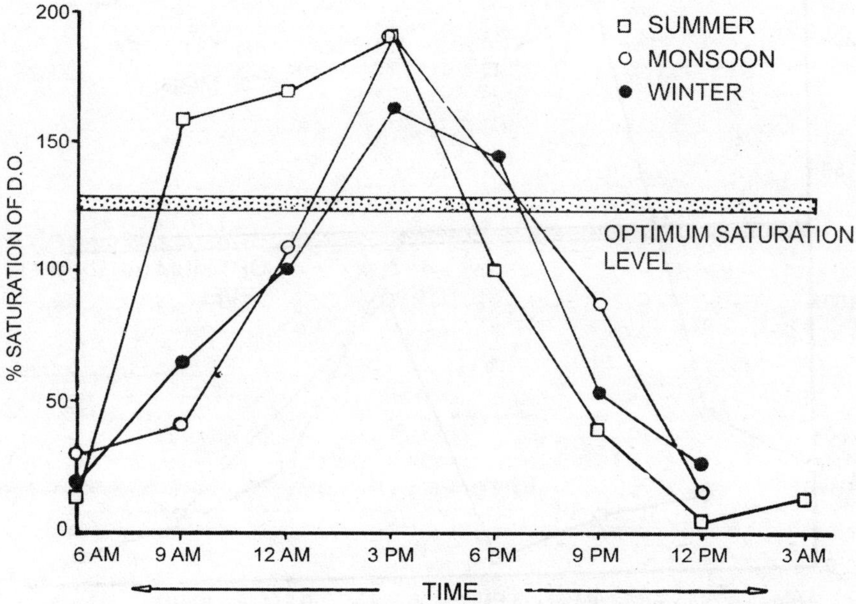

Figure 4 (B)

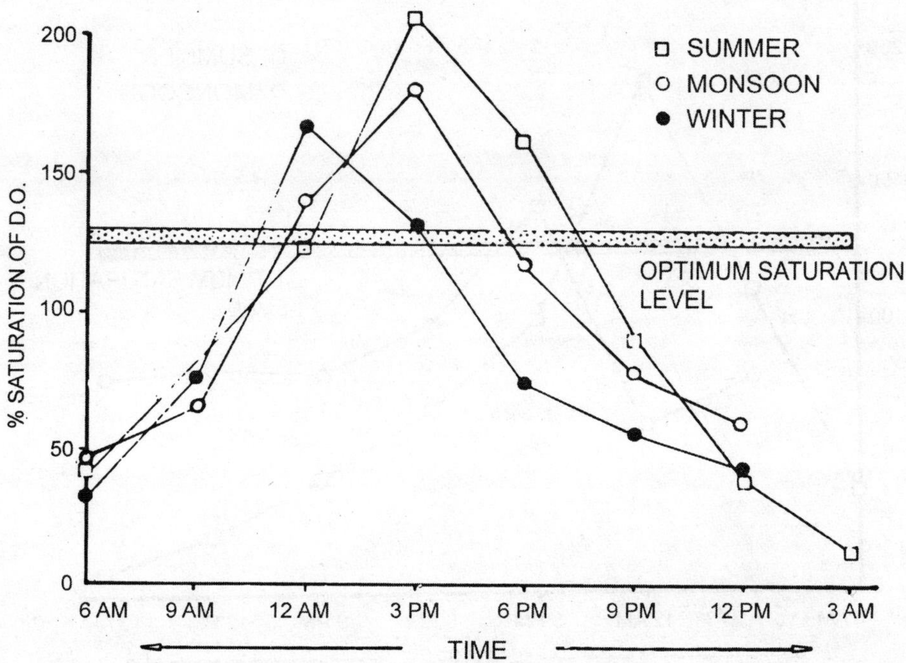

Figure 4 (C)

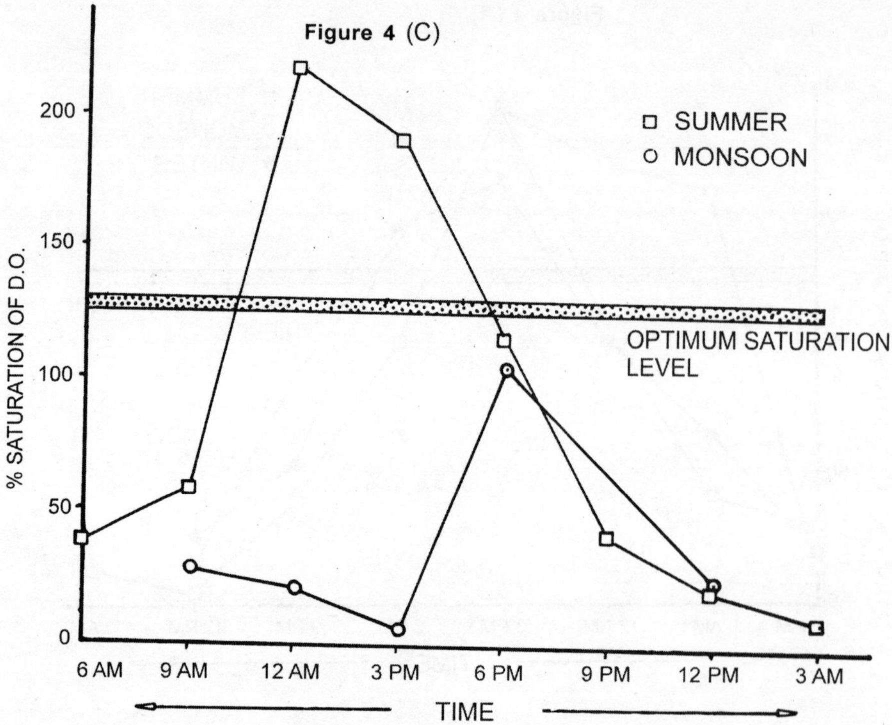

Figure 4 (D)

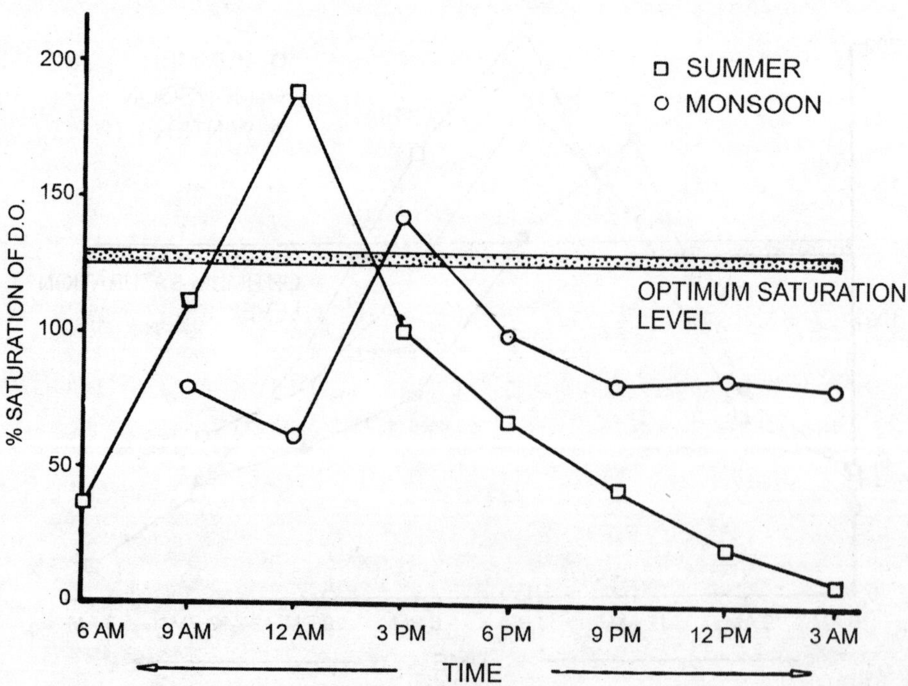

Figure 4 (E)

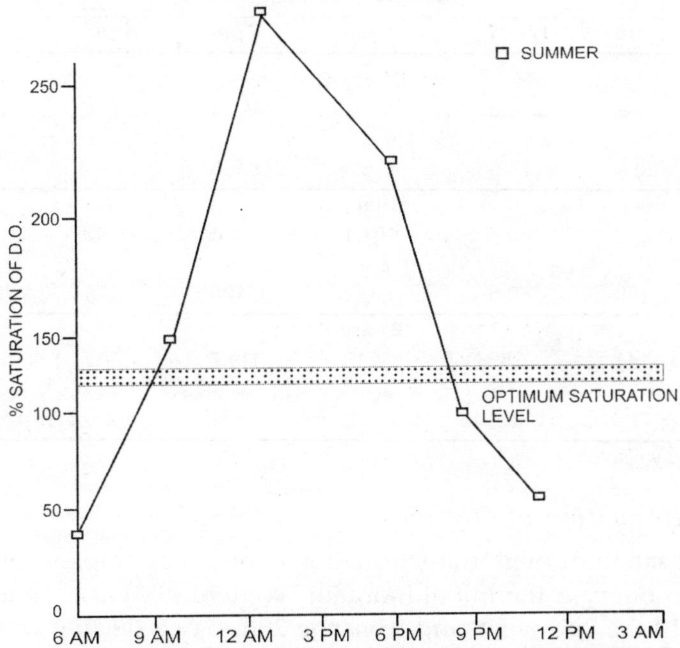

Fig. 4E Diurnal dissolved oxygen saturation in bheries.

the above order. In bhery B, the concentrations were Trace-0.1 ppm; 0.07-0.2 ppm; 0.05-0.1 pppm and trace-0.25 ppm in water, and in soil phase they were 53-100 ppm; 119-218.4 ppm; 215-410 ppm and 220-480 ppm respectively. Likewise in bhery C, the concentrations ranged Trace-0.1; 0.07-0.08 ppm; Trace-0.15 ppm and Trace-0.55 ppm in water, when their concentrations in soil phase were 1-4 ppm; 10.50-43.3 ppm; 15-90 ppm and 350-490 ppm respectively. In bhery D, the values were Trace-0.2 ppm; 0.08-0.1 ppm; 0.08-0.1 ppm; 0.05-0.25 ppm and 0.05-0.1 ppm in water and 1-3 ppm; 25.2.32.9 ppm; 20-185 ppm respectively in the soil phase. Lastly in bhery E, the concentrations were trace- 0.1 ppm; 0.06-0.1 ppm; trace- 0.1 ppm and trace 0.2 ppm in water when the corresponding values for the above parameters in the soil sediments were 1-17 ppm; 39.9-53.2 ppm; 60-425 ppm and 435-5500 ppm respectively. It is thus clear from the above that concentrations of metals in the soil sediments are much higher than the concentrations in the overlying water (Table 11B).

Table 12. Diurnal Dissolved Oxygen Saturations (%) in Five Bheries During Different Seasons

Seasons	*6 am*	*9 am*	*12 pm*	*3 pm*	*6 pm*	*9 pm*	*12 pm*	*3 am*
				Bhery-A				
Summer	14	159.7	171.2	192.2	101.4	39.7	5.1	14
Winter	18.8	65.3	102.4	164	144.8	54	26.9	-
Monsoon	30	41	109.5	193	16.2	88	16.2	-
				Bhery-B				
Summer	40.5	-	124.4	217.4	163.3	91.1	39	13.3
Winter	33.3	75.6	167.4	132.4	73.8	56	43	-
Monsoon	46	6	140	180	127	79	5	-

Table 12 – *contd...*

Seasons	*6 am*	*9 am*	*12 pm*	*3 pm*	*6 pm*	*9 pm*	*12 pm*	*3 am*
				Bhery-C				
Summer	37.8	58.9	218.3	192.4	115.5	41.6	20.5	10.5
Winter	-	-	-	-	-	-		
Monsoon	-	28.4	21.5	6.3	105	-	24.6	-
				Bhery-D				
Summer	36.5	110.9	188.7	101.1	67.3	43	20.5	7.7
Winter	-	-	-	-	-	-	-	-
Monsoon	-	80	60.5	142	100	76	77	74.3
				Bhery-E				
Summer	33.2	141	285	213	119.7	77.7	-	15.4
Winter	-	-	-	-	-	-	-	-
Monsoon	-	-	-	-	-	-	-	-

(Data for the blank are not available)

5.1.4 *Extent of Transformation of Organic Complex in Soil*

Extent of mineralisation through transformation of organic complex appears to be greatly affected (Table 14). In bhery A the initial inorganic content (9%) after incubation showed a steady rise (26%) until the 2nd week and from the 20th day a decline was noticed (23.6%). This is also reflected by moderate concentration of available nitrogen (4.36-8.36) and phosphorus value (4.4-08.2 mg/100 g of soil). In bhery B, the extent of mineralisation was slower than A. Mineralisation attained highest value (19%) from 0-day (7.5%) during the second week and degradation was noticed from the 3rd week (15.3%). Mineralisation rate in bhery C showed a steady rise (29%) from 0-day (8.5%) until the 3rd week unlike previous cases. This was followed by slow degradation. In bhery D although initial inorganic nitrogen value was followed by slow degradation. In bhery D although intial inorganic nitrogen value was a little high (10%) but extent of mineralisation was poor, the highest value attained during the 3rd week being 24.3%. Extent of mineralisation was appreciably low in bheri E. With initial nitrogen value (6.6%) the extent of increase from 7th to 20th day was between 19.6 and 22.3% only. This was followed by a fall showing low degradation. It is thus clear that mineralisation and its extent, in general, were slow and attained maximum value between 2nd and 3rd week after which a gradual fall was noticed.

Table 13: Metal Concentration in water and soil during January-December 1983

Bheri	*Heavy metals in ppm (ranges)*	*Water*	*Soil*
	Cu	Trace-1.5	15-18
	Mn	0.1-0.8	350-435
A	Fe	Trace-0.1	90-350
	Zn	0.07-0.2	29.4-56
	Cu	Trace-0.1	53-100
	Mn	Trace-0.25	220-480
B	Fe	0.05-0.1	215-470
	Zn	0.07-0.2	119-218.4

....Contd.

Table 13 – *contd...*

Bheri	*Heavy metals in ppm (ranges)*	*Water*	*Soil*
C	Cu	Trace-0.1	1-4
	Mn	Trace-0.55	345-490
	Fe	Trace-0.15	15-90
	Zn	0.07-0.08	10.5-43.4
D	Cu	Trace-0.2	1-3
	Mn	0.05-0.1	220-295
	Fe	0.05-0.25	20-185
	Zn	0.08-0.1	25.2-32.9
E	Cu	Trace-0.1	1-17
	Mn	Trace-0.2	435-500
	Fe	Trace-0.1	60-425
	Zn	0.06-0.1	39.9-53.2

Trace-0.02/0.03

5.2 *Biological Investigations*

There was hardly any information on the aquatic biomass of the Salt lake area around the 19th century. Biswas (1927) probably for the first time gave a detailed account of vegetation of the Salt lake in embankments and bunds in the lake proper and swamps (cf. Table 15a-c) in his important document '*Flora of the Salt Lake*' is exhaustive study on algae of the area is commendable and has immense practical value for researches. The various forms encountered under the prevailing meteorological conditions are indicated below:

Table 14: Extent of Mineralisation observed as Nitrate Nitrogen

	Number of Days				
	0	7	15	20	35
Bheri-A	8-10	15-21	23-30	18-23	10-15
Bheri-B	6-9	10-16	17-21	14-17	14-15
Bheri-C	7-10	10-16	21-25	27-31	14-17
Bheri-D	8-12	12-15	16-21	21-27	15-20
Bheri-E	6-8	10-18	14-21	20-27	16-19

Monthly mean of maximum and minimum temperature values and rainfall recoded during the period are given in Table 16.

A few years later Basu (1950) recorded *Typha* sp. *Pistia, Vallisneria, Hydrilla, Ceratophyllum, Eichhornia,* different species of *Cyperus* and *Lemna* as natural macro-vegetation while *Panicum* sp. *Alternenthera* and *Sweada* as terrestrial vegetations thereby, indicating that considerable change in vegetation has taken place. Plankton is considered as an important fish food organism. Although not quantitatively estimated Biswas (1927) has preferred an elaborate list of algae of the Salt Lake comprising different forms under Myxophycease, Chlorophyceae and Rhodophyceae groups (Table 17). This was followed by Sewell (1934) and Basu (1950) (Tables 19-21).

Table 15: *Vegetation of the Salt Lake*

(a) Vegetation of the embankments and bunds:

1. *Fimbristylis ferruginea*
2. *Suaeda maritima*
3. *Heliotropium curassavicum*
4. *Tamarix gallica* Var. *indica*
5. *Acanthus ilicifolius*
6. *Excoecaria agallocha*
7. *Avicennia officinalia*
8. *Clerodendron inarme*

(b) *Vegetation of the Salt lake:*

The vegetation of the Salt lake proper is chiefly composed of algae either floating as plankton or forming benthic layer or floating in thick blue-black or bluish green layers of considerable width attached to the haulms of grasses and other partly submerged smaller plants along the magin of lakes. The algae belong to cyanophyceae and most of them are members of the family Oscillatoriaceae.

(c) *Vegetation of swamps which are partly dry and partly flooded in summer is as follows:*

1. *Oscillatoria princeps*
2. *Oscillatoria limosa*
3. *Oscillatoria laetevirens*
4. *Euglena sp.*
5. *Pandorina morum*
6. *Arthrospira platensis*
7. *Oscillatoria sp.*
8. *Diatoms*
9. *Clathrocystis aeruginosa*
10. *Oscillatoria cf. subsalsa*
11. *Enteromorpha intestinalis*
12. *Enteromorpha prolifera*
13. *Chaetomorpha linum*
14. *Protococcus viridis*
15. *Avicennia officinalis*
16. *Excoecaria agallocha*
17. *Sonneratia apetala*
18. *Hellotropium curassavicum*
19. *Excoecaria agallocha*
20. *Avicennia Officinatis*
21. *Aegiceras magus*
22. *Sonneratia apetala*
23. *Sonneratia apetala*
24. *Malachra capitata*
25. *Cleoma viscosa*
26. *Cassia tora*
27. *Cassia Sophera*
28. *Cassia occidentalis*
29. *Acacia arabica*
30. *Euphorbia antiquorum*

Table 16: Meteorological Conditions

Month temp. (°F)	Maximum temp. (°F)	Minimum temp. (°F)	Rainfall (inches)
January	77.3	55.5	0.36
February	82	60	1.10
March	90.9	69.3	1.44
April	95.6	75.7	1.89
May	94.5	77.5	5.75
June	91.5	78.8	11.90
July	88.4	78.6	12.51
August	87.6	78.4	12.69
September	88	78	9.87
October	87.2	74.3	4.19
November	82	63.3	0.66
December	77	56	0.20

Table 17 : Algae of the Salt Lake

I. Myxophyceae

(i) Family: Chlorococcaceae

1. *Gloeocapsa aeruginosa* (carmich Kuetizing)
2. *Clathrocystis* aeruginosa (kuetz) Henfray

(ii) Family: Oscillatoriaceae

3. *Oscillatoria princeps* Vauch
4. *Oscillatoria limosa* (Roth) Ag.
5. *Oscillatoria limosa simplicissima* Gom
6. *Oscillatoria tenuis* Agardh
7. *Oscillatoria amphibia* Agardh
8. *Oscillatoria minnesotensis* Tiden
9. *Oscillatoria laetevirens* (crouan) Gom
10. *Oscillatoria salina* sp. nov.
11. *Oscillatoria formoso* Bory
12. *Oscillatoria cf Subsalsa* Agardh
13. *Spirulina major* Kuetzing
14. *Phormidium tenue* (Meregh) Gom
15. *Lyngbya crispa* Agardh
16. *Lyngbya cinerescens* Kuetz
17. *Hypheothris tenax* Wolle
18. *Symploca hydnoides* Kuetz
19. *Symploca muscorum* (Ag) Gom
20. *Sirocoleus Kurzil* (Zeller) Gom
21. *Microcoleus chthonoplaster* (Fl. Dan) Thuret
22. *Calothrix juliana* (Menegh) Bornet et Flah
23. *Leptothrix mamillosa* Menegh

(iii) Family: Scytonemaceae

24. *Scythonema* granulantum Martens

II. Chlorophyceae

(iv) Family: Volvocaceae

25. *Pandorina mormum* (Muell) Bory

(v) Family: Cladophoraceae

26. *Protococcus Viridis* Agardh

(vi) Family: Cladophoraceae

27. *Chaetomorpha linum* (Muell) Kuetz.
28. *Rhizoclonium hieroglyphicum* (Ag) Kuetz.

(vii) Family: Ulvaceae

29. *Enteromorpha prolifera* (Muek.) J.Ag.
30. *Enteromorpha intestinatia* (L) Link

(viii) Family: Mougeotiaceae

31. *Mougeotia affinis* Kuetz

....Contd.

Table 17 – *contd...*

III. Phaeophyceae

(ix) Family: Encoeliaceae

32. *Encoelium vesicatum* Kuetz

IV. Rbodophyceae

(x) Family: Compsopogonaceae

33. *Compsopogon lividus* (Hooker) De Toni.

(xi) Family: Rhodophyllidaceae

34. *Catenella opuntia* (Good. et Woodw.) Grev.

(xii) Family: Delesseriaceae

35. *Caloglossa Leprieuril.* J.Ag.
36. *Hypoglossum phgme aum* Mat.

(xiii) Family: Rhodemelaceae

37. *Polysiphon brodiaeae* (Dillow.) Grev
38. *Polysiphon augustissima* Kuetz.
39. *Bostrychia rivularis* Harv.

(xiv) Family: Ceramiaceae

40. *Ceramlium gracillimum* Griff et Harv.

The most interesting aspect Sewell (1934) has brought into light is the gradual changes in the zooplankton composition during the years, 1914, 1930 and 1932 which are indicated below in Table 18.

Table 18. Gradual changes in the Faunatic composition encountered during 1914, 1930 and 1932 (Sewell, 1934)

Species	Station 19 Oct. 1914	Chingrighatta 20 Oct. 1930	Dhappa Canal 29 Oct. 1932
Diaptomus controtus	+	+	+
Diaptomus cinctus	-	-	+
Pseudodiaptomus annandalet	+	+	-
Pseudodiaptomus binghami	+	+	
Acartia chilkensis	+	-	-
Acartia plumosus	+	+	-
Acartia (Acartiella) tortaniformis	+	-	-
Oithona brevicornis	+	-	-
Cyclopina longifurca	+	-	-
Halicyclops aequoreus	+	-	-
Halicyclops tenuispina	+	-	-
Mesocyclops (Mesocyclops) leuckarti	+	+	+
Mesocyclops (Thermocyclops) rylovi	+	-	-
Mesochra meridionalis	+	-	-
Laophonta bengalensis	+	-	-
Stenhelia longifurca	+	-	-

From the table it is evident that in 1914 out of total of 16 species of Copepod 14 definitely belonged to the group of brackish water forms and 2 to fresh water species, these latter being *Diaptomus contortus* and *Mesocyclops* (*Mesocyclops*) leuckerti. By 1930, 11 of the brackishwater species had disappeared from the locality and only 3 of the more adaptable forma namely *Psuedodiaptomus annandelei, P. binghami* and *Acartia Chilkaensis* remained. Finally in 1932 these three brackishwater forms also disappeared and an additional freshwater species of Copepod, namely *Diaptomus cinctus* made its appearance as well as enormous numbers of the Rotifer *Brachionus pala* appeared. Water was brackish in 1922, sp.Gr. was as high as 1006-1008.5 where as at the present time it is as low as 1000.9 (Salinity: 1.19).

Sewell (1934) has made an exhaustive qualitative study of fauna under the broad groups, Copepods and Cladocera in different canal system of the Salt Lake indicating the salinity in his valuable articles, "A study of the fauna of the salt lakes, Calcutta" Results obtained by him in different areas are also presented. In the same article Swell (1934) has also given such a list of species in the animal group indicated by other workers. (Table 19)

Table 19. Plankton Composition of Salt Lake (Sewell, 1934)

1. Tolly's Nullah; near Magrahat Railway Station 7 iii 33, Salinity 0-17 per mile.

Copepoda

Acrocalanus inermis Sewell.

Pseudodiaptomus binghami Sewell.

Acarlia (*Acarliella*) *tortaniformis* Sewell.

Mesocyclops (*Mesocyclops*) *leuckarti* (Claus).

Mesocyclops (*Thermocyclops*) *rylovi* Smirnov.

2. Tolly's Nullah; below Gariya Bridge 2. ii. 33. Salinity 1-87 per mile.

Copepoda

Diaptomus cinctus Gurney.

Pseudodiaptomus binghami Sewell.

Acartia Chilkaensis Sewell.

Acartia (*Acartiella*) *tortaniformis* Sewell.

Mesocyclops (*Mesocyclops*) *leuckarti* (Claus).

Cladocera

Diaphanosoma excisum var. *longiremis* Ekman.

In addition a few examples of a species of *Mysis. Mesopodopsis orientalis* Tattersall and the larvae of a species of mayfly, *Cloeon*, were present.

Among the Rotifera that were present were the following:-

Brachionus bakeri Ehrb.

Brachionus bakeri var. *brevispinus* Ehrb.

Brachionus rubens Ehrb.

3. Tolly's Nullah. Gangajoara. 2. ii. 33. Salinity 4-07 per mile.

Copepoda

Pseudodiaptomus binghami Sewell.

Acartia chilkaensis Sewell.

Acartia (*Acartiella*) *tortaniformis* Sewell.

....Contd.

Table 19 – *contd...*

Mesocyclops (*Mesocyclops*) *leuckarti* (Claus).

In addition examples of the Mysid, *Mesopodopsis orientalis* Tattersali were present.

Salt Lake Canal System.

4. Chingrighatta, in Canal. Surface tow-netting 19. x. 14.

Copepoda

Pseudodiaptomus annadalei Sewell.

Pseudodiaptomus binghami Sewell.

Pseudodiaptomus tollingeri Sewell.

Diaptomus contortus Gurney.

Acartia chilkaensis Sewell.

Acartia plumosa T. Scott.

Acartia spinicauda Giesbrecht.

Acartia (*Acartiella*) *tortaniformis* Sewell.

Oithona brevicornis Giesbrecht.

Cyclopina longifurca Sewell.

Halicyclops aequoreus (*Fisher*).

Halicyclop tenuispina Sewell.

Mesocyclops (*Mesocyclops*) *leuckarli* (Claus).

Mesocyclops (*Thermocyclops*) *rylovi* Smirnov.

Stenhelia longifurca, sp. nov.

Laophonte bengalensis, sp. nov.

Mesochra meridionalis Sars.

Cladocera

Daphnia lumholtzi Sars.

Diaphanosoma exeisum var. *longiremis* Ekman.

Diaphanosoma modiglianii Richard.

Moina dubia Richard.

Ceriodaphnia cornuta Sars.

Ceriodaphnia rigaudi Richard.

Alona quadrangularis Müller.

Chydorus sphaericus Müler.

In addtion the Mysid, *Mesopodopsis orientalis* Tattersall, was present in large numbers.

Numerous examples of the Rotifer *Brachionus urceolaris* Ehrb. were present.

The catch shown clearly that the reproductive season was in full swing; Nauplii larvae of several kinds and immature Copepodid stages of several species were present in large numbers and in addition females of *Pseudodiaptomus* and *Cyclopina* were carrying egg saes.

5. Salt Lake Canal System. Dhappa Lock; Inside Lock Gates. 20. x. 30.

Copepoda

Diaptomus contortus Gurney.

Pseudodiaptomus annandalei Sewell.

Pseudodiaptomus binghami Sewell.

Mesocyclops (*Mesocyclops*) *leuckarti* (Claus).

....Contd.

Table 19 – *contd...*

Cladocera

Diaphanosoma excisum var. *longiremis* Ekman.

Diaphanosoma modiglianii Richard.

Moina dubia Richard, in enormous numbers.

6. Salt Lake Canal System. Between Chingrighatta and Dhappa Lock. Surface tow-netting; 29. xii. 32. Salinity 1-19 per mille

Copepoda

Diaptomus cinctus Gurney.

Diaptomus contortus Gurney.

Mesocyclops (*Mesocyclops*) *leuckarti* (Claus).

In addition a species of Rotifer, *Brachionus pala* Ehrb., was present in enormous numbers.

7. Salt Lake Canal System. Kristopur Canal at Mile I. 28. xii. 32. Salinity 2.20 per mille.

Copepoda

Diaptomus cinctus Gurney.

Pseudodiaptomus annandalei Sewell.

Acartia chilkaensis Sewell.

Mesocyclops (*Thermocyclops*) *rylovi* Smirnov.

Cladocera

Moina dubia Richard, in large numbers.

In addition the following Rotifera were present:

Brachionus pala Ehrb.

Brachionus bakeri var. *latissimus* Schm.

Both species were present in large numbers. By March 20th, 1933, the Salinity had risen to 3-60 per mile.

8. Salt Lake Canal System. Kristopur Canal at Mile 4. 28. xii. 32. Salinity 2-41 per mille.

Copepoda

Diaptomus cinctus Gurney.

Pseudodiaptomus annandalei Sewell.

Acartia chilkaensis Sewell.

Acartia (*Acartiella*) *tortaniformis* Sewell.

Mesocyclops (*Thermocyclops*) *rylovi* Smirnov.

Cladocera

Moina dubia Richard.

In addition the same two species of Rotifera as were present at Station 15, namely *Brachionus pala* Ehrb. and *Brachionus bakeri* var. *lalissimus* Schm, were again present but in fewer numbers.

9. Salt Lake Canal System. Eastern Canal, just beyond junction with Kristopur Canal. 28. xii. 32. Salinity 2-03 per mille.

Copepoda

Pseudodiaptomus annandalei Sewell

Pseudodiaptomus tollingeri Sewell.

Acartia chilkaensis Sewell.

Acartia (*Acartiella*) *tortaniformis* Sewell.

Mesocyclops (*Thermocyclops*) *rylovi* Smirnov.

....*Contd.*

Table 19 – *contd...*

Cladocera

Moina dubia Richard.

10. Salt Lakes: Z.S. 1 Station 1. Large pond, connected with the canal of Landowne Jute Mill. 16-17 ii. 28. Salinity 12-74 per mille.

Pseudodiaptomus annandalei Sewell.

Mesocyclops (*Thermocyclops*) *rylovi* Smirnov.

11. Salt Lakes: Z.S. 1. Station 2. Near edge opposite Dakhindari. 16-17. ii. 28. Salinity (on 23. ii. 28.) 9-60 per mile.

Pseudodiaptomus annandalei Sewell.

12. Salt Lakes: Z.S. 1. Station 3. Northern end. 23-24 ii. 1928. Salinity 23-66 per mille.

Pseudodiaptomus annandalei Sewell.

Mesocyclops (*Thermocyclops*) *rylovi* Smirnov.

13. Salt Lakes; Z.S. 1 Station 3. Nothern end. 20. iii. 33. Salinity 1-474 per mille.

Copepoda

Mesocyclops (*Mesocyclops*) *leuckarli* (Claus).

Mesocyclops (*Thermocyclops*) *rylovi* Smirnov.

Cladocera

Moina dubia Richard.

Diaphanosoma excisum var. *longiremis* Ekman.

In addition a large number of examples of a Rotifer, *Asplanchna brightwelli* Hudson and Gosse, were present as well as a few examples of *Brachionus bakeri* Ehrb.

14. Salt Lakes; Southern end. Naoabad. 2. ii. 33. Salinity 5-54 per mille.

Copepoda

Pseudodiaptomus annandalei Sewell.

Pseudodiaptomus binghami Sewell.

Acartia chilkaensis Sewell.

In addition examples of the Mysid, *Mesopodopsis orientalis* Tattersall, were present.

15. Pool on Road-side near Milestone 4 on the way from Baruipur to Uttarbhag. 8. ii. 33. Salinity 9-24 per mille. A fortnight later the Salinity of the water had risen to 10-05.

Copepoda

Pseudodiaptomus tollingeri Sewell.

Acartia chilkaensis Sewell.

Cyclopina longifurca Sewell.

Halicyclops tenuispina Sewell.

Stenhelia longifurca, sp. nov.

Ectinosoma melaniceps Boeck.

Limnocletodes secundus, sp. nov.

Tachidius brevicornis.

16. Pools in a drying up "Paddy" field at Uttarbhag; 2. i. 33. Salinity 4-38 per mille.

Copepoda

Pseudodiaptomus binghami Sewell.

Acartia chilkaensis Sewell.

....Contd.

Table 19 – *contd...*

Acartia (*Acartiella*) *tortaniformis* Sewell.

Oithona horai, sp. nov.

Cyclopina longifurca Sewell.

Halicyclops tenuispina Sewell.

Mesocyclops (*Mesocyclops*) *leuckarti* (Claus).

17. Large connected pol at Uttarbhag. 2. 1. 33. Salinity 4-16 per mille.

Copepoda

Pseudodiaptomus tollingeri Sewell.

Acartia chilkaensis Sewell.

Acartia (*Acartiella*) *tortaniformis* Sewell.

Oithona horai, sp. nov.

Cyclopina longifurca Sewell.

Halicyclops aequorous (Fischer).

Halicyclops tenuispina Sewell.

Halicyclops propinquus Sars.

Mesocyclops (*Mesocyclops*) *leuckarti* (Claus).

Mesocyclops (*Thermocyclops*) rylovi Smirnov.

Ldophonte bengalensis, sp. nov.

Stenhelia longifurca, sp. nov.

Parategastes sphaericus var. *similis* Sewell.

18. Piali River at Uttarbhag. 13. iii. 33. 1st phase of High tide, after a "bore". Salinity 18-08 per mille.

Copepoda

Paracalanus dubia Sewell.

Acrocalanus inermis Sewell.

Pseudodiaptomus tollingeri Sewell.

Labidocera gangetica, sp. nov. Juv.

Acartia chilkaensis Sewell.

Oithona brevicornis Giesbrechit.

Cyclopina longifurca Sewell.

Laophonte bengalensis, sp. nov.

Stenhelia longifurca, sp. nov.

Cletocamptus conflueus (Schmeil).

Several species of fresh water sponges appear to have been able to accustom to life in brackish-water and Annandale (1911) mentions the following as having found the brackish-water pools at Port Canning and in the Salt Lakes:-

Spongilla lacustris subsp. *reticulata.*

Spongilla alba var. *bengalensis.*

Spongilla crateriformis.

In addition *Spongilla carteri* is at the present day common in those parts of the Salt Lakes that lie to the north and east of the Kristopur canal. Annandale (1922) has also recorded the presence of "the cosmopolitan parasite *Cliona vastifica,* which is not uncommon in oyster-shells in the lower part of the Delta".

Among the Coelenterata Annandale (1915, p. 69; 1922, p. 147) records the occurrence of the following forms:-

Dicyclocoryne filamentata.

....Contd.

Table 19 – *contd...*

Annulella gemmata.

Asenathia piscatoris.

Bimeria flumnialis.

Diadumene schillerianum.

Pelocoetes exul.

Phytocoetes gangeticus.

Campanulina ceylonensis.

Acromitus rabanchatu.

Virgularia sp.

The last species has since been identified by Abdul Hamid (1931) as *V. gracillima.* In the great majority of species the evidence indicates that they are able to withstand changes of specific gravity that have a range from 1000 to 1025-75; *Pelocoetes exul* has been taken in water having a specific gravity ranging from 1005 to 1010 and *Dicyclocoryne filamentata* in water having a specific gravity of 1015.

In the Polyzoa Annandale (1915) has recorded the occurrence of seven species from brackish-water in the Gangetic Delta, namely:

Membranipora bengalensis.

Membranipora hippopus.

Victorella bengalensis.

Bowerbankia caudata.

Loxosomatoides colonialis.

Barentsia gracilis.

Alcyonidium mytili.

This last species was taken attacked to a Sea-snake (*Enhydrina valakadien*) at Diamond Harbour on the Hooghly river.

The Annelida have been reported on by Southern (1921) and Fauvel (1932), who have dealt with the Polychaeta; and Harding (1920) and Kaburaki (1921) have described the Leeches. In the former group, seventeen species are now known to inhabit the delta, namely:-

Matla bengalensis.

Spio bengalensis.

Eteone barantollde.

Lycastis indica.

Nereis glandicincta.

Nereis chingrighattensis.

Nereis cricognatha.

Nereis cavifrons.

Dendronereis aestuarina.

Dendronereis heteropoda.

Polydora (*Carazzia*) *kempi.*

Nephthys oligobranchia.

Barantolla sculpta.

Sabellaria spinulosa var. *alcocki.*

Sabellaria pectinala.

Mastobranchus indicus.

Potamilla leptochaeta.

....Contd.

Table 19 – *contd...*

In the Oligochaeta the only reference that I can find to any aquatic forms is to *Branchiura sowerbyi*, that was taken together with *Lycastis indica* in the Belliaghatta Canal near Calcutta.

A single Echiuroid, namely, *Thalassema branchiorhynchus*[1] has been recorded by Annandale and Kemp (1915) from the Canal System of the Salt Lakes. Voa Linstow (1907) has described a Nematode. *Oncholaimus indicus,* from brackish-water in Port Canning.

In the Entomostraca Gurney (1906, 1907) has recorded the presence of the following species:

Copepoda

Mesocyclops (*Mesocyclops*) *leuckarti.*

Cladocera.

Ceriodaphnia rigaudi.

and Annandale has reported the presence in brackish-water tanks at Port Canning of the Cirripedes:

Balanus amphitrite.

Balanus patellaris.

Chthamalus stellatus.

Two species of Amphipoda have been taken in the Ganges and the surrounding deltaic region, namely *Ampelisca pusilla* and *Quadrivisio bengalensis.* The presence of the former is a matter of some considerable surprise, since it was originally described from the deep waters of the Arctic seas (*vide* Chilton, 1921). Two species of Mysids have been recorded by Tattersall from the Salt Lakes and the Gangetic delta, namely, *Potamomysis assimilis* and *Mesopodopsis orientalis.*

As regards the Decapod Crustacea de Man (1908) has identified the following species from brackish-water pools at Port Canning;

Scylla serrata.

Tympanomerus stapletoni.

Pachygrapsus propinquus.

Varuna litterala.

Metaplax dentipes.

Leunder sp.

Palaemon (*Eupalaemon*) *lamurrei.*

Palaemon sp.

Caradina sp.

Caradina propinqua.

Caradina nilotica var. *begalcusis.*

Kemp (1915, 1917) reports that the following species are known from brackish-water of the Gangetic Delta:

Pachygrapsus propinquus.

Varuna lillerata.

Scylla serrata.

Neptunus pelagicus.

Clibanarius padavensis.

Palaemon mirabilis.

Penaeus carinatus.

Penacus indicus.

Penaeopsis monoceros.

....*Contd.*

Table 19 – *contd...*

Pendepsis brevicornis.

Parapenaeopsis sculptilis.

Examples of the genera *Metaplax* and *Hymenicus* are common in the Delta. Alcock (1895-1900) has recorded the occurrence of *Hymenicus wood-masoni* and *II. inachoides* from Port Canning and *Metaplax intermedia* and *M. crenulata* from the Sundarbans and the mouth of the Ganges. In addition he records the following species from the Gangetic Delta and the mouth of the Hooghly river:

Doclea ovis.

Doclea japonica.

Doclea gracilipes.

Doclea tetraptera.

Calappa lophos.

Matuta victor.

Matuta lunaris.

Leucosia craniolaris.

Philyra globosa.

Philyra globulosa.

Arcania septemspinosa.

Dorrippe facchino.

Scylla serrata.

Charybdis (*Goniohellenus*) *ornata.*

Conchoecetes artificiosus.

Pinnoteres mactricola (from *Mactra violacea*).

Metapograpsus messor.

Ryxidognathus fluviatilis.

Sesarma tetragonum.

Metasesarma rousseauxii.

Annandale and Kemp (1916) record the presence in the Gangetic Delta of the following species:-

Nassa denegabilis.

Nassa orissaensis.

Hydrobia (*Belgrandia*) *myliacca.*

Stenothyra blanfrodiana.

Modiola undulata.

Clementia annandalei.

Cuspidaria annandalei.

Preston (1915) has reported on a small collection of molluscs that was made in the Salt Lake Canal system near Chingrighatta, and in addition to the species mentioned above he records the occurrence of:-

Nassa fossae.

Tiara (*Striatella*) *tuberculata* (=*Melanoides tuberculatus*).

Tiara (*Tarebia*) *lineata* (=*Melanoides lineutus*).

Iravadia princeps.

Assiminea francesiae.

Septaria crepidularia.

Septaria depressa.

....Contd.

Table 19 – *contd...*

Brachydontes emarginata.
Sinodia jukes-browniana
Cyrena bengalensis.
Macoma gubernaculum.
Anatina induta.

Annandale and Prashad (1919) have still further increased our knowledge of the Molluscan fauna of the Delta and added the following species to the lists of those known from this region:

Dostia cornucopia.
Dostia depressa.
Dostia platyconcha.
Littorina melanostoma.
Littorina subintermedia.
Littorina delicatula.
Stenothyra echinata.
Sienothrya soluta.
Bithinella miliacea.
Assiminea brevicula.
Assiminea beddomiana.
Melania (*Mainwaringia*) *paludomidea.*
Nassa ennurensis.
Nassa ennurensis var. *depauperata.*
Ringicula caeca.
Auricularia auris-judae.
Auricularia gangetica.
Auricularia translucens.
Stenothyra deltae.

Among the truly fresh water species that have been taken in the deltaic area and the Salt Lake are the following:

Amnicola orcula.
Melanoides tuberculatus.
Melanoides scabrus.
Melanoides lineatus.
Pila globosa.
Viviparus bengalensis.
Limnae ovalis.
Limnaea luteola.
Indoplanorbis exustus.

As long ago as 1869 Stoliezka recorded the presence in the deltaic region of four species of *Onchidium*, namely:-

Onchidium typhae.
Onchidium pallidum.
Onchidium tigrinum.
Onchidium tenerum.

With considerable passage of time Basu (1950) in his study observed zoo and phytoplankton to be predominant in summer, moderate in monsoon and negligible in winter. Cladocera were prominent in summer, copepoda and rotifera in monsoon and *Tribonema* and *Oedogenium* in winter. Later Datta *et al.* (1983) reported from the diurnal variation study of two ponds that Cyanophyceae, Chlorophyceae and Bacillariophyceae constituted important groups of phytoplanton. In pond I. Cyanophyceae constituted the major group (53.2%) followed by Chlorphyceae (17.3%), while in the second, the percentages were 23.6% and 32.8% of the total phytoplankton. Rotifers, Copepods and Cladocerans were the major zooplankton group in the first pond while rotifers were totally absent from the second. Protozoans and ostracods were present in minimum number in pond I, but ostracoda were absent in pond 2. The author while studying the plankton in all the five bheries (cf Table) recorded that in general, Chlorphyceae was predominant among phytoplankton and Rotifer among zooplankton. Interestingly all the bheries exhibited their dominance in summer. In all 16 genera of algae and 11 genera of Rotifer, Copepoda and Cladocera recorded are indicated below:

Chlorophyceae

Pediastrum, *Scendsmus, Oedogonium, Closterium*

Diatomaceae: *Navicula, Amphora, Nitzschia, Cymbella, Synedra. Melosira*

Euglenophyceae: *Phacus, Euglena* sp.

Blue green algae: *Microcystis, Anabaena, Spirulina*

Rotifera: *Branchionus, Asplanchna, Filina* sp. *Polyarthra, Keratella*

Copepod: *Nauplius, Cyclops, Diaptomus*

Cladocera: *Moina, Diaphanosoma, Cariodaphnia*

Individually, in Bhery A the highest concentration of plankton by number (phytoplankton : 835 units/lit; zooplankton : 435 units/lit) was encountered during summer followed by in monsoon (phytoplankton: 225 units/lit; zooplankton : 375 units/lit) and in winter (Phytoplankton : 50 units/lit; zooplankton : 135 units/lit). *Oscillatoria* sp. and *Navicula* sp. remained dominant during summer, followed by *Amphora* sp. and *Navicula* sp. and *Brachionus* sp. during monson.

Primary productivity was low to moderately high, 143-450 mg/C/m^3/hr (of Table 11A).

In bhery 'B' phytoplankton were 7620, 435 and 145 units/liter during winter, summer and monsoon respectively. *Oscillatoria* sp. and *Brachionus* sp. were dominant forms.

Organic carbon production indicated low values 250-350 mg C/m^3/hr (of Table 11A).

In bhery 'C', however, phytoplankton was found to be highly abundant on account of *microcystis* sp. during water, when zooplankton concentration was 585 units/litre during monsoon, zooplankton indicated fairly high concentration 4340 units/litres followed by summer 1475 units/litre. The corresponding phytoplankton density was 45 and 235 units/litre. *Brachionus* sp. *Asplanchna* sp. *Cyclops* sp. and *Diaphanosoma* sp. were found in abundance during monsoon followed by *Brachionus* sp. and *Cyclops* sp. in summer and *Cyclops* sp. and *Brachionus* sp. during water.

Table 20: Distribution of Phytoplankton in units/litre in five Bheris during January-December 1983

Phytoplankton (Genera)	Bheri-A	Bheri-B	Bheri-C	Bheri-D	Bheri-E
Pediastrum	35	40-180	45-65	35	65
Navicula	15-40	25-150			35
Scenedesmus	25-40	40	125	100-bloom	45
Oscillatoria	625	125-625	-	35	120
Oedogonium	35	35	-	20	-
Phacus	15	-	-	-	-
Euglena	25	-	-	-	-
Microcystis	-	-	45-bloom	-	-
Anabaena	-	-	60	15	-
Spirulina	-	-	-	25	-
Chlorella	-	-	-	-	40
Arthrospira	-	-	-	-	40
Closterium	-	-	-	15	-
Amphora	20-40	20	-	-	-
Nitzschia	40	40	-	25	25
Cymbella	30	-	-	-	-
Synedra	15	-	-	60-125	30
Melosira	-	-	-	115	215

Table 21. Distribution of Zooplankton in units/litre in five Bheris during January-December 1983

Zooplankton (genera)	Bheri-A	Bheri-B	Bheri-C	Bheri-D	Bheri-E
Brachionus	100-290	110-3350	150-2925	150-200	60-855
Asplanchna	35	-	40-420	35-110	90
Filinia	45	-	10-50	35	40-190
Polyarthra	-	-	25	20	60
Keratella	-	-	35-50	10	-
Diaphanosoma	15	60	395	-	40
Ceriodaphnia	-	-	10	35	-
Moina	35	25-300	15-50	25-100	75
Cyclops	45-85	85-1830	200-425	10-40	75-330
Diaptomus	25	-	-	-	40-125
Nauplius larva	60-65	10-2080	100-125	100	35-300

Organic production indicated fairly high values 781 mg C/m^3/hr during water followed by summer 486 mg c/m^3/hr and monsoon 250 mg C/m^3/hr (of Table 12).

As in the previous case, phytoplankton abundance in bhery 'D' was high during winter and was represented by *Scenedesmus* sp. This was followed by monsoon (phytoplankton: 475 units/litre) dominated by *Synedra* sp., *Melosira* sp. and *Scenedesmus* sp. The corresponding zooplankton concentration in winter was low (70 units/litre) with appreciable rise during monsoon (415 units/litre); while in summer the values were 35 and 430 units/litre *Brachionus* sp. dominated both during summer and monsoon.

Production of organic carbon was fairly high during water (912 mg C/m^3/hr) followed by monsoon (373 mg C/m^3/hr) and summer, 150 mg C/m^3/hr. (cf. Table 12).

In bhery 'E' plankton concentration was moderately high (Phytoplankton: 335 units/litre and zooplankton: 465 units/litre) during summer followed by monsoon (Phytoplankton: 255 units/litre and zooplankton: 1695 units/litre). *Oscillatoria* sp. in summer and *Melosira granulator* in monsoon were the dominant forms in phytoplankton, while among zooplankton, *Cyclops* sp. and *Nauplius* in water, *Brachionus* sp. in summer and *Brachionus* sp. and *Cyclops* sp. during monsoon dominated.

Primary production estimated as mg C/m^3/hr was highest during summer (472.5 units/litre), followed by monsoon (978 units/litre) and winter (250 units/litre) (cf. Table 12). The rotifer concentration in Bantala Bhery has been discussed by Siddiqi, Srivastava and Ray (1984).

5.3 *Benthic Fauna and Flora*

Preston (1915), as recorded by Sewell (1934), observed that a number of species of benthos collected from the salt lake canal system near Chingrighata were rather small, but they were represented by countless numbers. The lakes were rich feeding grounds and appeared to preserves of a comparatively few species which had succeeded in establishing themselves in trying and changing conditions. A list of benthos indicating their abundance is given below:

Table 22. Benthos of Salt Lake (After Preston, 1915)

Molluscs:	*Nassa fossac*
	Tiara (Striatella)
	Tuberculeta (=Melanoides
	teuberculatus
	Tiara (tarebia) lineata
	(=Melanoides) lineatus)
	Iravadia princeps
	Asminea francesiae
	Septaria crepidularia
	Septaria depressa
	Brachydontes emarginata
	Sinodia jukes-browniana
	Cyrena bengalensis
	Macoma gubernaculum
	Anatina induta
	Nassa denegabilis
	Nassa orissaensis
	Hydrobia (belgrandia)
	myliacea
	Stenothyra blanfordiana
	Modiola undulata
	Clementia annandalei
	Cuspidaria annandalei

Sewell (1934) has studied the benthos of the Salt Lake the bottom of which consisted of a black soft mud in places forming a bed of considerable thickness. His study was a little more informative as evident from the enumeration below:

Gastropods:	*Melanoides* sp: Empty shells of gastropods scattered about on the shores every where in large quantities. In canal they were very few.
	Stenothyra sp: Fairly in large numbers along with *Melanoides* in the bottom mud.
	Nassa sp: Small number and in canal.
	Polychaet worms: Common in canal.
Graposid Crab:	*Varuna littorate*- Extremely common; living under half dried mud in holes. Could also be seen on the soft shores at many places.
	Palaemon lammarrei Common in lake
	Caridina nilotica Predominant
Portunid	*Scylla serrata* Scarce
Prawn	*Palaemon lamarrei* Common
Mysid	*Mesopodopsis orientalis* (Tattersall) In vast shoals in lake.
	Potamomysis assimilis Occasional.
Actinian	*Phyocoetes qangeticus*: Large numbers.
Bivalve	*Modiola striatulus*: In vast numbers.
Lamellibranch	*Cuspidaria* sp. Mixed with *Cuspidaria* sp. in hard stick clay.

After the valuable information furnished by Sewell (1934) there has been no follow up of such studies untill the detailed study made recently by the author in randomly selected 5 bheries. Besides, Chironomid larvae and oligochaetes, which were distributed in all the bheries, *Digoni, ostoma, pulchella* of the molluscan group was also common. This was followed by *T. tuberculeta* and *Bellamya bengalensis.* Benthic flora was almost absent. Benthic fauna (expressed in number per 30 × 30 cm area) have suffered a great loss in all the bheries investigated. They were either dead shells or present in poor numbers. However, both flora and fauna encountered are discussed below: (cf Table 23).

In bhery 'A' benthic fauna was represented by dead shells of gastropods (*Thiara lineata, T. tuberculata, G. covexiculus, Digoniostoma pulchalla* and *Bellamya bengalensis*). Chironomids and oligochaetes were observed in low concentrations, 47 and 17.2; 13.6 and 52; and 122 and 5.4 units respectively during winter, summer and monsoon. Benthic flora were almost absent in the bottom sediments.

Benthic fauna in Bhery 'B' was poorly represented and concentrations of Chironomids and oligochaetes were 22.8 and 20.8; 33.8 and 10; and 108.2 and 7.4 units during winter, summer and monsoon respectively *Thiara tuberculata, Digoniostoma pulchalla* and *Bellamya bengalensis* of the molluscan group were found dead. As in 'A' benthic flora was absent in the soil sediments.

Table 23. Numerical count of benthic flora and fauna in five bheries during January-December 1983.

Organisms	*Bheri-A*	*Bheri-B*	*Bheri-C*	*Bheri-D*	*Bheri-E*
Thiara lineata	1.6	0.8	1-2.4	1.8-2.8	10.4-42.2
Thiara tuberculata	1-9.2	0.6-13	1-15.2	0.8-2.6	9.2
Gyrrulus convexiculus	-	0.2	1-1.4	0.4-2.2	0.6-0.8
Digoniostoma pulchella	2.4-12.6	2.4-19	2.6-38.8	0.4-6.5	30.4-72.4
Bellamya bengalensis	0.4-2.2	0.4-2.4	0.6-4.4	0.2	2-19.2
Chironomids	13.6-122	22.8-108.2	0.8-156.8	1-74.6	1.6-15.4
Oligochaetes	5.5-52	7.4-20.8	2.8-9	2-5.2	4.4-19.2

The values indicated above are the averages of 5 dredges of 30 × 30 cm area.

In Bhery 'C' chironomids and oligochaetes were found in poor numbers, 17 and 3; 156.8 and 9; and 0.8 and 2.8 units during summer, monsoon and winter respectively. The molluscan group consisting of *Thiara lineata, T. tuberculata, Gyrrulus convexiculus. Digoniostoma pulchella* and *Bellgalensis* was found in dead condition. Benthic flora was negligible.

In author bhery 'D' benthic fauna seem to have suffered equally, *Thiara lineata. T. tuberculata, Gyrrulus convexiculas, Digoniostama pulchella* and *Bellamya bengalensis* of the molluscan group were found all in dead condition except during winter. Oligochaetes and Chironomids were showing appreciable low concentrations in all the seasons, their respective number being 74.6 and 5.2; 1 and 2; and 23.8 and 4.2 units respectively during the above order of seasons. As usual benthic flora was almost absent in the soil sediment.

In bhery 'E' all 5 species of gastropod (*T. lineata. T. tubuculata, Gyrrulus convexiculus, Digoniostoma pulchella* and *Bellemya bengalensis*) were recorded in dead condition during summer. During winter and monsoon some were recoded alive (mostly young ones). Chironomids and oligochaetes were found in poor concentrations, 8.8 and 19.2 units in summer. 1.6 and 10.8 units during rainy season and 15.4 and 4.4 units in winter. Benthic flora was almost absent.

5.4 *Microbial Load*

Bacteriological tests in water, soil and in different organs of fish of Salt lake bheries were for the first time initiated by the author to determine the microbiol load in terms of *TPC, E. coil,* Coliform Gr. and *S. aureus*/g of material (cf. Table 23).

In bhery 'A' the total count (TPC) in water was highest during winter (9×10^5) followed by summer (6×10^5) and monsoon (10×10^4). Soil sediments also exhibited a similar trend-Winter; 4×10^6; Summer: 2.7×10^6 and Monsoon, 3×10^5. In fish samples total count was highest during summer in the liver of mrigal (*C.mrigala*) 10×10^5, intestine of rohu (*L. rohita*) (25×10^5) and in kidney of Tilapia (*T. mosambica*) (25×10^5).

In bhery 'B' microbal loads (TPC) in the water were 12×10^6, 7×10^5 and 50×10^4 during winter, summer and monsoon respectively against 5×10^6, 3.5×10^6 and 6×10^5 in the soil. In fish, TPC in kidney, liver and intestine of mrigal has shown values of 5×10^5, 4×10^5 and 15×10^5 against 5×10^5, 6×10^5 and 12×10^5 in rohu during summer. During winter, the total load in intestine was high (3×10^6) in mrigal but infection was high in intestine (40×10^5) and liver (50×10^4) of rohu during monsoon.

High TPC was encountered in the water of bhery 'C' during winter (3×10^6) followed by monsoon (5×10^4) and summer (1.5×10^4). In the bottom sediments, TPC was observed to be highest in winter (7×10^6) followed by summer (1.9×10^6) and monsoon (10×10^5). In fish, however, the bacteriological load (TPC) was observed to be very high both in rohu (4×10^6, 2×10^6 and 4×10^6) and Tilapia (1×10^6, 18×10^6 and 9×10^6) in kidney, liver and intestine during summer. In case, (*Catla catla*) these values were 1.5×10^5, 2×10^5 and 8×10^5 during winter. During monsoon the infection was comparatively less (Rohu: 10×10^4, 25×10^4 and 40×10^4) in intestine, liver and kidney respectively.

Table 24. Total bacterial load in soil, water & fish (Intestine, liver, kidney) during January-December 1983.

Bheri	Source	TPC	*E. Coli*	Coliform Gr.	*S. aureus*
	Water	10×10^4-9×10^5	nil-2	3.5×10^2-1.2×10^5	3×10^1-1×10^5
A	Soil	3×10^5-4×10^6	nil	12×10^3-1×10^5	12×10^2-2×10^5
	L. rohita	3×10^4-25×10^5	nil	nil-40×10^3	nil-10
	C. mrigala	14×10^4-20×10^5	nil	nil-40×10^4	5-5×10^2
	C. catla	1×10^4-20×10^4	nil	100-600	13×34
	T. Mossambica	2×10^4-24×10^5	nil-100	nil-1×10^4	nil-30
	Water	50×10^4-1.2×10^6	nil	nil-2×10^5	1.4×10^2-1.2×10^5
B	Soil	6×10^5-5×10^6	nil	3×10^4-2.6×10^5	nil-2×10^5
	L. rohita	10×10^2-40×10^5	nil-40	nil-8×10^5	nil-40×10^2
	C. mrigala	4×10^5-15×10^5	nil-5	30-2×10^3	nil-140
	Water	1.5×10^4-3×10^6	nil	200-15×10^3	30×100
C	Soil	10×10^5-7×10^6	nil-3	70-90×10^4	nil-2×10^2
	L. rohita	10×10^4-6×10^6	nil-1×10^2	nil-10×10^5	nil-2.8×10^4
	C. catla	2×10^5-1.5×10^6	nil	3×10^2-1×10^5	6×10^2-1.2×10^4
	T. Mossambica	1×10^5-18×10^6	nil-7×10^2	4×10^2-3×10^3	nil-4×10^2
	Water	4×10^5-2×10^6	nil	75×10^1-30×10^4	·93-8×10^2
D	Soil	5×10^5-6×10^6	nil	6-40×10^4	nil-10×10^2
	L. rohita	40×10^4-12×10^5	nil-40	nil-10×10^5	nil-10×10^3
	C. mrigala	3×10^4-15×10^5	nil-30	nil-15×10^4	nil-4×10^5
	C. catla	0.1×10^4-3×10^4	nil	nil-290	1-18
	T. Mossambica	0.5×10^4-4×10^4	nil	nil-180	2-20
	Water	10×10^3-2×10^6	nil-20	nil-2.5-10^5	nil-1×10^5
E	Soil	44×10^5-5×10^6	nil-10	nil-1.6×10^5	nil-1.2×10^5
	L. rohita	4×10^5-18×10^5	nil	35-2×10^3	5-200
	C. mrigala	6×10^5-12×10^5	nil	40-6×10^3	nil-30
	T. Mossambica	10×10^5-12×10^5	nil	15-2×10^3	nil-120

In bhery 'D' TPC registered highest value in water during winter (2×10^6) followed by summer (5×10^5) and monsoon (4×10^5). Coliform bacteria exhibited a high value during monsoon (30×10^4). In the soil phase, TPC showed highest value during water (6×10^6) followed by summer (2.7×10^6) and monsoon (5×10^5). As in water, coliform bacteria showed appreciably high value in soil during m onsoon 40×10^4. Bacterial load (TPC) was encountered highest value in kidney, liver and intestine in mrigal and rohu during summer. The corresponding values were 5×10^5, 10×10^5 and 15×10^5; and 5×10^5, 6×10^5 and 12×10^5. This was followed by monsoon when TPC in mrigal was 10×10^5, 3×10^4 and 40×10^3 and in rohu 5×10^5, 50×10^4 and 40×10^4 in intestine, liver and kidney respectively.

The TPC count was found to be highest in the water of bhery 'E' during summer (2×10^6) followed by winter (1×10^6) and monsoon (10×10^3). In soil, however, TPC continued to show high values in winter and summer (5×10^6). During monsoon the value was 44×10^5. Coliform bacteria, however, showed abundance during winter both in water and soil sediments. In fish, bacterial load (TPC) was high during summer 8×10^5, 6×10^5 and 12×10^5; and 8×10^5, 4×10^5 and 18×10^5, 12×10^5 and 12×10^5 in mrigal rohu and tilapia respectively in kidney, liver and intestine. This was followed by coliform bacteria, especially in intestine of mrigal in winter, and liver and intestine of mrigal, rohu and tilapia during summer.

6. BIOACCUMULATION OF METALS

Information on bioaccumulation of metals in different organs of fish is of vital importance as this helps in understanding the health condition of fish. Bioaccumulation of metals may lead to physiological damage which may ultimately affect the reproductive system. This is more important particularly at a time when the industrial complexes are rapidly growing in Calcutta. Unfortunately, sor far there is no information on record. Ray and Bandyopadhyay (1984) for the first time have presented valuable information on this aspect in five bheries investigated by them (cf. Table 25).

In bhery 'A' concentration of metals *e.g.* Copper, Zinc and lead estimated in fish tissue of mrigal (*C. mrigala*) indicated 1.25-6.468; 15.0-27.25 and nil-0.03 ppm respectively. Nickel was not recorded. In Catla (*C. Catla*) the concentrations of copper and zinc in tissue were 1.25 and 15.0 ppm respectively. Kidney, liver and ovary of mrigal indicated accumulation of 111, 93 and 94 ppm of Zinc, against 4.4, 3.4 and 5.5 ppm of Copper. In catla the accumulation of zinc was 72 and 145 ppm and copper 5.8 ppm respectively in kidney and liver. Nickel and lead were not detected.

In bhery 'B', the estimated heavy metal concentration in mrigal tissue was, copper 4.464 ppm, zinc 17.36 ppm and lead 0.031 ppm.

In bhery 'C' metallic bioaccumulation of zinc, copper and lead in mrigal tissue was at the levels 12.865-21.25, 1.25-3.215 and nil-0.021 ppm respectively.

In bhery 'D' copper, zinc and lead were found to accumulate in concentrations in the tissue of mrigal at 1.25-5.914, 22.176-25.0 and nil-0.02 ppm respectively. These observations indicate appreciable concentration of these metals.

In bhery 'E', however, of the three metals estimated (Copper, Zinc and Lead) in mrigal, lead indicated trace values (0.025 ppm) while concentrations of copper and zinc were 1.25-3.1 and 16.53-21.25 ppm levels.

Table 25. Accumulation of heavy metals in fish Tissue/Organ

Bheri	Test Fish	Tissue/organ	Heavy metal	Concentration in ppm
A	*C. mrigala*	Tissue	Cu	1.25-6.468
	C. mrigala	Tissue	Zn	15.0-27.25
	C. mrigala	Tissue	Pb	ND-0.03
	C. mrigala	Kidney	Cu	4.4
	C. mrigala	Liver	Cu	3.4
	C. mrigala	Ovary	Cu	5.5
	C. mrigala	Kidney	Zn	111
	C. mrigala	Liver	Zn	93
	C. mrigala	Ovary	Zn	94
	C.catla	Tissue	Cu	1.25
	C.catla	Kidney	Cu	4.8
	C.catla	Liver	Cu	5.8
	C.catla	Tissue	Zn	15.0
	C.catla	Kidney	Zn	72
	C.catla	Liver	Zn	145
	C.catla	Tissue	Pb	ND
	C.catla	Tissue	Ni	ND
B.	*C. mrigala*	Tissue	Cu	4.464
	C. mrigala	Tissue	Zn	17.361
	C. mrigala	Tissue	Pb	0.031
C.	*C. mrigala*	Tissue	Cu	1.25-3.216
	C. mrigala	Tissue	Zn	12.865-21.25
	C. mrigala	Tissue	Pb	ND-0.021
D	*C. mrigala*	Tissue	Cu	1.25-5.914
	C. mrigala	Tissue	Zn	22.176-25
	C. mrigala	Tissue	Pb	ND-0.02
E	*C. mrigala*	Tissue	Cu	1.25-3.1
	C. mrigala	Tissue	Zn	16.53-21.25
	C. mrigala	Tissue	Pb	ND-0.025

ND: Not Detectable

*Tissue is considered as fish muscle.

It is interesting to note that in some cases concentrations of metals, particularly zinc in all the organs of fish indicated higher accumulation although in water it was at low level. This may probably indicate cumulative accumulation over the period of life.

7. TOXICITY TESTS BY BIOASSAYS

To find out the toxic levels of the effluent using major carp fry and fish food organisms as test animals 96 hr. static bioassays were conducted. Toxicity experiments in 10 lit jars with adequate control were conducted with settled raw sewage using 10 fry (1.8-5.2 cm) each of rohu, catla and mrigal as test animals. Dissolved oxygen was maintained at 4.8-5.2 ppm level. Result shows only 10% mortality in case of rohu and mrigal in 100% concentration. This shows oxygenated raw waste may not have immediate harmful effect on test fish. Regarding fish food organisms *Daphnia* and oligochaete worms the 96 hr LC_{50} values with zinc sulphate were 0.061 and 157.8 ppm respectively. Oligochaete, however, are found to be more sensitive to copper chloride as evidenced from LC_{50} value, 0.263 ppm.

8. IMPACT OF WASTES ON BIOTIC AND ABIOTIC CONDITIONS OF WATER AROUND THE INLET POINT

Physicochemical and biological studies around the effluent receiving point in the bhery are important in determining the extent of pollution. The effluent introduced into the bhery has a variable composition and includes besides suspensoids, many other toxic substance of non-biodegradable and highly biodegradable nature. In this regard Basu (1950) ans Saha *et al.* (1958) made some studies worth mentioning. Their work related to physicochemical qualities and biological conditions of Calcutta sewage around the sewage linlet point from the viewpoint of pisciculture and the danger of feeding raw sewage to confined fishes (Fig. 5) Basu (1950) observed sewage having offensive smell and black colour with 100-115 ppm BOD, pH: 6.8-7.0' DO: nil-1.0 ppm; CO_2 32.5-37.5 ppm when introduced in the bhery showed changes in the polluted zone (Messosaprobic zone) as : DO nil; pH: 7.1-7.3; CO_2: 10-12 ppm; BOD: 70-90.5 ppm and increased turbidity. The aquatic life was absent. In the recovery zone the conditions were DO: 5.6-6.1 ppm; pH 8.0-8.2; CO_2: 10-12 ppm; BOD: 20-30 ppm. The concentration of plankton and aquatic life increased. Diurnal fluctuations of DO and pH indicated range of 6-14.6 ppm and 8.0-9.8 respectively.

He further observed that in summer BOD decreased to normal after 10-12 days which in winter took 15-20 days. Besides, increases in phosphate and nitrate from 0.25 to 0.5 ppm and 0.1 to 0.15 ppm respectively was observed after sewage feeding. At different places of the bhery pH fluctuation ranged 7.6-9.0 depending upon the submerged aquatic vegetation. After sewage feeding bacteria increased on the first day, followed by increase in protozoa and phytoplankton on the 2nd and 3rd day. Rotifer and crustacea increased on the fourth and fifth day respectively. Decrease in phytoplankton was followed by the increase in zooplankton.

Saha *et al.* (1958) in their study delineated the zone of pollution and the magnitude of recovery of chemical constituents from sewage inlet point of (Fig. 4). The observations made by these authors were base on analysis of the water samples at different points form and among the sewage inlet gate during introduction of raw sewage into a private fishery area. The study revealed that up to a distance of 15.24 m pH increased from 7.2 to 8.2; Dissolved oxygen increased from 0-7.8 ppm; Carbon dioxide value dropped from 56 ppm to 0 ppm; alkalinity and Oxygen consumption values dropped from 400 to 200 ppm and 37.6 to 20.28 ppm respectively. About 80% of the suspended solids were found to settle down within a distance of 3.04 m (10 ft) and about 95% got deposited within 12, 19 m (40 ft). Quantitative study on plankton population indicated that there was no plankton within 3.04 m (10 ft) from the inlet but beyond this point plankton appeared in increasing concentration. It thus appears that even by judicious regulation of does during intake, toxic effect of raw sewage in present up to a certain distance from the inlet end. This distance, however, is dependent on the depth of the bhery, its area, rate of intake, frequency of introduction of sewage etc. (Saha *et al.*, 1958).

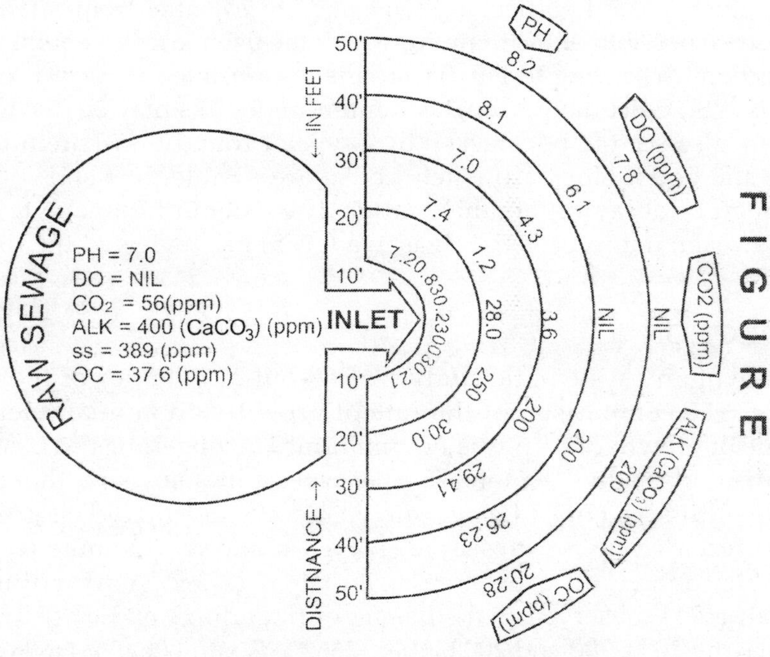

Fig. 5 Impact of waste on abiotic conditions of a bhery water around the inlet point
Source : Saha et al. (1958).

Regular introduction of raw sewage at any rate be encouraged. This may raise the bed level by gradual accumulation of suspended solids and such suspended solids mostly being highly oxidisable in character may undergo decomposition under favourable temperature and may cause an unhealthy ecological condition by creating negative oxygen balance in the bhery.

9. EPIZOOTICS IN BHERIES

Fish diseases in the freshwater bhery have not been studied extensively. Hardly, there is any published account except preliminay work of Pal (1982) and unpublished record by C.B. Srivastava (Personal Communication).

During winter when carps loose their appetite they get affected by a disease in the gill region, body, dorsal and ventral fins. The symptoms of this disease in minute white spots or patches spread over these parts. The affected fish when survive show retarded growth. Pal (1982) claims, single celled *Myxobolus, Thelohenelus,* etc., are responsible for ths diseases. Besides, overstocking of fish, inadequate supply of fish food are other managements measures which create an unhealthy environment. Carp fingerlings and *Cyprinus carpio* suffer from *Gyrodactylus* and *Dactylogyrus* infection during winter which may sometime cause fish mortality.

Some black spots representing strigeid infection are also seen on the body. This makes the fish weaker. Besides, when there is enough of accumulated bottom ooze in the bhery during summer, major carp suffers from a diseases which cause accumulation of water wither in the fish or under the scales. This is known as 'dropsy'. Excess of water accumulation causes bulging of the belly and particularly *Catla* and *Tilapia* suffer from cataract. Tilapia also suffers from a disease possibly of the brain for which the fish looses its equilibrium and move in a circular direction. According to Pal, bacteria is responsible for such disease. According to Srivastava, C.B., ZSI, Calcutta (personal communication) this may be due to *Myxobolus* sp. infection. *Catla* are also seen to be attacked by a type of round transparent louse (Argulus) that move over the body. Normally such Louse are visible on fishes of ponds which accumulate enough of decaying organic matter. The extent of damage to the entire fish population is, however, not indicated. Srivastava (1975) has reviewed the fish pathological studies in India.

10. FISH PRODUCTION

In the absence of any systematic study carried out of fisheries of the past it may be difficult to undertake comparison of the rate of growth and annual yields. Sen (1941) observed that rate of growth of carp was not uniform. The fingerlings of *L. rohita, C. mrigala* and *C. catla* attained normally ½ sr (468.75 g) in weight in one year although instances of carps growing to ¾th seer (703.13 g) in one year were not rare. The growth of Bhetki (*L. calcarifer*) was recorded up to ¼th seer (234.38 g) in one year. Minute fry of this species reared on suitable food could grow to a size of 1-3" (2.2-7.5 cm) within 3 months of introduction. Nair (1944) observed that stocking of fingerlings of 3-6" (7.5-15 cm) in three bheries of each having 2000, 700 and 200 bighas (266.6, 93.3 and 26.6 ha) in areas. Catla, rohu and mrigal attained weights of 1 ½, 1 and ½ seer (1406.25 g, 937 g and 468.75 g) respectively in 5 to 6 months time in water mixed with sewage. Along with this *Mystus gulio, Mugil parsia* and *p. sophore* contributed sizable fishery. Basu (1950) recorded *Channa*, Siluroidae and Anabantidae and minnows to form a natural byproduct in the fisheries in addition to carps cultured therein. He observed that in course of one year 1 cm fry of catla, rahu and mrigal attained a length of 40, 30 and 25 cm with corresponding weights 1 ¾ lb (794.5 g), 1 ½ lb (681 g) and 1 lb (454 g) respectively. Extensive survey made by the Secretary, Fish Producer's Association, in 1959 in the North and South Salt lakes (cf. Appendices I and II) indicated the nature of production in individual bheries prior to sixties when industrial development was at a low ebb. More recently Ghosh (1980) recorded an average annual yield from bheries not exceeding 4000 tonnes. According to him the average annual production is 1 t/acre. The catch contained 60-65% lower grade major carps, 20% *Cyprinus carpio* and 10-15% *Tilapia*. The range of average weight of major carps was 150-200 g, *Cyprinus* 250-400 g and *Tilapia*, 25-50 g. Average weight of *Tilapia* 10 year's before was 100-150 g. Ray *et al.* (1981) stated production in freshwater bheries even with 4-5 relay stocking got reduced to 1000 kg/ha from 1100 kg/ha obtained prior to 1969. The entire observation is presented in Table 27. In a personal communication Secretary, Fish Producer's Association, 24-Parganas, has given data on fish production for 10 years between 1959 and 1985 on a random basis along with the percentages of carps, *Tilapia* and other fishes in the catches (Table 26).

11. FISHES OF THE SALT LAKE

Sen (1941) prepared a list of fishes and Crustacea recorded in the Salt Lakes during 1937-40 which is given in Table 28. It is interesting to note that most of species recorded are of freshwater. This clearly indicates the change in ecology of these lakes from brackishwater to freshwater condition (Table 28).

Table 26. Fish Production of Salt Lake* (1959-85)

Year	Approximate area (acre) used for fish culture	Total Production (t)	Production t/ha	Carps (%)	Cyprinus carpio (%)	Tilapia (%)	Others (Forage, Murrels etc. (%)	Silver (%)
1959	18,000	8,940	1.24	96	-	-	4	-
1968	16,800	9,670	1.44	95	2	2	1	-
1969	16,200	Production could not be assessed due to some lands were acquired for irrigation purposes						
1970	356	320	2.24	5	15	80	-	-
1971	1350	1,500	2.77	15	5	80	-	-
1976	11,000	6,600	1.5	50	30	20	-	-
1978	9,000	7,500	2.08	60	25	13	2	-
1980	8,500	5,000	1.47	60	25	12	3	-
1983	8,000	7,500	2.34	75	15	7	3	-
1985	7,500	4,000	1.0	80	10	2	2	6

* Courtesy, Secretary, Fish producers Assocn. (24-Pgs.) Calcutta.

12. OVERALL IMPACT ON BIOTA

Investigations carried out so far since 19th Century in Salt Lake under different ecological conditions *viz.* (1) Brackishwater, (2) Transitional phase between brackish and fresh water condition, (3) Purely fresh water condition, and finally (4) freshwater associated with metropolitan wastes have brought out interesting results. Most of the studies made were, however, not comprehensive and were based mainly on aspects like, geographical conditions, information of fishery matters, flora, fauna, cultural aspect and of course more recently the sewage characteristics etc.

From the above findings some lacuna has been noticed during the process of culture of fry being practiced so far and they are :

1. Absence of proper analysis of waste water prior to their use.
2. In adequate knowledge of the quantity of use of highly rich organic wasted during the culture of duplicate fry/fingerlings in bheries.
3. Adverse effect of use of untreated sewage waste enriched with complex industrial wastes (rich in metallic contaminants) during fish culture resulting poor growth.
4. Lack of adequate scientific knowledge about the proper size of fry to be cultured to get beneficial results, stocking density and quantity of sewage to be used during different stages of fry fingerling culture.
5. Treatment of industrial waste before their use to get healthy growth of fishes.

It is for this reason the results obtained by different workers very. A sort of standardisation of size of fry/fingerlings, character of waste water, ratio of dilution (This may vary from season to season), stocking density, effect from use of complex industrial waste before and after their treatment etc. parameters have their own importance but do not reflect an overall picture of the ecology and bionomics of the salt lakes. The author has made an attempt to provide more valuable information on chemistry of water and soil, qualitative and quantitative study of plankton, their behaviour, biota, microbial load, presence of heavy metals in water, soil and their acumulation in different organs of fish etc. and the overall impact on abiotic characteristics of certain selected bheries.

Table 27. Fish Production in Salt Lake Bheries

	Species	*Size attained*	*Period*	*Remarks*
I. Sen (1941)	(1) *Catla catla*	Average size		
	L. rohita	227 gr.	1 yr.	681 gr. size not uncommon.
	C. mrigala			
	(2) Fry of major carp	2.5-7.5 cm	3 months	Not encouraging due to deficient sewage supply.
	(3) Fingerlings:			
	L. rohita			
	C. mrigala	Normally attains	1 yr.	Instances of corp growing up to 703.13 gr. was also there.
	C. Catla	234.4 gm.		
	L. Calcarifer	234.38 gr.	1 yr.	
II. Nair (1944)	(1) *Fry stocked : 7.5-15 cm. size*			
	C. Catla	1362 gr.	5-6 months	When adequate sewage waste was introduced.
	L. rohita	908 gr.		
	C. mrigala	454 gr.		
	Average	908 gr.		
	(2) *Fry. stocked : 1 cm. size*			
	C. Catla	794 gr.	1 yr.	
	L. rohita	681 gr.		
	C. mrigala	454 gr.		
	Avegage	643 gr.		
	(3) *Fingerlings : 7.5-15 cm. size*			
	C. Catla	1406.25 gr.		Water mixed with sewage.
	L. rohita	937. gr.	5-6 months	Along with this
	C. mrigala	468.75 gr.		*M. gulio*
				M. persia Contributed
				P. sophore sizable fishery.
III. Basu (1950)	*Fry 1 cm. size*			
	C. Catla	794.5 gr. (40 cm)		This is in addition to natural byproduct Channa, Siluroidae, Anabantidae and minnows.
	L. rohita	681. gr. (30 cm)	1 yr.	
	C. mrigala	454 gr. (25 cm)		
IV. Ghosh (1959)	Production of fish shown in appendixes I and II	—	—	When industrial development was a low ebb.
V. Ghosh (1980)	Average annual yield not exceeding major corps	4000 t (1 t/acre) 150-200 gr. (60-65%)		
	Cyprinus	250-400 gr. (20%)		
	Tilapia	25-50 gr. (10-15%)		
VI. Ray *et al.* (1981)	*Production of freshwater fisheries.* Reduced to 1000 kg./ha from 1100 kg./ha obtained prior to 1969.			Production from 4-5 relay stocking.

Note : Sewage effluents prior to 1963 was almost free or had negligible concentration of metallic contamination.

Table 28. List of fishes and crustacea in the Salt lakes during 1937-40

Scientific name	*Local name*
Fish	
Far. Megalopidae	
1. *Megalops cyprinoides* (bronssonet)	Amlet
Fam. Clupeidae	
2. *Engraulis tetera* (Ham)	Fresha-Phansa*
Fam. Notopteridae	
3. *Notopterus notopterus* (Pallas)	Fol ui (Phola)-Pholui*
Fam. Cyprinidae	
Sub. Fam. Abramidinae	
4. *Chela bacaila* (Ham)	Chela-Chela*
Sub. Fam. Rasborinae	
5. *Esomus danricus* (Ham)	Danrke-(Danrica) Danrika*
Sub. Fam. Cyprininae	
6. *Amblypharyngodon mola* (Ham.)	Maurala Mawrala*
7. *Barbus* (*Puntius*) *Sarana* (Ham.)	Swarna Punti (Sarana) Punti*
8. *Barbus* (*Puntius*) *Sophore* (Ham.)	Safari Punit Punti*
9. *Barbus* (*Puntius*) *ticto* (Ham.)	Tit punti Tita Punti*
10. *Catla catla* (Ham.)	Katla (catla) Katla*
11. *Cirrhina mrigala* (Ham.)	Mrigal-(Mrigala) Mrigala*
12. *Cirrhina reba* (Ham.)	Kharke bata (Batta)-Kharke batta
13. *Labeo bata* (Ham.)	Bata - (Dommarci batta)-Bhangna*
14. *Labeo calbasu* (Ham.)	Kalbose-(Kalbasu)- Kalbasu*
15. *Labeo gonius* (Ham.)	Kurchi bata (Karchi)-Karchi*
16. *Labeo rohita* (Ham.)	Rui-(Ruee)-Rui*
Fam. Claridae	
17. *Clarias batrachus* (Linn)	Magur (Maghur)-Magur*
Fam. Heteropneustidae	
18. *Heteropneustes fossilis* (Bloch)	Singi-(Singee)-Singhi*
Fam. Siluridae	
19. *Wallago attu* (Bloch)	Boal (Boyari)-Boal*
Fam. Bagridae	
20. *Mystus tengara* (Ham)	Tangra
Fam. Cyprinodontidae	
21. *Aplocheilus panchax* (Ham.)	Techoke-(Panchasi)
22. *Oryzias melanostigma* (Meclelland)	Chuno
Fam. Hemirhamphidae	
23. *Hemirhamphus limbatus* C.V.	Bogu

....Contd.

Table 28 – *contd...*

Scientific name	*Local name*
Fam. Ambassdae	
24. *Ambassis nama* (Ham)	Katchanda-Nama chanda*
25. *Ambassis ranga* (Ham.)	Ranga Chanda-(Chandee)-Ranga chanda*
Fam. Ophicephalidae	
26. *Ophicephalus gachua* (Ham)	Chang-Cheng*
27. *Ophicephalus punctatus* (Bloch)	Lata-Taki*
28. *Ophicephalus striatus* (Bloch)	Sole (Sol) Shol*
Fam. Osphrone midae	
29. *Colisa fasciata* (Bloch)	Khalisa-Khalisha*
Fam. Anabantidae	
30. *Anabas testudineus* (Bloch)	Koi-(Coi)-Koi*
Fam. Nandidae	
31. *Nandus nadus* (Ham)	Nandas-(Latha)-Nandus*
Fam. Gobiidae	
32. *Glossogobius giuris* (Ham)	Bele-Beley*
Fam. Mugilidae	
33. *Mugil corsula* (Ham)	Khorsola (Corsula)
34. *Mugil parsia* (Ham.)	Parse-(Tarui)
35. *Mugil tade* (Forsk)	Bhangan (Bangon)
Fam. Scatophagidae	
36. *Scatophagus argus* (Bloch)	Pyra chanda
Fam. Mastacembelidae	
37. *Mastacembelus pancolus* (Ham)	Pankal-(Pangkal)
Crustacea	
Fam. Palacmonidae	
1. *Palaemon lamarrei* M. Edw	Kucho chingri
2. *Paleamon carcinus* (Fabr.)	Mocha or Golda Chignri
Fam. Penaeidae	
3. *Penaeus carinatus* (Dona)	Bagda chingri
4. *Metapenaeus monoceros* (Fabr)	Honye chingri
Fam. Portunidae	
5. *Scylla serrata* (Forsk)	Nona Kankra
Fam. Grapsidae	
6. *Varuna litterata* (Fabr.)	Chiti Kankra

N.B.: The local names in brackets and those marked with asterisks are according to Day (1876-78) and Shaw and Shebbeare (1937) respectively. Species names are as given by Sen (1941).

The studies made so far reveal that around 19th century the water in the Salt lake was purely brackish in character as shown by the occurrence of brackish water flora and fauna (cf. Tables 17, 18 and 19), Biswas (1927) and Sewell (1934). The species cultured during that period were *Lates calcarifer* (Bloch), *Mugil parsia* (Ham) and *Mugil tade*. Due to heavy silting the river Bidyadhari lost most of its characteristics in about the year 1904. Although 14 out of 16 species of Copepoda pertaining to brackishwater continued to be present until 1914, they disappeared completely by 1932. The number of species of benthos present in the Salt lake was rather small through represented by countless number. The list of various species of fish and common crustacea prepared by Sen (1941) for the period 1937-40 clearly indicates that majority of the group belonged to freshwater, while the rest were Euryhyaline species. The observations made by Sewell (1934) clearly indicated that freshwater condition was established by 1932. This was supported by the species composition prepared by Sen (1941) during 1937-40.

With the regular introduction of Calcutta Municipal sewage into the bhery ecosystem, improvement in the growth rate of fish was noted. According to Sen (1941) average normal growth rate for major carps catla, rohu and mrigal was 1/2 sr (227 g) in one year although 3/4 sr (681) was not uncommon. The fry could grow upto 2.5-7.5 cm within 3 months of introduction. Such growth rate was undoubtedly not encouraging (Perhaps on account of deficient flow of sewage) against the growth of catlà, 1362 g, Rohu 908 g and Mrigal 454 g attained in 5-6 months, as observed by Nair (1944), when stocked with 7.5-15 cm fingerlings and 794 (40 cm), 681 g (30 cm) and 454 g (25 cm) respectively from 1 cm size of the above species in 1 year time.

Domestic sewage and effluents from various industrial complexes like acids, various chemicals, paints and varnishes, battery, cosmetics, perfumery, drugs and pharmaceuticals, electroplating, fertilizer, match factory, galvanizing, pulp and paper, soap detergents, photography and development of X-ray plate and a good number of tanneries contributing, besides metals (cf. Tables 9&29), conservative organic material along with highly biodegradable organic material. These affect the ecological conditions in the bhery ecosystem by altering pH. DO, CO_2 alkalinity, suspended solids, H_2S, ammonia etc. Such environment may affect the survial, growth and reproduction of the aquatic biomass like plankton, fish and dish food organisms. The organisms concerned or fish may not show any sign of distress for their survival, but being exposed for prolonged period, in an unfavourable ecological condition may suffer slowly.

In randomly selected five bheries of different sizes it was noticed that diurnal variation of pH was not that significant as expected in an organically enriched water where the variation was from around the neutral level or below to highly alkaline level. One of the reasons may possibly be comparatively low concentration of phytoplankton encountered in almost all the bheries. This results in moderately high primary productivity. Diurnal variations of DO in all the bheries showed that saturation of DO required for a healthy growth of smaller sized fish (optimum level 125-130% saturation) persisted for a shorter time (Fig. 4 A-E).

Metallic contamination, especially copper and zinc, might have some role in destroying the phyto and zooplankton (evidenced by LC_{50}) value of zinc as zinc sulphate to Daphinia). Individully, zinc and copper are toxic to fish in 8 ppm and 0.2 ppm concentration respectively, but when put together their toxic effect comes down to lower level (1 ppm zinc and 0.025 ppm copper). Likewise, it is experimentally shown that cadmium, chromium

(K_2CrO_4 $CrCl_3$, $6H_2O$), copper, lead, mercury, nickel, silver and zinc, if present in 1 ppm concentration in water can reduce BOD by 16, 18, and 35, 44, 23, 72, 25, 81 and 8 per cent (Stones, 1976).

Venogopalan (1977) from his experiments conducted to study the effects of copper and zinc on growth and photosyntheis of *S. constatum* and *N. longissima* (a marine species) remarked that copper concentration up to 10 µg cu/ 1 (0.01 mg/l) increased the grówth and photosyntheis in both the species. Although the cell multiplication was reduced at copper concentration above 25µg/l (0.025 mg/l) the photosynthesis was not affected even up to a concentration of 50 µg/l (0.0.5 mg/l). He further observed that unlike copper, relatively high concerntration of zinc of 75µg/l (0.75 mg/l) was tolerated by *N. longissima* without any dertrimental effect on growth and photosynthesis. On the otherland, *S. costatum* was able to tolerate only up to 10 µg/l of (0.01 mg/l) zinc.

TABLE 29 : Heavy Metals Employed in Industries around Calcutta

Industry	Cd	Cr	Cu	Fe	Hg	Mn	Pb	Ni	Zn	As	B	Ag
Piants and varnish	—	—					—					
Perfumery drug.												
Pharmaceutical, chemicals	—	—		—.	—		—					
Electroplating	—	—	—		—			—				
Fertilizer	—	—	—	—	—	—	—	—	—			
Galvanizing	—											
Pulp and Paper		—	—		—		—	—	—			
Tanneries		—								—		
Photography and Development of X-ray plate												—*
Cosmetics									—			
Battery	—				—		—		—		—**	
Soap detergents												

* High Toxic to bacteria, soil flora and plants.

** Powerful herbicide.

Since the metallic concentration in the bhery water is not analysed round the clock or with a greater frequency round the year it is diffifcult to make any statement on its average presence in the sewage and finally in bhery water. There are, however, reasons to suspect that concentration of metals in the bhery water is not always low. The higher concentration of metals has been recorded in the soil sediments.

Such high low concentrations of metal present in the water perhaps keep the growth of biomass controlled. BOD_5 either in raw in raw sewage or in bhery water, is undoubtedly indicaing low values, Metals estimated in certain organs (kidney, liver and ovary) are also much higher than in fish tissue. Bioaccumulation of metals in fish may be higher than its toxic level if fish is exposed for a longer period in water having such toxicants presents in sublethal level. Whether bioaccumulation of such toxicants estimated in fish tissue or other organs during the period of their culture has any adverse effect on the physiological system requires further investigation. Under these water quality parameters fish did not show any adverse symptom affecting their survival. The growth rate of fish as reported by Ghosh (1980) has undoubtedly gone down as disussed earlier. The soil sediments, concentration of copper and zinc if much higher than the overlying water. Iron and manganese are serveral times more. As toxicant copper in combination with zinc may check the growth of any algal matter over the soil sediments (algal growth found negligible) which could have been used as food for bottom dwelling fauna. Probably it might have caused restricted biodegradation of highly non-conservative organic material otherwise soil pH should have come down to near neutral or acidic and released more of nutrients (phosphorus and nitrogen) both to soil and the overlying water. But this is not the case. Calcium bound solid wastes from tannery also help in raising the soil pH. The range of fluctuation of soil pH (7.5-8.1) in the bheri is considered unfavourable for chemical and biological activities. Further, rate of mineralisation of organic complex in soil was not high enough and remained with in a level of 26-20% only in an incubation period of 15-20 days with 0-day value around 8% from 7th day a gradual rise took place in the two bheries until the 15th day and was followed by a decline. In case of other three bheries the increase was highest around 20th day and was followed by a decline. Rate or mineralisation was undoubtedly poor. All these tend to show that, pehaps native microbes could not proliferate poor. All these tend to show that, perhaps native microbes could not proliferate on the non-conservative organic material either in presence of individual toxic material or in combination. The effect is also reflected on CN ratio, the average value of which remained around 20. Likewise benthic fauna and flora also suffered. Pesticidal harmful levels were not detected.

Bacterial loas (TPC) in water and soil encountered during water, sunner and rains indicated ranges 9×10^5 –3×10^6, 1.5×10^4 – 2×10^6 and 10×10^{10} – 5×10^5 and 4×10^6 – 7×10^6, 1.9×10^6 - 5×10^6 and 3×10^5 – 4×10^5 respectively. In the raw sewage, the values were 4×10^5, 7.6×10^5 and 10×10^4 respectively. In fish maximun TPC values were observed in the investine of mrigal (3×10^6), rohu (6×10^6) and Tilpia (9×10^6) during winter. In catla however, TPC values remained 1.5×10^6 during winter in kidney. In summer, the bacterial load was maximum in kidney and intestine of rohu (4×10^6 each) and in liver of mrigal (20×10^5). During monsooon TPC value was higher in kidney of mrigal (20×10^5) and rohu (25×10^5). Coliform bacteria showed fairly high values during monsoon in mrigal (40×10^4) and in winter in catla (1×10^4). In general, concentration of coliform was indicating much lower value while *E. coli* was negligible. Concentration of such microbial load evern in raw effluent is much lower than what is normally observed in domestic wastes (Coliform bacteria and *E. coli* can go as high as 1.9×10^6 and 2.6×10^7/100 ml under tropical environment).

13. COST-BENEFIT OF FISH CULTURE OF A SEWAGE-FED BHERY

The cost-benefit structure of a privately-owned sewage-fed bhery near Bantala is given below with break-up of expenditure of each item involved in Fish Culture.

Type of Bhery	:	Privately owned
Area	:	7.7 ha
Average depth	:	60 cm

Break-up Statement of Expenditure Involved

A. Input

Labour

Manpower and their mode of duty	No. of employees	Working days/year	Total man day/year	Daily Rate (Rs.)	Total expenditure (Rs.)
1. Monthly paid employees	7	365	2555	10	25550
2. Night watchman	2	365	730	9	6570
3. Netting					
a. Dragging for fish capture	6	120	720		
b. Dragging for raking	6	18	108 908	7	6356
c. Fish trans-stocking	6	12-15	80		
4. To carry the catch to the assembly centre	8	120	960	3	2880
5. Removal of water Hyacinth	5	90	450	5	2250
6. Dyke repair	25	40	1000	12	12000
7. Net mending	3	25	75	12	900
8. Repair of sheds and watch tower	5	30	150	12	1800
9. Smearing th earthen floor	2	24	48	5	280
10. Boat repair	2	20	40	20	800

Sum up statement of expenditure:

(a)	Total salary of employees for 6910 man day	= Rs.	59,386
(b)	Bonus of the empolyees (9) @ Rs. 500	= Rs.	4,500
(c)	Contingencies:		
	1. Cost of spawn/fry		25,000
	2. Bamboo and ancillary material		5,000
	3. Rope, thread etc. for mending nets		2,500
	4. Wood, coal tar etc. for boat repair		500
	5. Artificial feed for fish, lime etc .		5,000
	6. Panchyat tax		200
	7. Cess for the effluent		300
	8. Tax for the water area @ Rs. 3 00-bigha		15,000
	9. Ancillary expenditure		2,000
	Total contingent expenditure		55,500

Total expenditure involved on a, b, & c Rs. 119386

B. Out put

Total yield of fish 3000 kg/ha/yr i.e. 22.5 t.

Cost of 22.5 t fish @ Rs. 7.50/kg	= Rs.	1,68,750
Net Annual profit (B-A) (l,68,750-11, 9386)	= Rs.	49,354

Production Proportionately increases if depth of bheries increases (90-120 cm)

Source : Based on the information given by Secretary, Fish producer's Association, 24 Parganas, W.B.

14. MANURIAL POTENTIALITY OF SOLID WASTE AND AGRICULTURAL PRODUCE

Vegetables belonging to different species are cultivated in an area of about 5,000 bighas (666.6 ha) of land around Dhapa, near Bantala, by manuring the land with the solid waste of the metropolitan city and irrigating them with the raw sewage. Such solid waste has high potential nutrients like nitrogen, phosphorus and potassium. This is evidenced from NEERI'Z report (stated by Samaddar, 1970). Analyes of samples of refuge drawn from the major dumping sites of Calcutta indicate C/N ratio and other chemical constituents as follows.

TABLE 30 : Chemical Constituents of Calcutta Refuge

A Parameters	SITE Dhapa	Bantala	Average
Carbon %	18.89	18.38	18.63
Nitrogen%	0.55	0.54	0.55
C/N Ratio	35.90	33.85	34.87
Phosphorus (P_2O_5)%	0.57	0.62	0.59
Potassium (K_2O)%	0.38	0.42	0.40

Thirteen years later, Banerjee and Srinivasan (1983) have shown that chemical characteristics of Calcutta city refuse maintained almost similar characteristics. Replication of 5 sampels indicate ranges as:

B Parameters	Ranges	Average Value
Moisture	39.378-46.03%	43.753%
pH	7.2-7.4	7.08
Nitrogen	0.429-0.584%	0.521%
Phosphorus (P_2O_5)	0.428-0.588%	0.504%
Potash (K_2O)	0.264-0.40%	0.320
Organic Carbon	17.63-22.16%	19.68%
C/N Ratio	36.49-41.09	38..23

Such potentially rich nutrients of N, P and K undoubtedly produce enough of vegetables. Though not being done very scientifically, it is reported that 5000 bighas (666.6 ha) of land around Bantala is being used for producing vegetables in large scale to feed the city markets regularly. Besides, maize, cauliflower cabbage, raddish, brinjal, bottle gourd, pumkin potato, mustard seed, variety of kitchen vegetables of bitter taste (karela), different types of edible herbs and varieties of flowers are largely produced. It is reported that Dhapa area is supplying, at present, 70-75 lakh tons of vegetables annually to Calcutta City aginst 1.7 lakh tons beig supplied 20 years back. That fall in the produce is mainly for two reasons (1) That area of cultivable land, 5000 bigha (666.6 ha) has been reduced to 3000 bigha (400 ha) due to the growing township in the Salt lake area and (2) the city garbage is being diverted elsewhere for filling the low lying swams to be used for urbanisation.

Out of 6000 cu ft of sludge production per day, Calcutta itself produces 2600 t of solid waste material. Based on the analysed data of sludge waste, the calculated annual nutrient potential come to 2710 , 2907 and 1971 t of N, P and K respectively while for silid waste material the value comes to 4745, 4745 and 2847 t respectively. On considering such high nutrient value the Agro Industries Corporation of Government of West Bengal has established a mechanical compost plant at Bantala. The plant was commissioned on 23.10.78 with a capacity process 125/150 t of city garbage daily in two shifts. Finished product was approximately 33% of the input. Due to frequent load shedding full capacity production was never achieved. Available production, as reported in the Annual Reports 1980-81; 1981-82 of Agro-Industries Corporation Limited, Govt. of West Bengal, is shown in the table below. No data is available after, 1981-82 and, as reported, the plant is no longer in operation. The fertilizers were being used by the agriculturists in different parts of Sunderbans.

1978	:	600 t (By enquiry)
1979-80	:	2065.56 t
1980-81	:	3262.4 t
1981-82	:	1377.430 t

15. PROBLEMS ASSOCIATED WITH PISCICULTURE AND AGRICULTURE – THEIR REMEDIAL MEASURES

Sewage, in liquid, sludge of solid (garbage) from, has its own potential nutrient value - volume of such liquid cannot increase overnight to meet the growing demand of the pisciculturists and agriculturists. The existing practice is to use the raw sewage indiscriminately round the clock in most of the bheries without considering its actual requirement. The principle behind is to overstock fry/fingerlings with a consideration that the food for the young fish always remains adequate. By doing so dissolved oxygen saturation remains below the optimum level for most of the time and at times causes asphyxiation of even death of fish. Oxygen production from photosynthetic activity cannot cope up with the high demand. Controlled use of sewage may avert this oxygen crisis.

Analysis of water reveals that there is hardly any difference in chemical and biological composition in the bhery water and that at the outlet point. The aim to introduce sewage it to enrich the bhery water to a desired level. In reality water that comes out of the bhery can be used in another bhery or even a series of bheries till it retains its nutrient potentiality. This can make the utilization of sewage economical. This is not normally practical and the bhery water is drained out. In this manner the bery owners face scarcity of sewage supply from time to time more particularly during dry seasons when the agriculturists compete for the sewage. Continuous use of raw sewage is a traditional practice. Whether the raw effluent can be periodically used by keeping the outlet closed for some days and lowering the stocking density is a problem which needs investigation. Plankton are likely to grow quicker in static shallow nutrient - enriched water exposed to direct sunlight and wind action.

The increasing demand for sewage for agriculture purposes has posed a constraint on pisciculture. This can be overcome by digging shallow wells around the garbage dumping sites and also near the sewage carrying drains. It has been investigated that such ground water contains enough of rich nutrients which could be an alternate arrangement to feed the bheries with sewage as and when demand arises. Some experimental work done by Subramanium and Kulasekaran (1985) and Handa (1985) in Jaipur in this line justify such a suggestion. The former authors observed chemical parameters of groundwater above and below tannery disposal points and the adjoining Palar river of Tamil Nadu. The results of this investigation are indicated below.

From the Tables 31 & 32 it is clear that by seepage, chemical characteristics of ground-water can be enriched appeciably and will be suitable for use either in pisciculture or in agriculture. Handa (1985) on the otherhand observed chemical composition of ground-water in shallow wells located at various distances from garbage disposing point (Table 25).

The above findings indicate that the water around 300-450 m maintained pH 7.8-8.5 Alkalinity: 300-360 mg/l; COD: 16 mg/l, Nitrate – N (6.0-8.0 mg/l) which are ideally suited both for pisciculture as well as agriculture. Thus, shallow wells can prove to be immense benefit particularly at a time of the year when sewage supply is short.

TABLE 31 : Chemical parameters of Groundwater around Tannery Disposal Points and the Adjoining Palar River.

	Above outfall (1.5-3.0 km)		*Below outfall (1.5-3.0 km)*	
Parameters studied	*1968*	*1981*	*1968*	*1981*
Electrical Conductivity (mho/cm)	850-1150	1100-1500	1000-3000	2000-4000
Dissolved solids (mg/l)	600-800	700-1100	700-2000	1500-2500
Chloride (mg/l)	90-180	150-270	150-600	350-700

TABLE 32 : Characteristics of Palar River Water

Parameters	*Above outfall*	*Heart of polluted zone*	*Below outfall (Stretch 17 km)*
Electrical conductivity (μmhos/cm)	765	6800	1955-5200
Total dissolved solids: (mg/l)	530-540	5300	1280-3710
Total hardness (mg/l)	162-172	1025	454-598
Choloride: (mg/l)	61-66	2100	411-1511

TABLE 33 : Characteristics of Groundwater at various points of garbage disposing centre

Location	*pH*	*Total ALK (mg/l)*	*Hardness (mg/l)*	*NH_3-N (mg/l)*	*NO_3-N (mg/l)*	*Fe (mg/l)*	*COD (mg/l)*	*SO_4 (mg/l)*
20 m	8.4	1240	980	0.8	0.8	3.0	200	680
150 m	8.4	463	548	0.4	6.0	0.8	40	160
300 m	7.5	300	360	-	6.0	0.6	16	160
450 m	8.5	360	310	-	8.0	0.4	16	96

Certain findings in the course of investigation viz., moderate organic production moderate mineralisation rate and low bacterial load, high pH level of soil, high accumulation of metals in some organs of fish are some of the parameters of concern and require proper attention for further investigation. It is suspected that metallic contamination prevents the bheries from deriving maximum benefit from organically enriched sewage by controlling plankton growth and organic decomposition, the process by which nutrients become available to the overlying water. Also bioaccumulation of metals prevents healty growth and reproduction of fish and other crustacean animals. Fish diseases are imminent. This is also supported by occasional mortality of fish more particularly of *Tilapia* which suffers in growth and heavy mortality due to some unidentified factors.

16. WATER HYACINTH – A MEASURE OF POLLUTION CONTROL

Recent investigation has shown that use of water hyacinth plant for control of water pollution is highly economical and effective method. This is being used even in developed countries (Wolverton, 1981). Controlled use of water hyacinth in conjunction with waste stabilisation ponds not only increases the BOD removal from sewage but also reduces the high total suspended solids in lagoon effluent by reducing density of algae which otherwise make up a major portion of the suspended solids. Nitrogen, Phosphorus, Potassium, Sulphur, Calcium and other minerals can also be removed from domestic sewage by harvesting the plant biomass. This harvested plant material is also a source of high quality protein, energy, fertilizer and other products. Further, the water hyacinth can readily absorb concentrated heavy metals such as lead, cadmium, mercury and nickel (Wolverton and McDonald, 1978b; Wolverton, Barlow and McDonald, 1976). It has been experimentally observed that water hyacinth were found to contain 594 ppm Copper, 297 ppm lead, 113 ppm silver, 164 ppm cadmium and 286 ppm Chromium, yet the plants exhibited no visible sign of toxicity (Wolverton and McDonald, 1977). It is interesting to note that water hyacinth can also rapidly concentrate cadmium from polluted water containing concentration as low as 0.001 mg/l (Wolverton and McDonald, 1978a). Further, in case of organic chemicals, Wolverton and McKown (1976) demonstrated that the water hyacinth can remove 36.4 g phenol per kg of plant in 72 hrs. Smith *et al.* (1977) monitored the water hyacinth absorption rate of toxaphene and found that 281 mg toxaphane per kg of plant could be removed in 48 hrs. Smith and Shore (1978) found that 258 mg mirex per kg of plant could be removed in 7 days. By this system of using water hyacinth in chemically polluted water comprised of photographic, general and pesticides even up to 95000 litre of waste per day can be treated. The overall BOD and COD can be reduced to 97% and 88% respectively. Thus its effectiveness is proved beyond doubt. Perhaps it is for this reason pollutants get absorbed by the water hyacinth abundantly growing near Bantala Lock gate and their presence is acting as blessing to the bhery owners. It is, hence, suggested that every bhery owner should arrange some space for use as settling tank where water hyacinth can profusely grow and control pollutants.

The saturated water hyacinth plants are to be periodically harvested and disposed off in an adjacent pit designed in a manner that contamination of groundwater is prevented from any of the heary metals leaching out from decaying plant material.

Water hyacinth, besides its use in the polluted water treatment, can also be used as fertilizer, Wolverton and McDonald (1980) have shown that the plant contains the following chemical constituents:

Crude	:	22.9 per cent
Fat	:	2.12 per cent
Fibre	:	17.8 per cent
Kjelhal Nitrogen	:	3.65 per cent
Phosphorus	:	0.87 per cent

Futher, it has been experimently shown in NSTL (National Space Technology Laboratories) in Mississipi that water hyacinth digested at 36°C can produce 350 to 411 litres of biogas per kg of dry plant material (Wolverton, McDonald and Gordon, 1975; Wolverton and McDonald, 1978c). Such biogas contained 60 per cent methane. Their estimate stated that one hectare of water hyacinth could produce 58,400 m^3 of biogas containing 35,100 m^3 of methane.

17. CONSERVATION AND MANAGEMENT OF BHERY AND ADJOINING FIELDS USED FOR AGRICULTURE

Besides appreciable silting of Kistopur canal, and DWF and storm water flow canals there is no other source of any diluent to dilute the raw waste except the seasonal rain water. The harmful effect of using raw waste indiscriminately is already discussed. It is, hence, necessary to recycle the waste (liquid, sludge and solid) appropriately using modern technology to achieve maximum benefit. There are already indications of shortage of waste water due to increase in demand for its use both in aquaculture and agriculture.

Saha (1970) showed a production of 1258 kg of fish/ha comprising *L. rohita*, *C. mrigala* and *C. catla* which was subsequently raised to 3237 kg/ha by addition of extra 0.0337 m m^3/ha of sewage round the year. The sewage used by him contained 330 kg of nitrogen and 52.5 kg of phosphorus (P_2O_5). But the prescribed dose suggested by C.I.F.R.I. Barrackpore (Jhingran, 1974) for achieving a high fish production up to a level of 10,000 kg/ha was 90 kg of nitrogen and 40 kg of phosphorus. This shows that in Saha's recommendation nitrogen content was 3 times more than suggested by C.I.F.R.I., Barrackpore. Thus, the higher dose may result in hazardous condition in the ecosystem by demand of high oxygen and/or by producing excessive growth of algae; hence may be avoided. The excess waste can thus be used somewhere else.

For effective use of wastes, integrated system of waste recycling may be followed. Sewage contains, besides large quantities of useful water, potentially useful substances such as proteins, fats, animal feed supplements and fine chemicals such as vitamins and nucleic acids. Similarly, a number of useful byproducts including gas, organic acids, aminoacids, vitamins, proteins, etc. can indirectly be reclaimed by fermentation and stabilization ponds. Zain-ul-Abedin (1977) has shown a schematic diagram (cf. Fig. 5) of an integrated system for algal biomass production and waste treatment which although appears to be feasible is slightly modified (Fig. 6). In this system sewage effluents are fed into oxidation ponds in series and parallel; with or without pretreatment. The clear water thus obtained is used for fish farming, agrihorticultural practices and industrial recycling. The fishes and algae derive valuable proteins. The sludge is either digested anaerobically to give biogas (methane + CO_2) and farm manure, or it is chemically processed to yield various chemicals and biochemicals which are recycled. In this way, waste is very much reduced giving rise to low waste (or no waste) technology and in terms of economics, low cost (or no cost) technology compared to high cost technology on treatment of effluents.

Foster (1973) has suggested that the sludge produced in the settling tanks could be converted to valuable commodities like fine chemicals, gas, oil, paper, plastics and animal feed. The Wolfsen laboratory Cardiff and Interplan Ltd., Croyoton, have recently produced a sponsored report claiming that protein production from a range of wastes, including liquid organic wastes and sewage sludge, should prove technically and economically feasible (Surveyor, 1974) for fish culture.

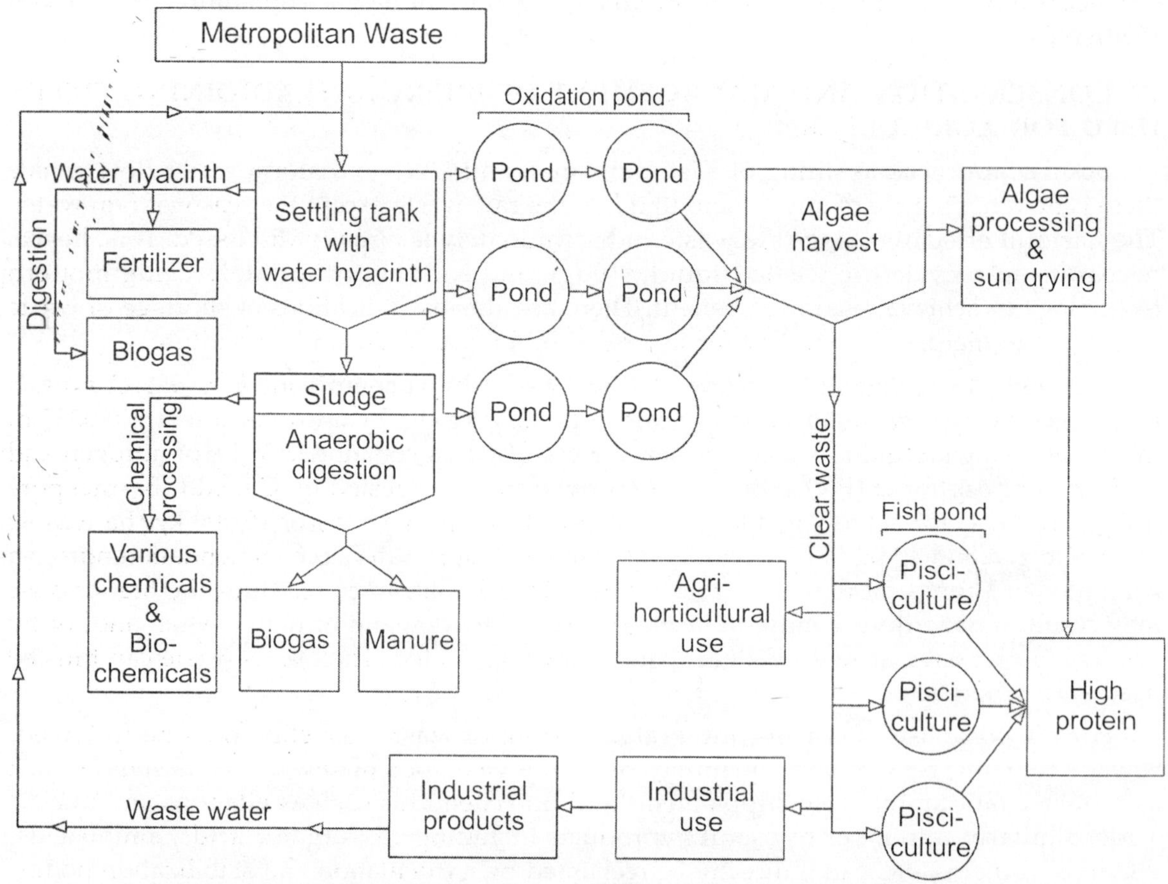

Fig. 6 : Schematic diagram showing integrated system of waste recycling. Modified from Abedin (1977)

In the field of agriculture the existing practice of using raw garbage is far from scientific hence there is enough scope to undertake improvement and increase production. Since sewage dumping grounds are important perennial source of supply of vegetables to the metropolitan city scientific application of measures NPK (produced from garbage) in the field supplemented by primary treated sewage can definitely enhance the production of vegetables. The fear of metallic interference in the crop may be remote as in agriculture, tolerance limit is high. There might be some health hazards as vegetables may get contaminated with parasitic eggs and larvae present in garbage.

18. SUGGESTION FOR FUTURE INVESTIGATION

Systematic monitoring of the metropolitan waste at certain points along the DWF canal must be carried out to collect the baseline data on physical, chemical, biological and bacteriological characteristics, since it is being utilized regularly in the bheries. A similar study is essential to find out the extent of changes that take place in the bhery water after the introduction of the waste. Such study will be extremely useful in estimating the quantum of effluents required to be allowed in the bhery for getting optimum production without creating any ill effect.

The presence of pesticides and heavy metals in the waste, water and soil of the bhery, and their effect on biota (fish of different sizes) need thorough investigation. Synergistic and antagonistic effect and Chelation of metal ions with organic compound in relation to their effects on aquatic life remain to be studied.

Bioaccumulation tests of metals in different organs of various fishes of different size groups should be carried out to determine index of heavy metal accumulation. This will help to ascertain whether growth and maturity of fish are affected and at which age group.

Toxicity tests by bioassays are well known to provied valuable information of the danger due to toxic material in excessive volume of waste water that are released in the aquatic ecosystem. Bioassay is a measure of toxicity effect without necessarily knowing the particular factor responsible for the same.

The basic bioassay consists of a 96 hr exposure of an appropriate organism in number adequate to assure statistical validity to an array of concentration of the substance, or mixture of substances that will reveal the level of pollution. It will cause (1) irreversible damage to 50 per cent of the test organisms (2) the maximum concentration causing no apparent effect on the test organisms in 96 hrs. Tests should be conducted in a flow-through system so that the organisms are exposed continuously to a fresh solution of the test material appropriately diluted with water of the same quality as that at the site of the proposed discharge.

Appropriate test must be undertaken to demonstrate the relative sensitivity of economically important species and substitute species to the test material so that meaningful interpretation of the data can be made. P. Doudoroff and his associates are the pioneers in this field and readers can refer to the paper published in 1951 in *Sewage and Industrial Wastes*, 23 (11): 1380-1397. *APHA* (current edition) also has provided basic principles of conducting bioassay experiments. The author has however, exhibited a simple device for continuous flow apparatus that can be used for bioassay studies.

19. CONTINUOUS FLOW APPARATUS FOR BIOASSAY EXPERIMENT

Effluent or test material in different concentrations (earlier determined by exploratory tests) in a selected dilution water are allowed to run in the mixing chambers for proper mixing and released into the test jars where test fish (at least 10 Nos. in 10 lit. jars) or organism are kept with adequate control (Fig. 7). Close observations on the reaction of the test fishes are to be made during the first few hours since the introduction of the fishes. This will give an indication of the nature of toxicant and serve as guide for futher tests. The number of fishes which die during 24, 48 and 96 hours of the test should be recorded. Test animal should be deemed dead if the respiratory and other movements cannot be readily detected by mechanical stimulation during an observation period of 5 mts. Dead fish and any other accumulated suspensoids collected at the bottom should be removed. The most simple method of determining LC_{50} value (median tolerance limit) is by plotting the experimental data on semilogarithmic co-ordinated paper with test concentrations laid off on the logarithmic scale (ordinate) and survival percentages on the arithmetic scale. A straight line is drawn between the two points representing survival percentages at two successive concentrations of the test series which are lethal to more than half and to less than half of the test animals. The perpendicular drawn from 50% survival line gives the LC_{50} value.

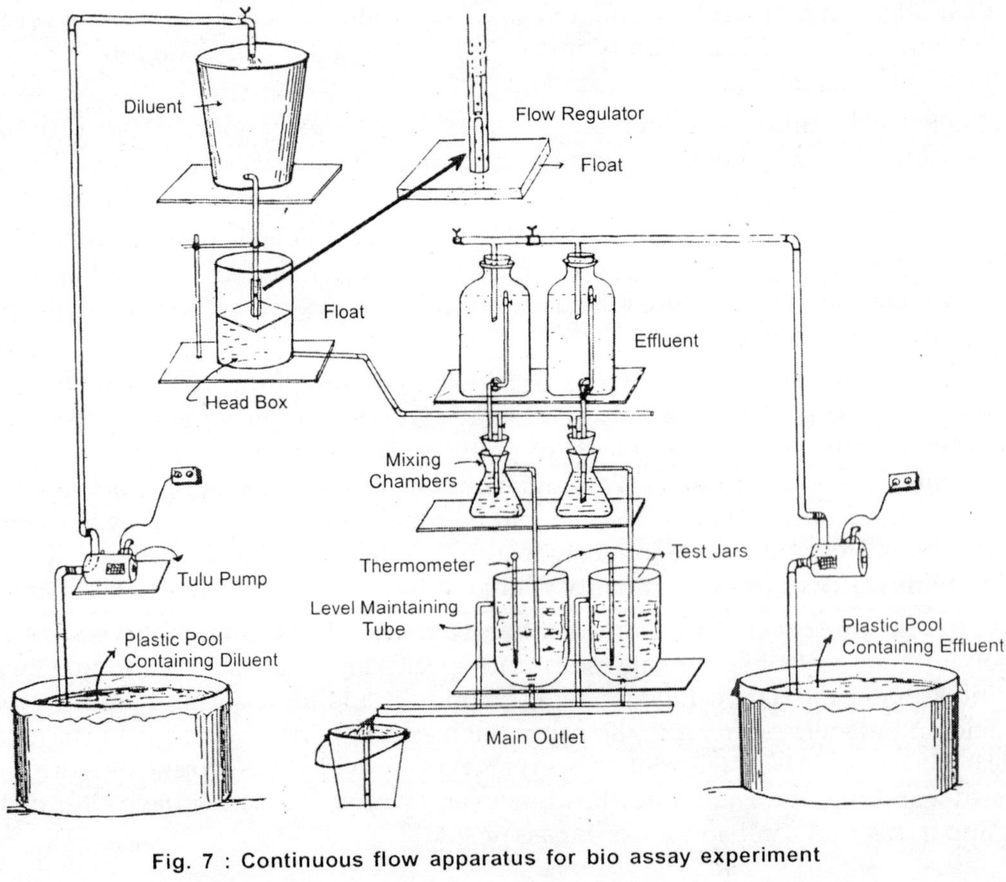

Fig. 7 : Continuous flow apparatus for bio assay experiment

An application factor of 0.1-0.1 is multiplied with the LC_{50} value to have a safe level for the discharge of the test material to the water body. During the entire period of experiment dissolved oxygen in the jars should be maintained at 4-5 mg/l. This can be done by maintaining the D.O. of the diluent at a higher level.

Since fish can avoid area contaminated by metals, even in low concentration, their avoidance reaction must be determined by testing effluents under different, dilution by a special device followed by Benjamin *et al.* (1956). The test is very sensitive and will help to find out if fish of any size group has a tendency to avoid any particular dilution. Such dilution level cannot be determined by any other method. A simple and modified form of the device followed by Benjamin *et al.* (1956) is indicated in Fig. 8.

Detailed investigation on epizootics in bheries, an essential aspect in pisciculture, is needed.

Standardisation of stocking density both in nursery and rearing ponds requires to be evolved.

20. AVOIDANCE TESTING APPARATUS

A number of devices are in use to study the avoidance reaction. The one described by Benjamin *et al.* (1956) is quite simple and is used in the present study with slight modifications.

The apparatus consists of a transparent plastic tank of the size 220 cm × 33.5 cm with 20 cm height. At one end it is divided into 4 channels, each 65 cm long and of equal width, formed by glass partition. The water input in each channel is controlled by separate valves. Each drain is slotted at its outlet in the remaining part of the main tank at a level 7.5 cm above the tank base. An oblique baffle prevents water from passing the drains and the water level is adjusted so that the tops of the drain are 6 cm beneath the water surface (Fig. 8). When the outflow from a channel is adjusted to equal the inflow to the channel, it appears that almost no water moving down the channel passes beyond the drain. Thus a fairly sharp boundary between the water flowing through the channel and standing in the tank beyond the end of the channel is maintained.

The waste to be studied is mixed in separate large plastic tanks with diluents in required dilutions (obtained from exploratory tests) and the mixture is introduced with sufficient force in three channels while the fourth channel is attached to a tank kept as control. Adequate dissolved oxygen in each of these tanks is ensured (by oxygenating the diluent, if required). Initially fish are left undisturbed in the tank for a short time for acclimatization. Observations made before the introduction of diluted waste indicate that the fish do not show any marked preference for any particular channel. Both the glass baffles were then slowly lifted to a height and the test water is introduced with adequate force giving chance to fish for selecting a particular channel and congregate in large numbers in preferred channel. One hour or less is usually sufficient for recordings fish entires after introduction of test water. The system of flow may be reversed to confirm the tests made.

The tank can be enclosed in a separate white walled compartment in which two flurorescent tubes are fixed from the ceiling to provide uniform light throughout the tank.

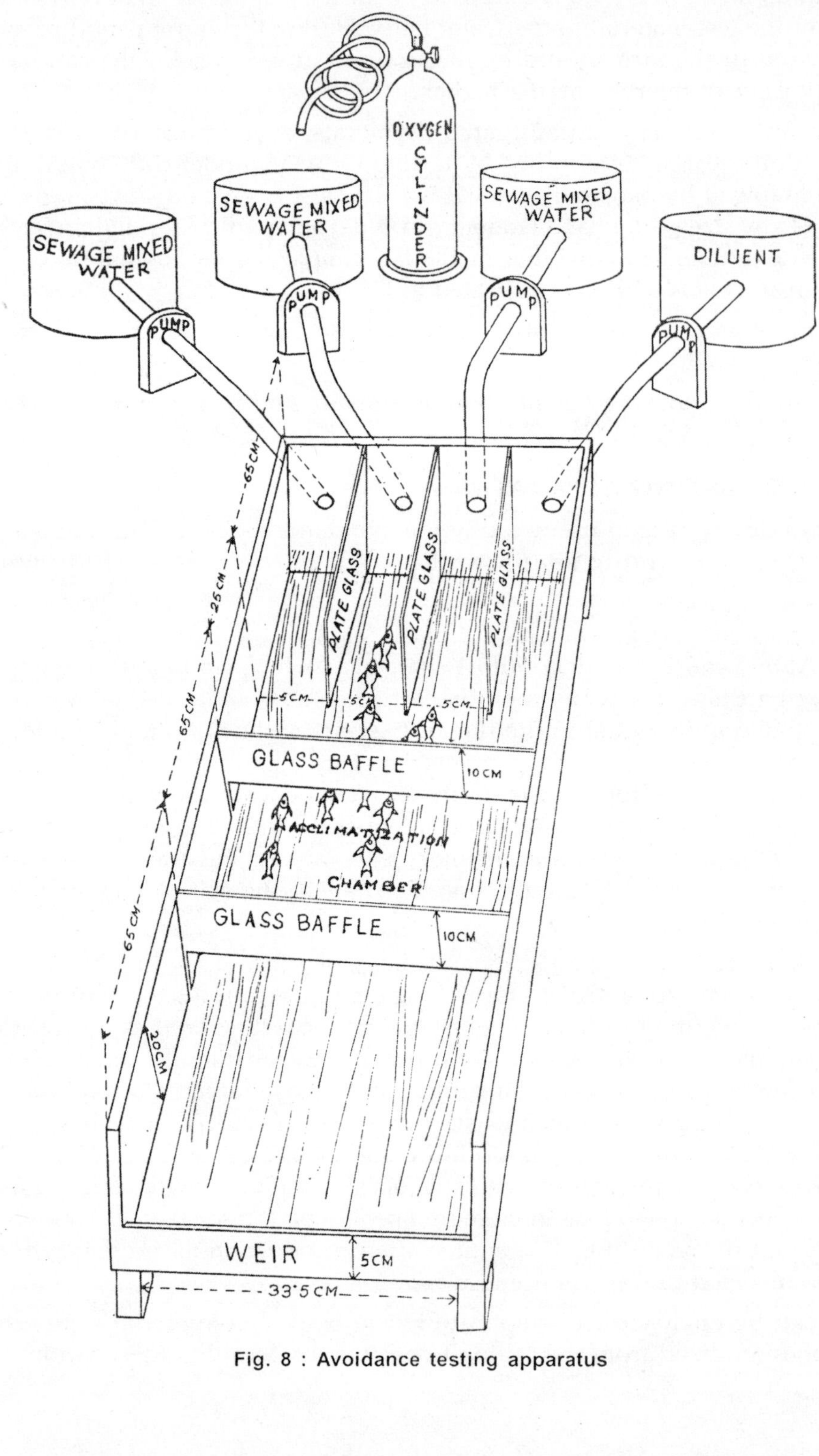

Fig. 8 : Avoidance testing apparatus

The percentage of avoidance is computed by the formula:

Per cent avoidance = 100 (E-A)/E,

where A = Total number of entries of test fish in the channel containing effluents, and

E = Total number of entries of test fish in the control.

Addendum - A (some tit bits) On freshwater fishes

Scientists have opened a new horizon in the field of fishery science by,(a) Genetically enhancing the growth of fish Catla (*Catla catla*) and Rohu (*Labeo rohita*) to nearly four times and (b) discovering a new cell in the fish body catla (*catla catla*) which can excrete insulins, (c) In hybrid fish culture system fishes (Rohu, Catla and mrigal) get infected by microscopic organisms of the mexogoa group. This affects the mother carp also. These are briefed below:

(A) Scientists Fish in Genetic Water

– Stanley Theodore (Hyderabad). *The Statesman*, 5.5.2003.

1. A silent revolution is taking place at the Centre for cellular and molecular Biology here which augur well for the quintessential Bengali foodie. Scientists are working on means to make fish grow faster and larger in size and buffaloes yield more milk with the help of genetic science.
2. The idea is to create genetically enhanced animals that will reach the optimum marketable size in greatly accelerated growing periods thus increasing their commercial value.
3. Scientists are working on two types of edible fish - the Rohu (*Labeo rohita*) and Catla (*Catla catla*) considered the testiest from amongst the repertoire that graces the Indian palate.
4. A growth hormone gene found in almost all living is extracted from one fish and grafted into the target. This becomes the transgenic fish which grows to nearly four times the normal size as stated by the scientists. Thus an ongoing genetic project in Hyderabad may bring about a revolution in the local market.

(B) Macher Anya Koseo Insulin – Abiskar Kolkatar

(Insulin is detected in other cells also of fish - Discovered in Kolkata).

– (Dabdut Ghosh, Thakur, Ananda Bazar Patrika 7.3.2003).

Quite silently revolution has taken place by the Indian Institute of Chemical Biology (IICB) Jadavpur (Kolkata).

The Scientists have traced a cell in fish body that can excrete insuin, jut at a time when Scientists around the world are trying their best to find out some preventive medicines for diabites. Such discovery is a worth mentioning step - claimed by Samir Bhattacharya, head of IICA. A special type of adipose tissue in fish which helps forming fat in the body has been traced which creates insulin. This helps reducing the quantity of glucose. The fact that chemical insulin when used for a long time develops adverse reaction is now established. Samir Bhattacharya has further stated that adipocytes cell obtained from the fat tissue of catla fish (*Catla catla*) has been transplanted in the body of hamstar (a kind of animal of money group). It has been noticed that adipocyte cell is already growing insulin and simultaneously reducing the glucose content. The experiment performed on hamstar is quite encouraging. Such experiment will now be tried in the human body. The research findings of scientist which claims international credit has been welcomed for publication in the well known journal "Endocrinology" which means its recognition in the "Endocrine society of America". The highest organisation in the world which recognises only top research work on hormone.

The entire work has been done jointly by the scientists of "Guromar Institute for molecular and cellur regulation of Japan and School of Life Science, Vishavharati" and jointly discovered the cell in fish.

(C) Bangaleer Mach Khower Din Sesh Hote Choleche

(Bangalees' fish eating days are being numbered)

– Nirmal Ghosh

Evening edition of daily Newspaper "PRATIDIN" (Kolkata) 12.11.2003.

To satisfy the increasing demand of fish the hybrid system of fish culture has been in practice. This has come as a worning for fish culturist.

Dr. Saugata Basu of Kalyani University (West Bengal) while conducting the research on this problem has shown some interesting findings that the hybrid fish rapidly get infected by microscopic organisms of the mexojion group. Such infection takes place in the mother carp also. Dr. Basu further observed that of the infection resulted from 19 mexojion organisms 11 are new species. From a survey it is revealed that when the percentage of infection in the mother corp Rohu, Catla and mrigal are 18.62, 20.37 and 26.91 their corresponding infection in hybrid fishes are 61.37, 58.20 and 78.10 percent respectively which is quite alarming. Dr. Basu also recorded the size and weight of fishes.

The above interesting findings are published in the journal "PROTOJOLOGICA" of Poland and also in different news papers of the country.

REFERENCES

Abedin, Zain-Ul, 1977, Domestic sewage and sewage treatment. Lectures presented at the 5th FAO/ SIDA workshop on "Aquatic Pollution in relation to protection of living resources" Manila, Philippines, January & February.

Banerjee, A.N. 1931, History and Hydraulics of the river system near Calcutta, Calcutta Corporation.

Banerjee, A.K. 1964, *Forest of Sundarbans*, Centenary commemoration volume, Writers Buildings, Calcutta, India. pp-166-75.

Basu, S.P. 1950, A study of the ecology and bionomics of Indian Carps, *Catla catla* (Ham), *Laboo rohita* (Ham) *and Cirrhina mrigala* (Ham) cultured in the sewage irrigated fish farms near Calcutta. *Proc. Nat. Indian Sc. Cong.* ; p 3. Abstracts : 252-253.

Biswas, K.P. 1927, Flora of the Salt lakes, University of Calcutta, Science college.

Bose, P.C. 1944, Calcutta Sewage and Fish culture *Proc. Nat. Inst. Sci. India.* 10 : 443-454.

Chatterji, G.C. 1934, *Biological Control of malaria in the rural areas of the lower Bengal.* Central Co-operative Antimalaria Society Calcutta : 15-16.

Chandra, S. 1977, An Ecofloristic Survey of the Mangroves of Sundarbans, West Bengal. *Trans. Bose. Res. Inst.* 40 (1) : 5-14

Chandra, S. and B.B. Mukherjee 1969, Radio carbon Dating of two Micro-fosiliferous Quaternary Deposits in and around Calcutta. *Sci & Cult.* 35 : 275-6

De, K.C. 1910, *Report of the fisheries of Eastern Bengal and Assam.* Eastern Bengal and Assam secretariat, shillong, p. 33.

Dutta, N.C., B.K. Bandopadhaya, N. Mondal and C. Maity 1983, Diurnal variation of some abiotic factors and plankton communities in two fresh water ponds at Calcutta. *Environ Ecol.* 1 : 129-133.

Dutta, N.C. and S. Chaudhury 1984, Primary productivity and its correlation with the physico-chemical parameters of a freshwater sewage-fed impoundment near Calcutta. *Environ. Ecol.* 2 : 65-68.

Foster, C.F. 1973, Eff 1. *Water treat J. 13* (11) 697. In : *Notes on water pollution.* No. 67, Dec. 1976, Water Research Centre, Stevenage lab. Herts, U.K.

Gangopadhaya, Naresh Chandra 1983, Deficiencies in development planning in Agriculture of West Bengal, Souvenir, 1983, organised by Institute of Applied Humanities and Socio-Economic Development consortium, Calcutta-78.

Ghosh, A.K. 1941, Submerged Forests in Calcutta, *Sci & Cult.* 6 : 669-70.

Ghosh, D. 1983, Anterjatic Jaler Dasak Moilajaler Samasaya O Amader Partibesh. (Water decade of International Polluted water, its problem and our environment) In : Subhadut : March, 1983.

Ghosh, P. 1980, Saptarathir Akromone Jarj Arita Mithe Jaler Bheri Samuha Muktir Path Khoje. (The sweet water bheries due to their sufferings from numerous opponents are seeking their way of rescue). Observation of Fish culture day and Seminar, December. Convened jointly by Department of Fisheries, West Bengal and Fish Producers' Association, Calcutta.

Gupta, K.G. 1908, Results of enquiry into the fisheries of Bengal and into the fishery matters. Bengal Secretariat Book Depot, Calcutta, pp. 65-109.

Gupta, A.C. 1957, The Sundarbans, its problem, its possibilities. *Ind. For.* 83 : 481-7

Handa, B.K. 1985, Ground water quality Management with reference to anthoropogenic activities. *National seminar on pollution Control and environmental management.* Vol. 3, 16-18, March 1985. National Environment Engineering Res. Institute, Nagpur.

Jhingran, V.G. 1974, *Fish and Fisheries of India.* Hindustan Publishing Corporation, Delhi 110007.

Joues, Benjamin, F., Warreu, Charles E., Bond, Carol E, and Peter, Doudoroff. 1956, Avoidance reaction of salmonid fishes to pulp mill effluents. Robert A. Taff sanitary engineering centre USPHS. Cincinnati Ohio, Sewage and Industrial wastes. Vol. 28. No. 11 November.

Maity, J. 1976, *Mahatirtha Ganga Sagar.* Progressive Book Forum, Calcutta.

Mitra, A. 1954, Census, 1951, West Bengal, District Hand Books-24 Parganas, West Bengal Government, Press, Alipore, W.B.

Mukherjee, H.K. 1938, Report of the scheme for the investigation of the life history, bionomics of fresh water fishes of Bengal, Simla, pp. 1-33.

Mukherjee G.D. 1959, Drainage in Greater Calcutta area. *The Journal of Institution of Engineers* Vol. XXXIX, July No. 11 pp. 6.1, 1061-1083.

Mukherjee B.B. 1972, Quatermary pollen Analysis as a possible indicator prehistoric agriculture in the Deltaic part of West Bengal. *J. Palynol.* 8 : 144-51.

Nair, K.K. 1944, The effects of Calcutta sewage on fish life. *Proc. Nah. Inst. Sci. India* 10 : 455-458.

Naskar, K.R. 1985, Halophytes and their unique adaptation on the Sundarbans Mangrove swamps in India. *J. Indian Soc. Coastal agri. Res.* 1 (2) 91-105.

Naskar, K.R. and D.N. Guha Bakshi 1987, Mangrove Swamps of the Sundarban. An ecological perspective. Naya Prakash, 206, Bidhan Sarani, Calcutta-6.

National Environmental Engineering Research Institute (NEERI) 1978, *Water Quality Survey of the Hooghly Estuary* (1974-76) Part III. *Survey of major Pumping stations of Calcutta including their disposal system* sponsored by CMDA.

Preston, H.B. 1915, Report on a collection of mollusca from the outskirts of Calcutta. *Rec. Ind. Mus.* Vol. XI.

Pal, R. 1982, Paschimbanglar Chabbish Pargana Jaleer Mechogherite Macher Rog/Tar Pratikar (Infection in fish in beheires of 24 Parganas (West Bengal) and their remedy). Convened by MPDEA. Department of fisheries West Bengal and Business Association-Prawn Export, Calcutta.

Raha, Atanu 1999, Sundarbone Misti Jal Komaya Sonokote Gach, Bagh. Reduction of fresh water ingress has caused forests and Tigers in trouble - Debdoot Thakur. In : Ananda Bazar Patrika : Nov. 17, 1999.

Rao, V.S. 1959, Mangrove forests and the problem of reclaiming Saline banks. *Proc. Mangrove symp.* Faridabad, pp. 58-65.

Ray, P. 1993, Aquaculture in Sundarban Delta-Its perspective-(An assessment) International Books and Periodical supply service, Delhi.

Roy, P. and S. Bandopadhyay, 1984, Utilization of Municipal wastes of Calcutta Metropolis in Freshwater Bheries for Aquaculture Programmes and Problems associated with its recycling. Final Report. 1982-84, ICAR Research Project. CIFRI, Barrackpore, West Bengal.

Ray, P., S.B. Saha and R.K. Banerjee, 1981, A case study of use of Calcutta Metropolitan, wastes for fish culture in the Bidyadhari-Kulti complex, West Bengal. International symposium on wastes Resources conservation, Pollution and abatement. Dec. 11-13, Dept. of Civil Engineering, University of Roorkee, Roorkee.

Saha, K.C., D.P. Sen, P.C. Mukherjee and S.K. Chakraborty, 1958, Physico-Chemical qualities of Calcutta sewage from the viewpoint of pisciculture and the danger of feeding the raw sewage to confined fisheries. *Indian J. Fish,* 5 (1) :

K.C. Saha, 1970, Fisheries of West Bengal Govt. Press, Calcutta, pp. 138.

Sen Purnendu 1941, Observations on the method of carp culture in the so called salt lakes near Calcutta with a note on the fish fauna of the lakes. *Jour Royal Asiat Soc.* Bengal Science, 8 : 7-13.

Sen Gupta, Bimal Kumar, 1980, Kolkatar Pasei Laban Radh : Mahanagarer Sathi Ebong Bandhu : Laban Hrader Cromo-Bibartaner Etihas. Matsa Chasi Dibas Palan O'Seminar (Salt lake is by the side of Kolkata : A companion and friend of the great city. The history of progressive revolution of Salt Lake). Observation of Fish culture day centinary and seminar. Convened Jointly by Dept. of Fisheries, West Bengal and Fish producers' Association, Calcutta.

Sen, P. and M. Banerjee 1984, On the occurrence of reptition remnants in the Peat Bed of Barrackpore, 24-Parganas, West Bengal. *Paleobot. Soc. Ind.* 29 : 47-51.

Sewell, R.B. 1934, A study of the fauna of the Salt Lakes, Calcutta. *Rec. Ind. Mus.* 36 : 45-121.

Shaw, G.E. and E.O. Shebbeare 1937, The fishes of North Bengal. *Jour. Roy. As. Soc. Bengal.*

Smith, R.W. and F.L. Shore 1978, Water hyacinths for removal of mirex from water. *Mississippi Acad. Sci.* Supplement, pp. 22, 23.

Stones, T. 1976, A discourse on the biochemical Oxidation of sewage effluent. *Water Treatment Journal,* 16 : 352-355.

Subramonium, M. and A. Kulasekaran 1985, Tanneries and ground water pollution in polar river Basin, Tamilnadu. National seminar on pollution control and Environmental Management, 16-18 March. *National Environmental Engg. Res. Institute* Vol. 3. Calcutta.

Srivastava, C.B. 1975, Fish Pathological studies in India - a review. In : *Dr. B.S. Chauhan Commemoration Volume.* Ed. K.K. Tiwari and C.B. Srivastava. *Zool, Society of India.*

Thakur, N.K. 1975, Observations on the availability of brackish water fish seed in the Kulti estuary, West Bengal. *J. Inland Fish. Soc.* India, Published in March 1976.

Thakur, Kishalaya 1986, Gobarier Songkot Gavir Biparjayer Ingitbahi. (The problems of Gobario is an indication of great disaster). In : Ananda Bazar Patrika, June 1986.

Venugopalan, V.K. 1979, Effect of Pesticides and heavy metals on growth and primary production of a few species of marine algae. Expanded Abstract. *National Seminar on Protection of Marine Environment and related ecosystems.* National Institute of Oceanography Goa. No. 13-15.

Vishnu Mitre and H.P. Gupta 1972, Pollen Analysis Study of quaternary Deposit in the Bengal Basin. Palesbotanist. 19 : 297-306.

Wolverton, B.C. 1981, Water hyacinths for controlling water pollution. *Water Pollution and Management reviews.* South Asian Publishers Pvt. Ltd. New Delhi.

Wolverton, B.C. and R.C. Mc Donald 1977, Waste water treatment utilizing water hyacinths. *Proceedings of the 1977 National conference on treatment and disposal of Industrial waste water and residue,* University of Houseton, Texas : 205-207.

Wolverton, B.C. and R.C. Mc Donald 1978 (a), Bioaccumulation and detection of trace levels of Cadmium in aquatic systems of *Eichhornia crassipes. Environmental Health Perspectives. 27* : 161-164.

________________ 1978 (b), Water hyacinth sorption Rates of lead, mercury and Cadmium, NASA/ ERL Report No. 170 (Washington D.C.).

________________ 1978 (c), Energy from aquatic plant waste water treatment system NASA. *Technical Memorandum* TM-X-72730 (Washington D.C.).

_______________ 1980, Nutritional composition of water hyacinths grown on domestic sewage (In press).

Wolverton, B.C. and N.M. McKown, 1976, Water hyacinths for removal of phenols from polluted waters. *Aquatic Botany*, 2 (3) : 191-210.

Wolverton, B.C., R.C. Mc Donald and J. Gordon 1975, Bioconversion of water hyacinths into methane gas : Part I. NASA. *Technical Memorandum* TM-X-72725 (Washington, D.C.).

Wolverton, B.C., R.M. Barlow and R.C. McDonald. 1976, Application of vascular aquatic plants for pollution removal, energy and food production in a biological system. In : Tourbier J. and Pierson, R.W. (Jr.) (eds.) *Biological control of water pollution.* University of Pennsylvania Press Philadelphia. Pennoylvamid, pp. 141-149.

APPENDIX I

Reconnaissance Survey fo Freshwater Impoundments in The North Salt Lake Area*

Serial No.	Local name of the Fishery	Area of the Fishery (bigha)	Name and Address of Lessee/Owner	Daily average catch	Annual Return Md.	Annual Return Amount Rs.	P.S.
1	2	3	4	5	6	7	8
+1.(A)	—	305 bighas	Bidyadhari, Matsajibi Samabay Samity, Dattabad	5 mds.	1250	75,000/-	Dum Dum
+ (B)	—	160 bighas	-do-	2.5 mds.	625	37,500/-	-do-
+ (C)	—	784 bighas	-do-	10 mds.	2500	1,50,000/-	-do-
+ (D)	—	358 bighas	-do-	5 mds.	1250	75,000/-	-do-
+ (E)	—	365 bighas	-do-	5.5 mds.	1375	82,500/-	Rajarhat
+ 2.	Dipu Bhery	250 bighas	Sri Upendra Nath Pramanick Krishnapur Vill. Naya Patti	5.5 mds.	1375	82,500/-	-do-
+ 3.	Bara Bhery	1000 bighas	Shri Jugal Mondal Krishnapur. Vill. Mahisghot	30 mds.	7500	4,50,000/-	-do-
+ 4.	Choto Bhery	150 bighas	Shri Nilkanta Mondal and Bhusan Kr. Naskar -do-	5 mds.	1250	75,000/-	-do-
+ 5.	—	31 bighas	Shri Narendra Nath Naskar -do-	1/2 mds.	240	14,400/-	-do-
+ 6.	Kazir Bhery	700 bighas	Shri Kanailal Mondal, Nayyapatti Shri Narayan Mondal, Krishnapur Thakurdari	8 mds.	2000	1,20,222/-	-do-
+ 7.	Adhar Babur Bhery	50 bighas	Shri Adhar Chandra Pramanick Nayapatti, Krishnapur	1.5 mds.	375	22,500/-	-do-
+ (B)	Natun Bhery	200 bighas	Shri Gupi Nath Pramanick & Shri Upendra Nath Pramanick Nayapatti.	4 mds.	1000	60,000/-	-do-
+ 8.	Smitir Bhery	270 bighas	Shri Krishna Pada Pramanick Nayapatti	10 mds.	2500	1,50,000/-	Rajarhat
+ 9.	Akrur Bhery	300 bighas	Shri Harendra Nath Pramanic -do-				
+ 10.	Hauser Bhery No. 1	450 bighas	Shri Lalchand Pramanick Nayapatti	6 mds.	1500	90,000/-	-do-
+ 11.	Hauser Bhery No. 2	60 bighas	Shri Aswani Mondal -do-	1.5 mds.	375	22,500/-	-do-
+ 12.	—	105 bighas	Shri Kamal Ch. Mondal Mahisbathan	3 mds.	750	45,000/-	-do-
+ 13.	—	58 bighas	Shri Lakshni Kanta Pramanick -do-	1.5 mds.	375	22,500/-	-do-

....Contd.

Appendix I – *contd...*

1	2	3	4	5	6	7	8
14.	—	40 bighas	Shri Dhirendra Nath Naskar -do-	1.5 mds.	375	22,500/-	-do-
15.	—	9 bighas	Shri Chandan Mondal -do-	10 Srs.	62	3,720/-	-do-
16.	—	36 bighas	Shri Abodh Pramanick Thakurdari Krishnapur	1 md.	250	15,000/-	-do-
17.	—	30 bighas	Shri Khagen Mondal Mahisghot	30 srs.	187	11,220/-	-do-
+18.	—	778 bighas	Joint Stock Co. Dattabad	6 mds.	1500	90,000/-	Dum Dum
×19.	—	102 bighas (Tanks)	Fishery Deptt. West Bengal	0.5 md.	174	10,440/-	Dum Dum
×20.	Kashinath Bhery	85 bighas	Shri Narendra Nath Bar; Shri Nepal Bar & Shri Ratan Kanna Dattabad	3 mds.	750	45,000/-	Bhangur
×21.	Daser Bhery	200 bighas	Shri Harendra Nath Saha 128, Bow Bazer Street, Calcutta	4 mds.	1000	60,000/-	-do-
×22.	Gurer Bhery	650 bighas	Shri Gopal Ch. Bar Ghar-bhanga Paschim	15 mds.	3750	2,25,000/-	-do-
×23.	Kansari Bhery	200 bighas	Shri Balaram Mondal Mahisbathan	6 mds.	1500	90,000/-	-do-
24.	Choto Kansari Bhery	175 bighas	Shri Raj Banshi Singha 52, Chowl Patty Road, Calcutta	5 mds.	1250	75,000/-	-do-
×25.	Bager Bhery	583 bighas	Shri Jogendra Nath Naskar Beliaghata	5 mds.	1250	75,000/-	-do-
×26.	Choto Munshir Bhery	200 bighas	Shri Balaram Mondal Mahisbathan	10 mds.	2500	1,50,000/-	-do-
27.	Munshir	300 bighas	-do-				
×28.	Bhery	46 bighas	Shri Khagen Mondal Mahisghot	1.5 mds.	370	22,500/-	-do-
29.	Mondaler Bhery	385 bighas	Shri Tarak Ch. Mondal Hatgacha	6 mds.	1500	90,000/-	-do-
×30.	—	50 bighas -do-	Shri Hem Pramanick	1 md.	250	15,000/-	-do-
31.	Diller Bhery	127 bighas	Shri Lakshimi Kr. Pramanick, Mahisbathan	2.5 mds.	625	37,500/-	Bhangur
32.	Patrabad Bhery 1,2,3	950 bighas	Shri Bidhu Bhusan Sarkar and others Beliaghata	19 mds.	6350	3,81,000/-	-do-
33.	Narkeltala Bhery	540 bighas	Shri Nil Kanta Pramanick, Mahisbathan	8 mds.	2000	11,20,000/-	-do-
34.	Chinta Singha Bhery	550 bighas	Shri Sita Nath Bose Shri Janaki Nath Bose Shri Kalipada Bose Shri Hem Lata Bose Beliaghata	10 mds.	2500	1,50,000/-	-do-

....Contd.

Appendix I – *contd...*

1	2	3	4	5	6	7	8
35.	Nalban Bhery	1400 bighas	Shri Samir Sarkar and others. Beliaghata	25 mds.	8650	5,19,000/-	-do-
×36.	—	50 bighas	Shri Jew Prodhan -do-	1 md.	250	15,000/-	-do-
×37.	Saheb Bhery	232 bighas	Shri Naresh Ch. Ghosh -do-	5 mds.	1250	75,000/-	-do-
×38.	Rajar Bhery	312 bighas	Shri Probadh Maity Dattabad	10 mds.	2500	1,50,000/-	-do-
×39.	—	100 bighas	Shri Hiralal Mondal	—	—	Dry	-do-
×40.	Nurtala Bhery	208 bighas	Shri Sunil Mitra Ballyganj, Calcutta.	10 mds.	2500	1,50,000/-	-do-
×41.	Kudghata Bhery	196 bighas	Bidyadhari Matsajibi Samabay Samity, Dattabad	5 mds.	1250	75,000/-	-do-
42.	—	23 bighas	Shri Gopal Bar Dattabad	1 md.	250	15,000/-	-do-
43.	—	7 bighas	Shri Jatin Pali Syaratola, Gharbhanga Paschim	10 Srs.	62	3,722/-	Bhangur
+44.	—	20 bighas	Shri Balaram Kumar Naobhanga	0.5 md.	125	75,000/-	-do-
45.	No. 4 Bhery	700 bighas	Shri Samir Sarkar and others Beliaghata	15 mds.	5000	3,00,000/-	-do-
46.	Sarder Bhery	490 bighas	Shri Kamala Kanta Pramanick Mahisbathan	10 mds.	2500	1,50,000/-	-do-
47.	Chaker Bhery	220 bighas	Shri Santosh Pramanick Mahisbathan	5 mds.	1250	75,000/-	-do-
48.	Bara Paresh Bhery	775 bighas	Shri Hirendra Nath Bose 42/A, Grey Stree, Calcutta	15 mds.	3750	2,25,000/-	-do-
49.	Choto Paresh Bhery	300 bighas	-do-				
50.	Jhagra Shisha Bhery	600 bighas	Shri Upen Sapul Garfa	15 mds.	3750	2,25,000/-	-do-
51.	Nutan Bhery	115 bighas	Shri Noni Gopal Ghosh Dhakuria	3 mds.	750	45,000/-	-do-
52.	Gopeswar Bhery	265 bighas	Shri Khagen Mondal Mahisbathan	5 mds.	1250	75,000/-	-do-
53.	Gomokpota Bhery	500 bighas	Shri Samir Sarkar Beliaghata	10 mds.	2500	1,50,000/-	-do-
54.	Sahebamara Bhery	700 bighas	Shri Samir Sarkar 102, Beliaghata Main Rd.	20 mds.	5000	3,00,000/-	-do-
55.	Natar Bhery	1000 bighas	Shri Bidhubhusan Sarkar 6/4, Fern Road, Calcutta-10	15 mds.	3750	2,25,000/-	-do-
56.	Goltala Bhery	1250 bighas	Hasirani Sarkar. 69, Beliaghata Main Road Calcutta	5 mds.	5000	3,00,000/-	Bhangur

....Contd.

Appendix I – *contd...*

1	2	3	4	5	6	7	8
57.	Chhto Chhai-lavi Bhery	150 bighas	Shri Khagendra Nath Mondal and others 6/4 Fern Rd., Calcutta 19.	2.5 mds.	625	37,500/-	-do-
58.	Bara Chhailavi Bhery	285 bighas	Shri Bibhuti Ghosh 65/2, Hindustan Park, Ballyganj	3 mds.	1095	65,700/-	-do-
59.	Harhare Bhery	175 bighas	Shri Bhabatosh Ghosh & Others, 68/B, Belliaghata Main Road, Calcutta.	2.5 mds.	625	3,75,000/-	-do-
60.	Hara Singha Bhery	50 bighas	Shri Sukumar Datta 8/1/B Kirti Mitra Lane	7 mds.	1750	1,05,000/-	-do-
61.	Garumara Bhery	285 bighas	Shyambazar Area				
×62.	Durga Bhasan Bhery	2000 bighas	Shri Bibhuti Bhusan Ghosh 63/2, Hindusthan Park Ballyganj	30 mds.	7500	4,50,000/-	-do-
+63.	Haldia Bhery						
64.A	Charcharia No. 1	500 bighas	Shri Prakash Chandra Bose 56/G. Monohar Pukur Road	10 mds.	2500	1,50,000/-	-do-
B	-do- No. 2	18 bighas	Shri Binode Sardar Bamanghata	1 md.	250	15,000/-	-do-
C	-do- No. 3	40 bighas	Shri Gopal Mondal -do-	1.5 mds.	375	22,500/-	-do-
65.	Karunamayee Bhery	250 bighas	Shri Bihari Sardar Beliaghata	2.5 mds.	625	37,500/-	-do-
66.	Darir Bhery	120 bighas	-do-	3 mds.	750	45,000/-	-do-
67.	Buno Beel	110 bighas	Shri Ramanath Patra and others -do-	4 mds.	1000	60,000/-	-do-
68.	Bara Bhery	500 bighas	Shri Panch Mondal Bamanghata	8 mds.	2000	1,20,000/-	Bhangur
69.	Batar Bhery	68 bighas	Shri Lalit Mondal Hatgocah	5 mds.	1250	75,000/-	-do-
70.	Kangsha Chak	69 bighas	Shri Jatin Mondal Bamanghata	2 mds.	500	30,000/-	-do-
71. (A)	Bara Abad No. 1	250 bighas	Shri Abinash Chakraborty, Dhakuria	2 mds.	500	30,000/-	-do-
(B)	-do- No. 2	600 bighas	Shri Sushil Sapin. 161/1A Rash Behari Avenue, Calcutta	12 mds.	3000	1,80,000/-	-do-
72.	Hanakhali Bhery	1800 bighas	Shri Samir Sarkar Calcutta	30 mds.	7500	4,50,000/-	-do-
73.	Khal	15.5 bighas	Shri Jugal Pargre Ghar-shanga Paschim	15 Srs.	94	5,640/-	-do-
74.	Nalbon Khal	85 bighas	Shri Bangshi Singha, Sarkar Bazar, Beliaghata	2 mds.	500	30,000/-	-do-

....Contd.

Appendix I – *contd...*

1	2	3	4	5	6	7	8
+75.	Harhare Bhery	50 bighas	Shri Ganesh Ch Biswas and Indubala Biswas 21, New Tangra Road, Calcutta.	2 mds.	500	30,000/-	-do-
76.	Badyir Bhery	162 bighas	Shri Ashutosh Mondal Thakurdari	2.5 mds.	625	37,500/-	-do-
77.	Garumara Bhery	200 bighas	Shri Ganesh Ch Biswas and Kartik Chandra Biswas 21, New Tangra Road, Calcutta	3.5 mds.	875	52,500/-	-do-
78.	Chaparasi Khal	15 bighas	Shri Anil Kundu, 7, Champatala Sealdah	20 Srs.	125	7,500/-	-do-
79.	Chapra Khal	42 bighas	Shri Rakhal Sarder Dhapa Manpur	1 md.	250	15,000/-	-do-
80.	Bangheri Bhery	120 bighas	Shri Lalih Kr Pramanik Mahisbatan, Kistopur	2.5 mds.	625	37,500/-	-do-
81.	Chatar Bhery	18 bighas	Shri Subal Naskar Bamanghata	1 md.	250	15,000/-	-do-
82.	Ghosher Bhery	80 bighas	Shri Panchu Mandal Bamanghata	—	—	Paddy cultivated	-do-
×83.	—	15 bighas	Shri Ratan Kawra Gharbanganga Purba	15Srs.	94	5,640	Rajarhat
×84.	Khal	25 bighas	Shri Benoy Kr. Pramanick Krishnapur	1 md.	250	15,000	-do-
×85.	—	11 bighas	Shri Lalchand Pramanick Nayapatti	1.5 md.	125	7,500	-do-
×86.	—	9 bighas	Shri Khagendra Mondal Mahisbathan	1.5 md.	125	7,500	-do-
87.	—	9.5 bighas	Shri Lakshmi kr. Pramanick Mahisbathan	1.5 md.	125	7,500	-do-
×88.	—	47 bighas	Shri Tarapad Bar Gharbhanga Paschim	1.5 mds.	275	22,500	Dum Dum
×89.	—	39 bighas	-do-	1.5 mds.	375	22,500	-do-
×90.	—	94 bighas	Shri Gopal Bar Gharbhanga Paschim	2.5 mds.	625	37,500	-do-
+91.	—	25 bighas	Shri Atul Kr. Mondal Gharbhanga Paschim	20 Srs.	125	7,500	-do-

Source : * Secretary, Fish Producer's Association, 24, Parganas, W.B. in 1959.

+ Areas occupied or city extension.

Data not available for Blanks.

APPENDIX II

Reconnaissance Survey of Freshwater Impoundments in The South Salt Lake Area*

Serial No.	*Local name of the Fishery*	*Area of the Fishery (bighas)*	*Name and Address of Lessee/Owner*	*Daily average catch*	*Annual Return*		
					Md.	*Amount Rs.*	*P.S.*
1	*2*	*3*	*4*	*5*	*6*	*7*	*8*
1.	Nayabad Bhery	306	Sm. Haripriya Ray, Shri Pasupati Ray, Shri Umapada Ray, Shri Chhita Ranjan Ray. 46/9, Ballyganj	5 mds.	1250	75,000/-	Sonarpur
2.	Gaurapaja Bhery	200	Shri Khagen Mondal Bhagaban pur, Sonarpur	2.5 mds.for 7 months	525	31,500/-	-do-
3.	Keutemari Bhery	179	Shri Sarada Naskar Sonarpur	1 md for 7 months	210	12,600/-	-do-
4.	Bagdoba No. 1	75	Shri Protiva Ch. Roy 95, Picnic Garden, Calcutta	5 mds.	1250	75,000/-	-do-
5.	-do- No. 2	175	-do-	—	—	—	-do-
6.	-do- No. 3	83	-do-	—	—	—	-do-
7.	-do- No. 4	60	-do-	—	—	—	-do-
8.	Doltala Bhery	75	Shri Atul Kr. Ray 95, Picnic Garden, Calcutta	2 mds.	500	30,000/-	-do-
9.	Tharkurda Budhuk Bhery	171	Shri Protiva Ch. Ray, 95. Picnic Garden, Calcutta	2.5 mds.	625	37,500/-	-do-
10.	Gune nala Bhery	287	Shri Pranab Prasad Ray 106/A, Raja Basanta Roy Road (Lake Market)	5 mds.	1250	75,000/-	Sonarpur
11.	Belekhali Bhery	135	Shri Sushil Sapul; Shri Sashi Sapul & Shri Upen Sapui, Garfa 161/1A, Rashbihani Avenue	2.5 mds.	625	37,500/-	-do-
12.	Banshtala No. 1	70	Shri Nani Gopal Ghosh Deara	2 mds.	500	30,000/-	-do-
13.	-do- No. 2	47	Shri Nani Lal Mondal Deara	1 md.	250	15,000/-	-do-
14.	-do- No. 3	49	-do-	1 md.	250	15,000/-	-do-
15.	-do- No. 4	40	Shri Chhotulal, Kasba	1.5 mds.	370	22,500/-	-do-
16.	-do- No. 5	40	Shri Upendra Sapui, Garfa	1.5 mds.	370	22,500/-	-do-
17.	-do- No. 6	25	Shri Biswanath Mondal Deara	0.5 md.	125	7,500/-	-do-
18.	Sayarer No. 1	35	Shri Nishi Kr. Pramanick Deara	30 Srs.	240	14,400/-	-do-
19.	-do- No. 2	36	Shri Abinash Mondal Deara	30 Srs.	240	14,400/-	-do-
20.	Neurer Bhery	75	Shri Protiva Ch. Ray Bondel	2.5 mds.	625	37,500/-	Sonarpur
21.	Khorki No. 1	130	-do-	7 mds. for 3 months	630	37,800/-	-do-
22.	-do- No. 2	44	Shri Upen Sapul, Garfa	3 mds. for 3 months	270	16,200/-	-do-

....Contd.

Appendix II – *contd...*

1	2	3	4	5	6	7	8
23.	Bara Bhery	175	Shri Sushil Sapul, Garfa	6 mds. for 3 months	720	43,200/-	-do-
24.	Chhoto Bhery	69	Shri Jogendra Patra Shri Khagendra Bar Haltu, Kasba	1.5 mds.	375	22,500/-	-do-
25.	Jeler Bhery	233	Shri Khagen Mandal Ballyganj	5 mds.	1250	75,000/-	-do-
26.	Sunderer Bhery	156	Shri Pasupati Ray, Ballyganj	2.5 mds.	625	37,500/-	-do-
27.	Kajir Bhery	200	-do-	3 mds.	750	45,000/-	-do-
28.	Ghoser Bhery	165	Shri Ajit Kr. Naskar Ballyganj	2.5 mds.	625	37,500/-	-do-
29.	Majher Bhery	183	Sm. Kanak Prova Naskar; Sm. Neli Naskar; Sm. Kamala Mondal; Sm. Binapani Saha, Ballyganj	2.5 mds.	625	37,500/-	-do-
30.	Hogol Banda	233	Shri Mahananda Ray 46/9, Ballyganj Place	5 mds.	1250	75,000/-	-do-
31.	Hogol Banda	213	Shri Chittaranjan Ray -do-	4 mds.	1000	60,000/-	-do-
32.	-do- No. 2	90	Shri Jogen Bar, Phagabanpur	3 Srs.	188	11,288/-	-do-
33.	Singher Bhery	66	Shri Badal Singha, Bondel	20 Srs.	125	7,500/-	Sonarpur
34.	Chouddo Singh No. 1	279	Shri Nandu Ghosh, Bondel	5 Srs.	1250	75,000/-	-do-
35.	-do- No. 2	45	Shri Gopal Ch- Garai Chomurhal, Sonarpur	1 Sr.	270	16,200/-	-do-
36.	Uparer Bhery	350	Shri Ashuthosh Ghosh, Deara	3 Srs.	1095	65,700/-	-do-
37.	Nicher Bhery	150	Shri Bejoy Kr. Roy, Bondel 46/9 Ballyganj Place.	Not cultivated for one year			
38.	Sarash Danga Bhery	70	Shri Mohini Mohan Das Naskar & others, 229, P.G. Garden, Bondel	1 Sr.	250	15,000/-	-do-
39.	—	40	Shri Laiit Anra, Sonarpur	1 md.	250	15,000/-	-do-
40.	Muchir Bhery No. 1	33	Shri Amiya Bhusan Mondal 8, kali Tara Bose Lane	0.5 md.	125	7,500/-	-do-
41.	-do- No. 2	45	-do-	1 md.	250	15,000/-	-do-
42.	Badar Bhery	35	-do-	0.5 md.	125	7,500/-	-do-
43.	Atghara Bhery	63	Shri Atul Mondal, Garia	1.5 mds. for	130	7,800/-	-do-
44.	Muchir Bhery No. 3	33	Shri Debendra Nath Mondal & Others, Rana Bhutia	3 months			
45.	Nalban Bhery	75	-do-	7 mds. for	630	37,800/-	-do-
46.	Badyir Bhery	40	-do-	3 months			

....*Contd.*

Appendix II – *contd...*

1	2	3	4	5	6	7	8
47.	Muchir Bhery No. 4	99	Shri Abani Mondal & others -do-	4.5 mds. for 5 months	405	24,300/-	-do-
48.	—	48	Shri Sudhansu Bh. Gain Rana Bhutia	3 mds. for 3 months	207	16,200/-	Sonarpur
49.	Nalban Bhery No. 1	60	Shri Dhananjoy Mondal	1.5 mds. for 3 months			
50.	-do- No. 2	60	Shri Khagen Mondal -do-	1.5 mds. for 3 months	135	8,100/-	-do-
51.	-do- No. 3	60	Shri Kali Ch. Mondal -do-	1.5 mds. for 3 months	135	8,100/-	-do-
52.	-do- No. 4	26	Shri Kartic Ch. Mondal -do-	2.5 mds. for 3 months	135	8,100/-	-do-
53.	-do- No. 5	60	-do-		225	13,500/-	-do-
54.	-do- No. 6	190		4.5 mds for 3 months	405	24,300/-	-do-
55.	-do- No. 7	14	Shri Puten B. Mondal -do-	1 md. for 3 months	90	5,400/-	-do-
56.	-do- No. 8	10.5	Shri Niranjan Mondal -do-	0.5 md. for 3 months	45	2,700/-	-do-
57.	Atghara Bhery	63	Shri Atul Mandal, Garia	1.5 md. for 3 months	130	7,800/-	-do-
58.	Bagdir Bhery	364	Shri Charan Naskar, Nayabad	2.5 mds. for 10 months	750	45,000/-	-do-
59.	—	125	Shri Harendra Dey Sarkar Baishnab Ghat	2.5 mds.	625	37,500/-	-do-
60.	—	70	Shri Khagendra Dey Sarkar Baishnab Ghat	1.5 mds. for 3 mds.	130	7,800/-	-do-
61.	Heder Bhery No. 1	49	Shri Paresh Ch. Dey P. 10 C.I.T. Road, Calcutta	1 md.	250	15,000/-	-do-
62.	Heder Bhery No. 2	50	Shri Nagen Mondal (Malikapur)	1 md.	250	15,000/-	-do-
63.	Heder Bhery No. 3	90	Shri Atul Mondal & others Mahisghot	1.5 mds.	345	20,700/-	Sonarpur P.S.
64.	Jagatipota Bhery	170	Shri Himangsa Sapui, Garfa	4 mds.	1000	60,000/-	-do-
65.	Kalir Bhery	277	Shri Kali Pramanick, Kalikapur	—	—	Not cultured at present	
66.	Ghugur Bhery	70	Shri Samaresh Roy Chowdhury 91, Debendra Dey Road.	1 md.	250	15,000/-	-do-
67.	Ram Mondal Bhery	70	Shri Ramapada Mondal Chow Bhagha	1 md.	250	15,000/-	-do-

....*Contd.*

Appendix II – *contd...*

1	2	3	4	5	6	7	8
68.	Hare Krishna Bhery	50	Shri Jugal Pada Mondal -do-	0.5 md.	125	7,500/-	-do-
69.	Bara Bhery	50	Shri Nemal Ch. Mondal Chak Kolar Khal	0.5 md.	125	7,500/-	-do-
70.	Majher Bhery	29	Shri -do-	0.5 md	125	7,500/-	-do-
71.	Char Kanta	23	Sm. Alamoni Dasi -do-	0.5 md	125	7,500/-	-do-
72.	Hansar Bhery	15	Shri Suren Mondal -do-	10 Srs.	62.5	3,750/-	-do-
73.	Dhaner Bhery	12	Sm. Alamoni Dasi -do-	10 Srs.	62.5	3,750/-	-do-
74.	Ghar Bhery	12	Shri Bhola Nath Mondal; Shri Nani Mondal & Shri Phani Mondal -do-	10 Srs.	62.5	3,750/-	-do-
75.	Nutun Bhery	55	Shri Subal Mondal Chow Baga	1.5 Srs.	375	22,500/-	Tallygunj
76.	Corporation Bhery	62	Shri Santosh Ch.	1 Sr. for 7 months	210	12,600/-	-do-
77.	Choto Jeler Bhery	93	Shri Khagen Mondal Bhagabanpur	1 Sr. for 7 months	210	12,600/-	Sonarpur
78.	Jeller Bhery	489	-do-	4 Srs. for 8 months	960	57,600/-	-do-
79.	Maliker Bhery	130	-do-	2 mds. for 6 months	360	21,600/-	-do-
80.	Abader Bhery No. 1	200	Shri Ardhendu Sekhar Naskar and others, Beliaghata			Other Information not supplied.	-do-
81.	-do- No. 2	339	Shri Khagen Mondal Bhagabanpur	3 mds. for 7 months	630	37,800/-	-do-
82.	Pakhir Bhery	190	Shri Joy Krishna Mondal, Shri Hare Krishna Mondal Kalikapur	—		Not cultivated at present	Tallygunj
83.	—	150	Shri Kali Pramanick Kalikapur	—			
84.	Hossairpur Bhery	94	Shri Beraj Mondal; Shri Deben Monal & Shri Lalit Mondal, Madu Daha	— 1 md.	-do- 250	-do- 15,000/-	-do-
85.	Bogur Bhery	300	Shri Provat Ch. Roy, Protiva Roy, 95, P.G. Road.	6 mds.	1500	90,000/-	-do-
86.	Bunor Bhery	156	-do-				
87.	Nulor Bhery	55	Shri Pasupati Roy 46/9 Ballyganj Place	1 md.	250	15,000/-	Tallygunj

....Contd.

Appendix II – *contd...*

1	2	3	4	5	6	7	8
88.	Bara Bhery	25	Shri Nemal Mondal Chak Kolar Khal	0.5 md.	125	7,500/-	-do-
89.	Kalikapur No. 1	270	Shri Bejoy Kr. Roy 46/9, Ballyganj Place	3 mds.	750	45,000/-	-do-
90.	-do No. 2	150	Shri Hare Krishna Pramanick Shri Joy Krishna Pramanik, Kalikapur		Not cultivated at present		-do-
91.	Bara Heder Bhery	126	Shri Atul Mondal Krishnapur	3 mds.	750	45,000/-	-do-
92.	Kultala Bhery	240	Shri Anil Kr. Chowdhury Dhakuria	4 mds.	1000	60,000/-	-do-
93.	Bara Khola Bhery	540	Shri Ganga Das Pal 25, Tangra Road.	5 mds.	1250	75,000/-	-do-
94.	Rajapur Bhery	375	Shri Behai Mondal Garfa	1 md.	250	15,000/-	-do-
95.	Ghoser abad No. 1	42	-do-	—	300	18,000/-	-do-
96.	Forman Bhery	500		7 mds.	1750	1,05,000/-	-do-
97.	Badrak Bhery	350	-do-	6 mds.	1500	90,000/-	-do-
98.	Bara Bhery	175	Shri Sachin Kolay & Provat Mondal, 8/4A, Hatibagan	4 mds.	1000	60,000/-	-do-
99.	Musher Bhery	136	Shri Satish Baidy, Garia	4 mds.	1000	60,000/-	-do-
100.	Ghoser abad No. 2	35	Shri Bepin Sarkar Hatibagan	0.5 md.	375	22,500/-	Tallygunj
101.	Ghoser Bhery No. 3	25	Shri Debendra Baidya & Other Garia	1 md.	250	15,000/-	-do-
102.	9 No. Bhery No. 1	188	Sm. Ratna Prova Mondal 19, Rupchand Mukherjee Road.	2 mds.	500	30,000/-	-do-
103.	9 No. Bhery No. 2	259	Sm. Bandana Singha 398/1, Keatola Lane, Calcutta	3 mds.	750	45,000/-	-do-
104.	Bagdir & Jhang-hanir Bhery	315	Sm. Kanika Das W/o Suniti Das	3 mds.	750	45,000/-	-do-
105.	Bogdoba Bhery	415	Sm. Provita Naskar W/o Amiya Naskar EX-D.F.O.	6 mds.	1500	90,000/-	-do-
106.	Naturn Bhery	229	Atul Kr. Das 95, P.G. Road	3 mds.	750	45,000/-	-do-
107.	Nartola Bhery	419	-do-	6 mds.	1500	90,000/-	-do-
108.	Uchhepota Bhery	139	Shri Rajen Naskar Sonarpur	1.5 mds.	315	18,900/-	Sonarpur
109.	Bara Nemair Bhery	65	Shri Panchi Bar -do-	0.5 md. for 7 months	105	6,300/-	-do-
110.	Arjun Barer Bhery	50	Shri Upen Bar Bhagabanpur	0.5 md. for 7 months	105	6,300/-	-do-

Source : *Secretary, Fish Producers' Association, 24 Parganas West Bengal, 1959 Data not available for blanks.

2
Aquaculture in Brackishwater Impoundment

1. INTRODUCTION

With the diversion of the main Ganga flow through the Padma, destructive forces rapidly set in and shattered most of the active rivers which gave way to unstable tidal rivers of the Sundarban Delta particularly in the North 24-Parganas district (Fig. 1, Chapter 1).

Gobaria (Haroa PS) a flood prone Bhery of Bidyadhari river is one of the saline water bheries of Sundarbans. Once it was infested with dense forest and was known as 'Corwns land'. It was in the year 1840 when one foreigner Armanian Aga Bey by name took lease of 36000 bighas of forest land at the rate of two annas per bigha for the purpose of cultivation. But he could not do anything due to repeated flooding. Finally, in 1857 he left everything to one Mr. Umesh Chandra Mitra who in the name of the Company of Bhawanipur Zaminder tried to make of the land. In 1880 one Upen Basu by name together with few others became the lease holder of this company. The present Basu's of Gobaria are the decendents of Upen Biswas. The cultivated the land to some extent by reclaiming the forest. But unfortunately during 1922 the entire Sundarbans was flooded by a 22 ft. high flood water of Bidyadhari. This resulted creation of innumerable number of saline water behries. The lot No. 70 consisting of 4 *moujas* formed four major bheries like Gobaria, Sarbaria, Ramchak and Ramjoy. Each of them was a mile wide and 2-3 miles long. The area of Gobaria was about 7.5 thousand bighas (ha). In all the salt water bheries prawn (*P. monodan*) was mainly cultured. This was followed by bhetki (*L. Calcarifer*) and Persia (*L. persia*).

Between the large number of estuaries and rivers are innumerable streams and water courses, called khals/beels and lakes, forming a perfect network of channels and ending ultimately in little channels that serve to draw off water from each block of land. For each block is formed a saucer with high ground along the banks of the khals surrounding it, and with one or more depressions in the middle according to their size. The water collected in the depressions is drained off by the little khals into large ones and ultimately into rivers; conversely, when the water swells in the rivers, it floods the country through the same channels. Such lakes and depressions are formed more due to inflow of tidal water with imporper or no exit and natural calamities like floods and cyclone. From a list of natural calamites (cf. Table 1) it is clear that flood occurred on several occasions associated with storm or cyclone. Such floods are capable enough to change the topography and ecology of the Sundarban delta.

The 24-Parganas like other deltaic districts is studded with large number of marshes and swamps (beels) situated between the elevated tracts which mark the courses of the rivers. These marshes are chiefly situated in the east of the district. The principal of these are as follows:

1. The Dhapa or Salt water Lake, between the Hooghly and Bidyadhari is a low basin, commences from about 5 miles (8 km) east of Calcutta and has an area of 30 sq miles (77672 ha). The Salt lake is discussed in Part-1.
2. The Kulgachi Beel, situated on the west of Balinda in the centre of the insular protion surrounded by the Bidyadhari and the Katakhal.
3. The Bariti Beel, halfway between Balinda and Basirhat between the Bidyadhari and the Jamuna rivers.
4. Bayra Beel is the largest in the district comprising an area of 40 sq miles (103560 ha), the greater part of which is covered with reed jungles. It is situated on the east of the Jamuna and in the north-east of the district, and is being gradually drained and reclaimed.

5-6. The Balli and Dantbhanga Beels, the former has a spread of 10 and the latter 12 sq miles (2589 and 3106.8 ha), are also situated on the east of the Jamuna and on the north of the Bayra Beel.

7. The Barti Beel, on the east of Shyamnagar, has a station on the Sealdah Branch of the Eastern Railway.
8. The Dholkhera Beel, on the south-east of Barasat.
9. Ghazalmari Beel

10-11. The Nagarghata and Khaliskhali Beels on the east of the Bayra Beel.

Table 1 : Natural Calamities like floods and cyclones in Sundarbans

1582 - Storm	1896 - Cyclone
1737 - Severe, Earthquake, Rains, Storm	1897 - Earthquake
1770 - Storm	1898 - Cyclone
1823 - Storm	1899 - Storm
1830 - Storm	1900 - Flood; Storm
1832 - Storm	1901 - Cyclone
1833 - Storm	1904 - Flood; Storm
1834 - Storm	1907 - Flood; Storm
1839 - Storm	1909 - Cyclone
1840 - Storm	1911 - Flood
1841 - Storm	1913 - Flood; Cyclone
1844 - Storm	1916 - Cyclone
1844 - Storm	1917 - Storm
1848 - Storm	1919 - Storm
1850 - Storm	1921 - Cyclone
1852 - Storm	1922 - Flood; Storm
1856 - Flood; Storm	1927 - Storm

contd...

Table 1 – *contd...*

1858 - Cyclone	1928 - Storm
1859 - Cyclone	1929 - Storm
1864 - Flood, Cyclone	1932 - Severe Storm
1865 - Famine from	1933 - Flood
1866 - Oct. to Dec.	1934 - Earthquake
1867 - Cyclone	1934 - Flood, Severe storm or cyclone
1869 - Storms	1936 - Severe storm or cyclone
1871 - Flood	1937 - Severe storm or cyclone
1872 - Storm	1939 - Flood
1877 - Cyclone	1940 - Severe storm or cyclone
1878 - Storm	1941 - Flood, Storm
1880 - Storm	1942 - Storm
1881 - Cyclone	1943 - Famine
1882 - Cyclone	1948 - Storm
1883 - Storm	1956 - Storm
1884 - Storm	1960 - Storm
1887 - Cyclone	1962 - Storm
1888 - Cyclone	1970 -Severe storm or cyclone
1889 - Cyclone	1978 - Flood, Cyclone
1893 - Cyclone	1981 - Cyclone (A large number of new fisheries have sprung up).
1894 - Storm	1988 - Severe Storm

Sources

i) Report of the Fact Finding Committee, Land Record and Survey, W.B., 1984.

ii) Census, 1951, District Hand Book, W. Bengal - Ashok Mitra (1954).

Mukherjee (1959) from a record of 74 years rainfall (1880-1953) in the eastern track of the Hooghly established that:

(a) The total rain varied in a year between a maximum of 85.0" (2159 mm) to a minimum of 30.4" (772.16 mm) only, the average being 57.4 inches.

(b) The monthly distribution of rain for the same period was as:

(c) The deluge of 1990, which put the drainage problem of this region in the forefront, recorded a rainfall of 38.8" (985.52 mm) at Diamond Harbour between September 18 and 25.

(d) Fortunately, a rainfall of over 4" a day was hardly widespread. Old data evenced that on a particular day 12.3" (312.42 mm) of rain was recorded at Dum Dum, 6.6" (167.64 mm) at Alipore and 2.7" (68.58 mm) at Barrackpore. All these are situated within a radius of 30 miles (48 km) of Raj Bhavan, Calcutta.

It is thus clear that climatic (rainfall, temperature, humidity and radiation of solar energy for appreciable time) and ecological conditions (tidal flow and fertility of soil), being favourable, helped to afford a foothold for numerous marsh species like *Sphenoclea, Hydrolea, Ammania, Enhydra, Wedelia, Hygrophila* and *Sesbania;* while the numerous ponds and ditches are filled with submerged and floating water plants, like *Utribulari, Pistia, Hydrilla,*

Month	*Minimum rain (in inches)*	*Mean rain (in inches)*	*Maximum rain (in inches)*	*Mean of 1870-81 (in inches)*
May	1.0 (25.2 mm)	5.5 (139.7 mm)	17.1 (434.34 mm)	-
June	1.6 (40.64 mm)	11.7 (297.18 mm)	31.2 (792.48 mm)	10.2 (259.08 mm)
July	4.5 (114.3 mm)	13.1 (332.74 mm)	25.4 (645.16 mm)	12.6 (320.04 mm)
August	4.3 (109.22 mm)	13.3 (337.82 mm)	26.5 (673.1 mm)	13.5 (342.9 mm)
September	2.6 (66.04 mm)	10.4 (264.16 mm)	40.5 (1028.7 mm)	9.1 (231.14 mm)
October	Nil	4.6 (116.84 mm)	14.6 (370.84 mm)	5.0 (127 mm)

Lagarosiphos, Ceratohyllum, Ottelia and *Vallisneria*. Remarkable among these for rarity and interesting on account of distrubution in Europe on one hand and Australia on the other are the floating *Droserad, Aldrovanda.* The edges of sluggish creeks are tined with large sedges and bulrushes; the banks of river have a hedge like shrub jungle of *Derris, Dalbergia* and *Caesalpina,* with climbing *Ipomoea, Argyreia* and Menioperms and a few trees like *Pongamia glabra, Barringtonia acutangula* and *Thespesia pepulnea.* The sides of embankments and village sites, not occupied by habitations, are densely covered with village shrubberies of semi-spontaneous species like *Odian, Zizyphus, Acacia, Glycosmis* and *Trema,* often intersprersed with clumps of planted bamboos and groves af *Areca, Morina Mangifera* and *Anona,* waysides and waste places are filled with grasses and weeds, usually of little intrinsic interest, but often striking because of their distribution. A very proportion of the species of this class to be met with in the district has been inadvertently introduced by human agency; besides weeds that are indigenous in other parts of India, these include European or African species like *Senebrera pinnatifida,* and *Xanthiusa spesidsum,* and especially American species like *Ageratum conyeoides, Scoparia dulcis, Wissadula rostrata, Evolvulus nummulasifolus, Pepromia peilucida, Malachra capitata. Herpestris chamoedryoides, Croton sparsiflorus* and many others, which not only hold their own with, but spread plentifully than similar weeds of truly Indian origin, e.g., water hyacinth (Reproduced from Imperial Gazetteer by Late Lieut-Col. Sir, D. Prain, Quoted by Mitra, (1954). Further, information on vegetation of the embankments of Saltlake and swamps provided by Biswas (1927) may also hold true for the other part of deltaic Sundarbans being under same ecological condition and for this reference may be made in part I of this account. Similarly benthos of salt lake discussed by Sewell (1934) comprising Gastropods, Grapsid crap, Portunid, Prawn, Mysid, Actinian, Lamellibranchs etc., and abundantly present, are also likely to occur in this area of Sundarbans. These may be referred in part I of this account. Added to it information of fauna (Copepoda and Cladocera) will provide additional knowledge of this ecosystem.

Large areas of marsh land in the Sundarbans have been reclaimed and brought under tillage by means of raising embankments to keep out salt and brackishwater. Some of the large marshes in the heart of the district have also been drained and rendered fit for cultivation by constructing embankments and sluices.

Ecology and climatic condition of deltaic Sundarbans favourable natural fish stocking was also abundant and hence fish are plentiful and found at all times of the year Sir K.G. Gupta in his report predicted the future prospect of indefinite expansion and development of such fish farms as early as 1908. Fish farmers take the full advantage of tidal fluctuations as the ingress of floodwater carries/innumerable young ones of commercially important estuarine fishes. Fish culture practice in the brackishwater region came into existence almost at the same time in the salt lake area, but was restricted to low lying areas in Sundarbans (South of the Dempire Hedges Line of the deltaic West Bengal) and substantial portion of these lands were preferably utilized in rearing Saltwater fishes. According to Mitra (1954) the fish found in the waters of this district mostly belong to the carp and dog fish fishes families. The varieties commonly found in the Sundarbans are *Bhetki, Hilsa, Tapsi, Bhangan, Kanmagur, Rekha, Rucha, Chitra or Bisatara, Parse, Tengra, Bunspata, Datnia, Pangas* and *Bagda Chingri*. Crabs and shrimps are common. Sharks are also common in these parts.

Mitra (1954) further stated that *Koi* and *Magur* were common in the Beels and ponds outside the reserve forest. Reared *Rui, Catla* and *Mrigal* are also found in tanks and ponds in localities outside the reserved forest area. A classified list of such fishes are indicated below in Table 2.

Table 2 : Classified list of fishes (Mitra, 1954)

Order Clupciformes
Family Clupcidae
Hilsa illisha
Gonlolosa manmina
Family Engraulidae
Engraulius telara

Order Osteoglossiformes
Family Notopteridae
Notopterus chitala
Notopterus notopterus

Order Cypriniformes
Family Cyprinidae
Chela bacaila = Oxygaster bacalla
Esomus danicontus
Amblypharyngodon mola
Puntias tecto
Puntius conchonius
Puntius phutunio
Labeo calbasu
Labeo bata
Labeo rohita
Cirrhinus reba
Catla catla

Order Siluriformes
Family Siluridae
Wailago attu
Callichrous pabda + Ompak pabda
Family Bagridae
Mystus aor = Aorichthys aor
Mystus gullo

Order Atheriniformes
Family Belonidae
Xenentodon cancila

Order Channiformes
Family Channidae
Channa gachua
Channa punctatus
Channa striatus

Order perciformes
Family Nandidae
Badis badis
Nandus nandus

Family Gobiidae
Glossogobius giuris
Family Centropomidae
Lates calcarifer
Family Chandidae
Chanda nama = Ambaosis nama
Family Sciaenidae
Pama pama
Johnius coltor
Family Mugilidae
Liza parsia = Mugil parsia
Liza tade = Mugil tade
Liza corsula = Rhinomugil corsula
Family Polynemidae
Polynemus pamdiseus
Polynemus indicus
Eleutheronema tetradactylum
P. tetradactylus
Family Anabantidae
Anabas testudeneus

contd...

Table 2 – *contd...*

Mystus tengra
Rita rita
Family Aricidae
Tachysurus sienensis = S. sienensis Bagarius bagarius
Family Pangasiidae
Pangastus pangasiun
Family Schilbeidae
Eutropilchthys vacha
Silonia silondia
Clupisoma garua = Pseudotropius garua
Family Clariidae
Clarias batrachus
Family Heteropneustidae
Heteropneustes fossilis

Family Harpadontidae
Harpodon nehereus
Order Mastacembeliformes
Family Mastacembelidae
Mastacembelus pancalus
Order Tetraodontiformes
Family Tetraodontidae
Tetradon cuteutia

Family Belentidae
Colisa fasciata
Colisa lalius
Colisa chuna = Colisa sota
Family Stromateidae
Parastromateus niger = Stromateus niger
Family Carangidae
Caranx sp.

Family Trichiuridae
Trichiurus savala
Order Osteoglossiformes
Family Notopteridae
Notopterus notpterus
Notopterus notpterus
Notopterus chitala
Order Synbranchiformes
Family Synbranchidae
Monopterus cuchia = Amphipnous cuchia

With the increase in population of Calcutta and rising demand for fish there was a tendency among the lotdars (Lessees of Sundarbans lots) to increase the extent of fisheries. They did not even hesitate to inundate agricultural lands of their ten ants with saline water and resumed the lands on the grounds of abandonment. As for example, the history of lot No. 70 comprising Gabariabad illustrates the strategy employed by lotdars to convert agricultural lands into fishery. In the year 1922, the embankments around the lot collapsed through tidal pressure of the Jagannath Khal and was inundated with saline water. No serious attempt was made to repair the damage and in 1930 the area was converted into fishery. The history of lot nos. 71 and 72 is also the same.

Settlement records show that the conversion of agricultural lands into fishery took place in the Sundarbans on a large scale. It will be seen from the table below that as much as 53.3% of the area under fisheries in the region was agricultural land (Chosh, 1973).

Classification	*Area in acre*	*Percentage of the area under fisheries*
Marsh (Khal. bils etc)	2,607,66	16.4
Agricultural lands	8,573,23	53.3
Non-agricultural lands	4,643,61	29.1
Land accreted from rivers	197,09	1.2

From the report of 1958* it appears that there were 92 fisheries in the Sundarban area. The biggest being Bokaria Fishery in Haroa P.S. with an area of 2469.31 acres. These fisheries were classified in Table 3.

* Report of the each Finding committee set up by the West Bengal State Land use Board on the problems of Paddy-cum-Brackishwater Fish/Prawn Cultivation and forcible conversion of Paddy Lands into fisheries in parts of 24-Parganas District, 1984.

Table 3 : Trend of Fisheries in the Sundarban Area

	No.	Area
1. Fisheries in the marshy area	14	(3.6 Sq. Miles).
2. Fisheries created partly in marshy areas and partly in inundated agricultural lands	9	(2.15 Sq. miles)
3. Fisheries created partly in non-agricultural lands and partly in inundated agricultural lands.	26	(12 Sq. miles).
4. Fisheries created exclusively in agricultural lands but subsequently inundated	31	(6.5 Sq. miles).
5. Fisheries created exclusively in non-agricultural lands but subsequently inundated	6	(5 Sq. miles).
6. Fisheries created by constructed embankments in the low lying areas accreated from the rivers.	6	(3 Sq. miles).

Table 4. Indicating a list of fisheries constructed in the Sundarbans during different periods.

Year	Number of fisheries constructed	Year	Number of fisheries constructed
1915	1	1945	4
1925	1	1946	9
1926	8	1947	4
1932	1	1948	6
1934	3	1949	10
1935	5	1950	6
1938	1	1951	14
1939	4	1952	5
1940	1	1953	4
1942	3	1954	1
1943	1	-	-
	29		63

Total 29 + 63 = 92.

From the District settlement records it is seen that 53.3% of the total fishery area in the Sundarban area were fisheries where agricultural lands were converted. Such tendency persisted all throughout.

Of the 92 fisheries existing in 1958, 26 were in Canning P.S., 9 in Haroa, P.S., 16 in Mathurapur P.S., 10 in Jayanagar P.S., 19 in Sandesh Khali P.S., 6 in Hasnabad P.S., 4 in Baruipur P.S. and 1 each in Kakdwip and Sagar P.S.

Thus there has been an unprecidented proligeration of Paddy-cum-fisheries in the 24-Parganas district particularly in the Sundarban area. It has been noticed that the trend of conversion of paddy lands into fisheries started long ago and probably the present period is the peak period of such fishery activity in the area. This is further reflected from the survey report conducted in 1978 by the Department of Fisheries, West Bengal (1984) which indicates progressive increase of Brackish water fisheries in the Sundarban delta.

2. BRACKISHWATER IMPOUNDMENTS IN FOUR SUB-DIVISIONS OF DELTAIC WEST BENGAL

In the subsequent years construction of roads and other communications in the reclaimed areas of the Sundarbans have given impetus to further development of brackishwater fish farms and lately of bheries for fish culture and even fish-cum-paddy cum fish culture. Results of inventory survey of bheries in 4 Subdivisions of the district (24-Parganas) by Pakrasi (1965) are listed below (Table 5a and Fig. 1) indicating number of bheries and their nature in the four subdivisions of Basirhat, Barasat, Alipur and Diamond Harbour and reconnaisence survey of Brackishwater impoundments are indicated.

It is clear from the table that paddy-cum fish culture practices were already in vogue. But fish production being high and more paying, the system was changed to fish-cum-paddy culture where fish is the primary produce.

Table 5a : List of Brackishwater Impoundments in four Subdivisions of Deltaic Bengal (Based on the Survey Report of Pakrasi (1965)

	Particulars	Number
1. Basirhat Subdivisions		
i.	*Basirhat P.S.*	
	Total number of bheries	17
	Paddy-cum-fish farm	10
	Fallow or derelict fish farm	1
ii.	*Baduria P.S.*	
	Total number of bheires	2
	Paddy-cum-fish farm	2
	Fallow or derelict fish farm	Nil
iii.	*Haroa P.S.*	
	Total number of bheries	11
	Paddy-cum-fish farm	2
	Fallow or derelict fish farm	Nil
iv.	*Hasnabad P.S.*	
	Total number of bheries	14
	Paddy-cum-fish farm	2
	Fallow or derelict fish farm	2
v.	*Sandeshkhalt P.S.*	
	Total number of bheries	17
	Paddy-cum fish farm	1
	Fallow or derelict fish farm	7
2. Barasat Subdivision		
i.	*Barasat P.S.*	
	Total number of bheries	22
	Paddy-cum-fish farm	3
	Fallow or derelict fish farm	4
ii.	*Deganga P.S.*	
	Total number of bheries	4
	Paddy-cum-fish farm	3
	Fallow or derelict fish farm	Nil
3. Alipur Subdivision		
i.	*Canning P.S.*	
	Total number of bheries	31
	Paddy-cum-fish farm	Nil
	Fallow or derelict fish farm	2
ii.	*Jayanagar P.S.*	
	Total number of bheries	13
	Paddy-cum fish farm	4
	Fallow or derelict fish farm	2
iii.	*Sonarpur P.S.*	
	Total number of bheries	1
	Paddy-cum-fish farm	Nil
	Fallow or derelict fish farm	1
iv.	*Baruipur P.S.*	
	Total number of bheries	4
	Paddy-cum-fish farm	2
	Fallow or derelict fish farm	Nil
4. Diamond Harbour Subdivision		
i.	*Kakdwip P.S.*	
	Total number of bheries	2
	Paddy-cum-fish farm	Nil
	Fallow or derelict fish farm	Nil
ii.	*Sagar P.S.*	
	Total number of bheries	1
	Paddy-cum-fish farm	Nil
	Fallow of derelict fish farm	Nil
iii.	*Mathurapur P.S.*	
	Total Number of bheries	13
	Paddy-cum-fish farm	1
	Fallow or derelict fish farm	Nil

Other information like number of bheries, their areas, species cultured and production per acre are indicated in Appendix III. This is further supplemented by additional information on sizes, groups of area, type of bheries etc. indicated in the "Master Plan for Fisheries Development, West Bengal, Part I & III, 1974". The partial survey of 'Nona bheries' (24 Parganas) reveals that there were 99 nona bheries (brackishwater) out of which 70 were in Barasat area of North 24-Parganas and 29 in Canning of South 24-Parganas (Table 5B).

3. SURVEY OF NONA BHERIES OF 24 PARGANAS

Nona Bheries (Saline water bheries are broadly known as nona bheries) of sizes from 50 to 2000 bighas of 24 Parganas (North & South) were surveyed and classified under different categories. Those are further classified under different categories into two groups, one purely used for fish culture and the other used for paddy-cum-fish culture and are presented in Table 5 and 6 A & B.

These Nona bheries are of three different categories, viz.,

(a) Nonabhery fish farm (purely for fish culture)

(b) Nonabhery : Paddy-cum-fish culture (Paddy is primary produce)

(c) Nonabhery : Fish-cum-paddy culture (Fish is the primary produce).

The size group of Nonabheries given in table 5 consists of first two categories of farms. The size groups of these two categories of Nona bheries (a) and (b) are given in Table 6A & B.

Table 5B : Classification of Nona bheries

Size group of bheries (bigha)	*24 Parganas North (Basirhat area)*		*24 Parganas South (Canning Area)*		*Total*	
	No.	*Area (bigha)*	*No.*	*Area (bigha)*	*No.*	*Area (bigha)*
Up to 50	15	331	1	36	16	367
50 - 100	7	580	2	200	9	780
100 - 500	29	8,060	10	3.700	39	11.760
500 - 1000	12	11,200	10	7.206	22	18.408
1000 - 1500	2	2,700	2	2.600	4	5.300
1500 - 2000	2	3,900	1	2.000	3	5.900
Over 2000	3	16,900	3	17.000	6	33.900
Total	70	43,671	29	32.742	99	76.415
		(5822.8 ha)		(4365.67 ha)		(10188.66 ha)

1 ha = 7.5 bigha.

It is further stated that Paddy-cum-fish cultural nona bheries are mostly concentrated in the North 24-Parganas.

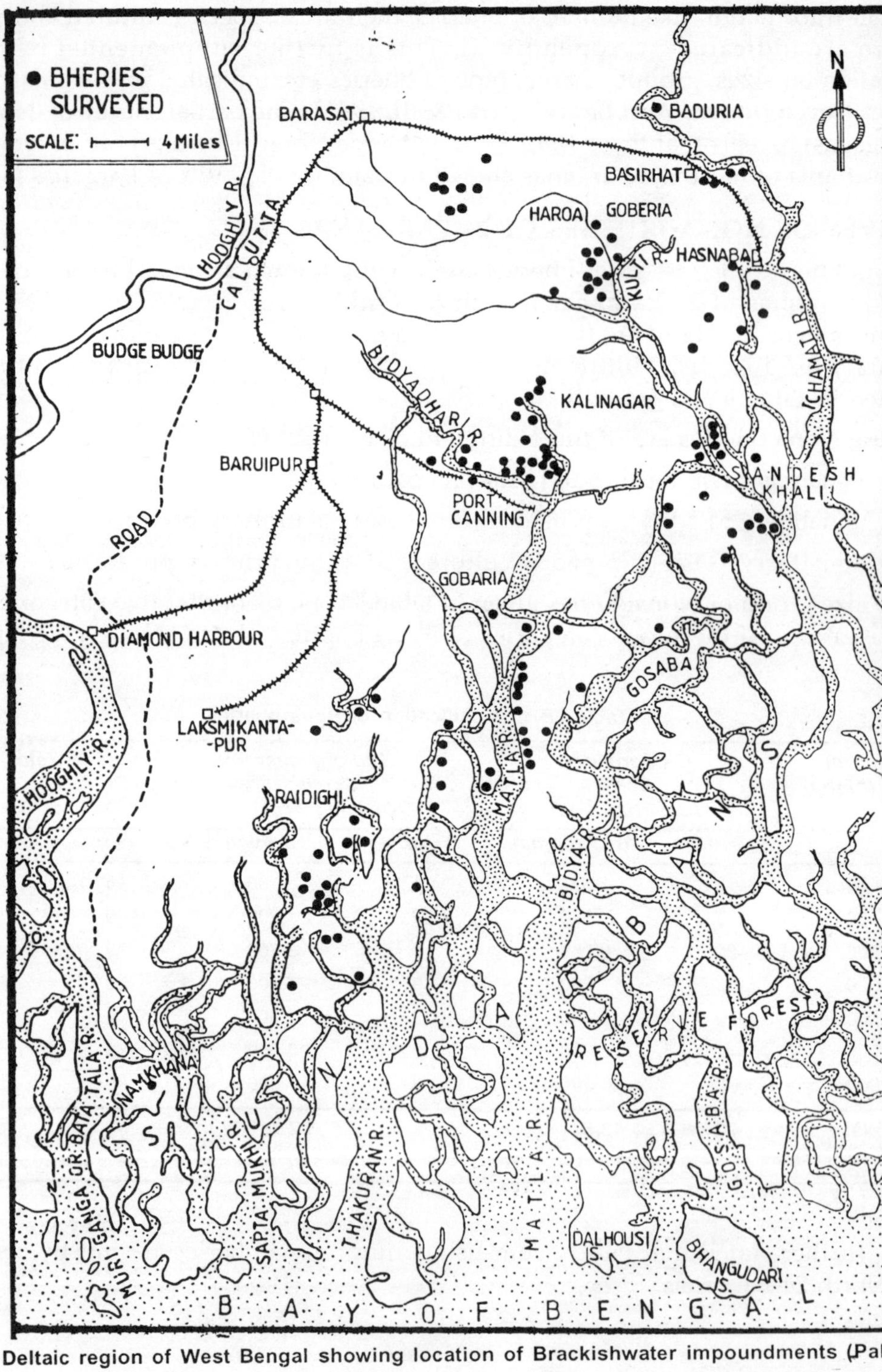

Fig. 1 : Deltaic region of West Bengal showing location of Brackishwater impoundments (Pakrasi, 196

Though paddy-cum-fish culture is an age old practice in Bengal of late fish-cum-paddy culture has been incrasing rapidly. It is worther mentioning here that brackishwater bheries, although in existence since sixties have increased rapidly around seventies and was as much as 420% (Ray *et al.*, 1981). In a most recent survey the culturable water of deltaic Sundarbans is classified as seasonal bhery in low saline area. These bheries are in existence in Basirhat, Baduria, Rajarhat, Swarupnagar and Ghusighata, seasonal and perennial bheries of medium saline zone at Haroa, Hasnabad, Hingelganj, Minakhan and Baduria; and seasonal and perennial zones of high saline zone at Mathurapur, Canning, Parthapratima, Gosaba and Jayanagar, (cf. Fig. 1). Gangopadhaya (1983) estimated area of Brackishwater bhery as 20000 acre (800 ha) and that of Bhasabada fisheries/perennial brackishwater bhery (Recorded fishery) as 0.20 lakh acre (0.08 lakh ha) (Master Plan Part I, Government of West Bengal 1974).

Table 6 : Survey of Nona Bheries of Different Categories

A. (North 24-Parganas)

Size group of bheries (bigha)	*Nona-bhery Purely Fish Culture*		*Nona-bhery Paddy-cum fish Culture*		*Total*	
	No.	*Area (bigha)*	*No.*	*Area (bigha)*	*No.*	*Area (bigha)*
Up to 50	1	12	14	319	15	331
50 - 100	—	—	7	580	7	580
100 - 500	1	250	28	7,810	29	8,060
500 - 1000	10	10,000	2	1,200	12	11,200
1000 - 1500	2	2,700	—	—	2	2,700
1500 - 2000	2	3,900	—	—	2	3,900
Over 2000	3	16,900	—	—	3	19,900
Total	19	33,762 (4501.6 ha)	51	9,909 (1321.2 ha)	70	46,671 (6222.8 ha)

B. (South 24-Parganas)

Size group of bheries (bigha)	*Nona-bhery Purely Fish Culture*		*Nona-bhery Paddy-cum fish Culture*		*Total*	
	No.	*Area (bigha)*	*No.*	*Area (bigha)*	*No.*	*Area (bigha)*
Up to 50	1	36	—	—	1	36
50 - 100	1	100	1	100	2	200
100 - 500	10	3,700	—	—	10	3,700
500 - 1000	10	7,208	—	—	10	7,208
1000 - 1500	2	2,600	—	—	2	2,600
1500 - 2000	1	2,000	—	—	1	2,000
Over 2000	3	17,000	—	—	3	17,000
Total	28	32,644 (4352.53 ha)	1	100 (13.3 ha)	29	32.744 (4365.86 ha)

1 ha = 7.5 bigha.

4. BHERI MANAGEMENT

With the passage of time the people of Sundarbans have gained more experience in fish and paddy culture by the judicious use of water and land. Besides culturing fish in perennial brakishwater bhery (Recorded fishery) they have also devised methods for successfully culturing both fish and paddy simultaneously in the same field. In the former type of bhery, fish is the dominant crop while in the other paddy dominates. Of course, the latter type is less common than the former. Besides, transplantation of fry, selective stocking with compatible species and rearing of fry in nurseries, which are in practice, are gradually gaining importance but not fully in the present system of culture. The use of sewage-mixed water of the Kulti estuary has undoubtedly enhanced the crop further. The types of brackishwater bheries that are in practice in the area are as follows :

1. Perennial Brackishwater bhery (recorded fishery)
2. Fish-cum-Paddy culture bhery (seasonal)
3. Paddy-cum-fish culture bhery.

4.1 Perennial Brackishwater Bhery (recorded fishery)

In early days, as evidenced from Mitra's (1954) census record, suitable areas, near tidal rivers or channels of their beds, were protected from being flushed out by monsoon or tidal floods by putting embankments around them keeping a water level of 5 feet (1.5 m). From March/April onwards during suitable flood tides salt water was let into the embanked areas from the adjoining areas through sluice boxes provided in the mouth of small and narrow canals. Larvae and fry of fishes *viz. Lates calcarifer, Mugil tade, M. parsia, Polynemus indicus, Scatophagus argus, P. monodan, Metapenaeus* sp. *M. gutio, Scylla* sp. and other brackish water species entered the protected areas. With the required amount of water taken when sluice boxes were closed entire area appeared as a vast tank of standing water teeming with millions of fish fry. The young ones thus entrapped were usually allowed to remain in these closed area for 8-10 months till they attained marketable size. For proper exploitation, segregation of fishes was done considering their sizes and market value. The bhetki (*L. calcarifer*) has been observed to grow to 5-6 seers (4.7-5.6 kg) at the end of the second year and fetch a very handsome price. Crabs are found throughout the district and are plentiful in the Barasat Sub-division.

Pillay (1954) gave a precise account of fish culture practices in the bheries. He observed, "the cultural practices commence in January-February when ponds are stocked with fry. The ponds which contain only little water at that time, as a result of dewatering for fishing in September-December, are filled with water to the maximum extent possible. Generally, each pond has only main sluice gate, but during this period emergency inlets are made to enhance intake of water. It is only during spring tides that the tidal water reaches the sluice gates. The main sluices are provided with shutters to regulate the water level. At high tide these doors are opened to allow tidal water to flow in and during low tide the gates are closed to present water from flowing out. In front of each main sluice gate is a V or W-shaped screen of bamboo-grating with gaps at the apex of V or epices of the W. In these gaps are placed 'atols'. In fixing them, the open side with the value-like arrangements of the bamboo spindles is kept facing the pond. The bamboo-gratings are kept in front of sluice gates to prevent the fishes in the ponds from escaping when the gates are open."

After the intensive stocking of ponds from February to April no attention is paid by the fishermen to improve the growth of fish crop, apart from letting in water into the ponds, attending to the occasional repairs of the embankments etc., and keeping watch over the stock. The fishes feed on the autochthonous food material of the ponds as well as those brought in by the tidal water. During the dry months, when the water level falls, several twigs are stuck into the bottom of the canals to prevent easy fishing by the poachers. Mainly to meet the farmer's needs the ponds are fished on a small scale during May-June. On one side of the main sluice gate and the canal leading from it, they dig a shallow canal. The mouth of this shallow canal is guarded by a low fence of bamboo grating and some distance away from it is another similar fencing with a gap in the middle. Tidal water from the main feeder canal leading from the river is allowed to flow into the pond through the canal. The fishes ascend the canal and gather below the bamboo-fencing. At the turn of tide or when a significant number of fish has gathered in the canal it is closed at the mouth either by means of shutters or by putting an earthen bund. The gap in the lower bamboo-fencing is also closed. After the water in the channel drains off, the fish that have been trapped in the channel are removed by hand. Cast nets are operated for the capture of all the fishes from the ponds during all seasons.

Large-scale fishing operations are carried out between September and November. As no tidal water is required in the ponds during this period as much water as possible is let out. The fishes move into the canals and deeper areas of the ponds where these are fished by means of dragnets, generally during the evening or night so that the catches could be sold in the morning. The catches are sometimes kept in the live wells or live cages known as 'Hapar'.

During the usual annual intensive fishing, large number of immature fish are also caught and sold. Formerly, it was the general practice to stock them in the subsidiary ponds where they were allowed to grow till the main pond was again filled with water and could gain access to the canal system and the pond proper, a practice which is now seldom followed.

Gangopadhya (1983) is more specific in his statement in bhery operation. According to him generally from February onward from 11th to 15th following full and new moon periods of spring tides water is drawn into the bheries through the sluice box along with this tidal water enough of spawn and fry of various brackishwater fishes enter the bhery. From 5th to 10th day during neep tides the excess of water is released back taking care that the spawn and fry do not escape from the bhery. Any fish of marketable size stocked earlier are trapped and disposed off. Such mode of operation is practised every month and the bheries are thus self-stocked adequately.

But sometime stocking is also done from outside sources. Besides, paddy cultivation in the exposed area of the bhery, where fertile aluvial soil is largely deposited, is also practised by the people of Sundarbans.

4.2 Fish-cum-Paddy Culture

This industry has largely developed in the upper Sundarbans. After initial drying of the bhery, water from the Kulti estuary is drawn sometime between November and December through sluice boxes and allowed to stabilize for 10-15 days to help in growing algal matter. A constant deth of 0.3-1 m water level is always maintained in the bhery. During each tidal phase of full moon and new moon periods certain volume of water released earlier is drawn back into the bhery from the Kulti. Such an operation lasts for 5-6 days to replenish the ecosystem as well as to maintain optimum salinity in the water, which, otherwise, would have shoot up during summer. During the slack period when the water level in the river remains at a lower level, water is lifted into the bhery through a pump. As far as possible direct feeding of water from the estuary is done in each bhery through canals. The advantage of brackishwater bheries over the freshwater ones is that always diluted sewage water in varying concentrations, depending upon the tidal phases, is drawn into them. Diluted sewage acts as a potential fertilizer and enhances the production in comparison to other brackishwater bheries. In freshwater bhery it is the raw sewage that maintains the level of water. There is no possibility of any dilution except by rainwater.

Paddy seedlings are lifted during August and September when water salinity is considerably low in the bheries. The bhery owners lease out the bheries to other parties interested for paddy cultivation temporarily during the wet months with the condition that no pesticides would be used in the bhery for protection of plant crop as this might harm the fish. Thus in the same plot two parties are benefitted. The average paddy yield so obtained varies between 6 and 8 md/bigha (1682-2243 kg/ha). Salinity being low major carp fingerlings, mainly Catla (*Catla catla*) of 70-100 g size, are brought normally from freshwater bheries (Bantala) and stocked. Paddy harvesting is completed by November-December. Prior to this, the bhery water is drained out to trap the fishes, as by that time fish (Catla) would grow 1 1/2-2 kg in size. After the removal of fish the bhery bed is dried for a short period, ploughed and between November and December water is drawn to stagnate and allow the water to enrich with plankton. Seedlings of prawn and other brackish water species are stocked heavily. Harvesting begins by April, with peak period in June-July, when prawns, *P. monodan* (tiger shrimps) attain weight between 70 and 100 g, and is completed before the onset of monsoon as sudden drop in salinity is not congenial for their survival.

Since the time of Commissioniñg the DWF canal in 1940 Calcutta Metropolitan wastes are regularly discharged into the Kulti estuary. Initially the local people were hesitant to use such polluted water as fish mortality in the river was a common phenomenon. Further, since fish could neither migrate nor survive in a certain stretch around the outfall, the estuary gradually became devoid of fish. It was around sixties some people initiated to use this water for pisciculture and paddy cultivation purposes. It proved highly beneficial because of its manurial value. As a result, within a few years fish and paddy culture practices in bheries rapidly increased and the area of use of such water in bheries flanked on both sides of the estuary extended from Haroa in the North to Chital-Halgachi (Minakhan) in the South. Relay stocking practices in bheries (more than once) are very common on account of high rate of mortality (70-75%).

4.3 *Paddy-cum-Fish Culture*

Simultaneous culture of paddy and fish at some places in the lower Sundarbans is being practised. By this process as Gangopadhyay (1983) explains, they get both paddy and fish, paddy being the primary produce. For such mixed cultivation, culturable area adjoining the tidal rivers, channels, creek, etc., of the Sundarbans are selected. Such areas are well protected by embankments on both the sides of the river, channels, creeks etc. from free incoming of highly saline water. Interested parties willing to culture fish group together, arrange a number of plots of large paddy fields covering an area of 300-400 bighas (40-53.3 ha) having gradual slope from the embankment. All the lands are properly dyked keeping the irrigation embankment. The depressed area on one side of this bhery is again dyked. The smaller bhery is connected to the inlet and outlet sluice boxes constructed on the irrigation embankments.

By January, when the salinity goes down to the minimum, sufficient water is drawn in the smaller bhery through the inlet sluice box during spring tide. From February onward, when salinity increases in the rivers and channels, the seeds of different species of brackishwater fish become available. Such saline water with seeds are drawn in the smaller bhery and salinity at that time practically has little effect on the saline base because of dilution. Since the breeding habit of different species is different, water is drawn in the bhery for a few months consecutively, so that there is maximum composition of fish in the bhery.

With the onset of monsoon from June onwards the high salinity of the water in all the rivers decreases. Enough of rainwater is then traped in the bigger bhery. After thorough ploughing paddy seeds are planted during July. The low saline water of the river and channel being supplemented by rainwater dilutes the salinity of the smaller bhery considerably. By this time paddy seeds get rooted. The smaller bundhs of different plots of lands together with the main embankment of the smaller bhery are levelled and the small fishes are allowed to move freely in the paddy field. Later on in the month of August paddy seeds are transplanted in the smaller bhery after a single ploughing. From December onwards, paddy harvesting begins, but prior to that the bhery water is let out and fishes are trapped and removed. No water is taken in during November and December.

From January onward again necessary arrangements (repair of dykes, sluice gates etc.) are made to start the operation for the next year. by adopting this system 8-9 mds/bigha (56.07 kg/ha) production of paddy can be obtained while average fish production is about 20 mds per 100 bigha (56.07 kg/ha) in a season. The production of fish is however, dependent on the size of the smaller bhery and there is scope for increased production in case the fish culture operations are done more scientifically. Lime in small does is used as prophylactic against infections and manuring by rice bran and oil cake 500-100 kg/ha at 3:1 ratio is a recent practice, low dose of urea is also tried.

4.4 *Hydrobiological Investigations*

While much importance is laid on the development of brackishwater bheries only in the recent past, prior to fifties there was hardly any systematic study except casual mention of the mode of operation of fish culture in shallow impoundments fed by tidal waters.

Information on some aspects of study on ecology of brackishwater bheries with reference to fish culture was first initiated by Pillay (1954) in an area which is free from any sewage contamination. The same author (1962) in a subsequent study has discussed the benthic flora in brackishwater bheries highlighting the floral production at different depths. Subsequently Ray *et al.* (1981) made some systematic studies in an area influenced by Calcutta metropolitan wastes. Both the investigations made were in medium saline zone. More recently Saha *et al.* (1986) made some investigations covering ecology and management of brackishwater bheries.

4.4.1 Perennial Brackishwater Bhery (recorded fishery)

The following observations are based on the study of Pillay (1954).

Pond topography : In the shallow portion of the pond where average depth is 6"-9" (15-22.5 cm) during rainy season and 1'-2' (30-660 cm) when a good supply of tide water is obtained.

Meteorological observation : Rainfall ranged between 0.2 and 36.5 cm. During actual wet months i.e. from June-September the normal rainfall varied between 24.43 and 36.5 cm. Highest rainfall (36.5 cm) recorded was in August.

Water characteristics : Temperature ranged between 14.4 and 35.0°C. During February to September, the period of 8 months, the ranges were between 27.5 and 35.0°C; October-December, between 24.5 and 26.5°C and in January the lowest temperature (14.4°C) was recorded.

Hydrogen-ion concentration : The water seems to have a moderately high buffering capacity as a result of which changes in the pH recorded are not very much (7.9-8.3), except on one occasion when it was 7.5. A gradual rise in pH (7.8) from August was distinctly observed until December, when the value reached as high as 8.3. Then again after a fall (pH 8.2) during January and in February (pH 7.5) a tendency of rise (pH 8.3) was noticed up to April.

Chlorinity : In general salinity indicated moderately high values (6.05-11.10‰) except during July to September (monsoon months) when due to high dilution there was a fall in the chlorinity (1.80-2.20‰) and also in March (1.60‰).

The hydrological changes discussed above were recorded at 8 a.m. in the morning for one year and are indicated in Fig. 2.

Hydrographical changes during inter-spring tide : Changes in the water during inter-spring tide is also an important aspects of study to know the behaviour of the biomass. In one such study Pillay (1954) observed that changes in depth and salinity of a bheri are inversely related. From the new moon onward there is a gradual fall in depth of water until the 13th day and from full moon a gradual rise is noticed. Likewise, with the fall in the water level there is a gradual increase in salinity until the 13th day and again from fullmoon day a fall is noticed. Water temperature in general ranged between 30.50 and 30.60°C, pH showed simply a marginal change (8.3-8.5). (cf. Fig. 3).

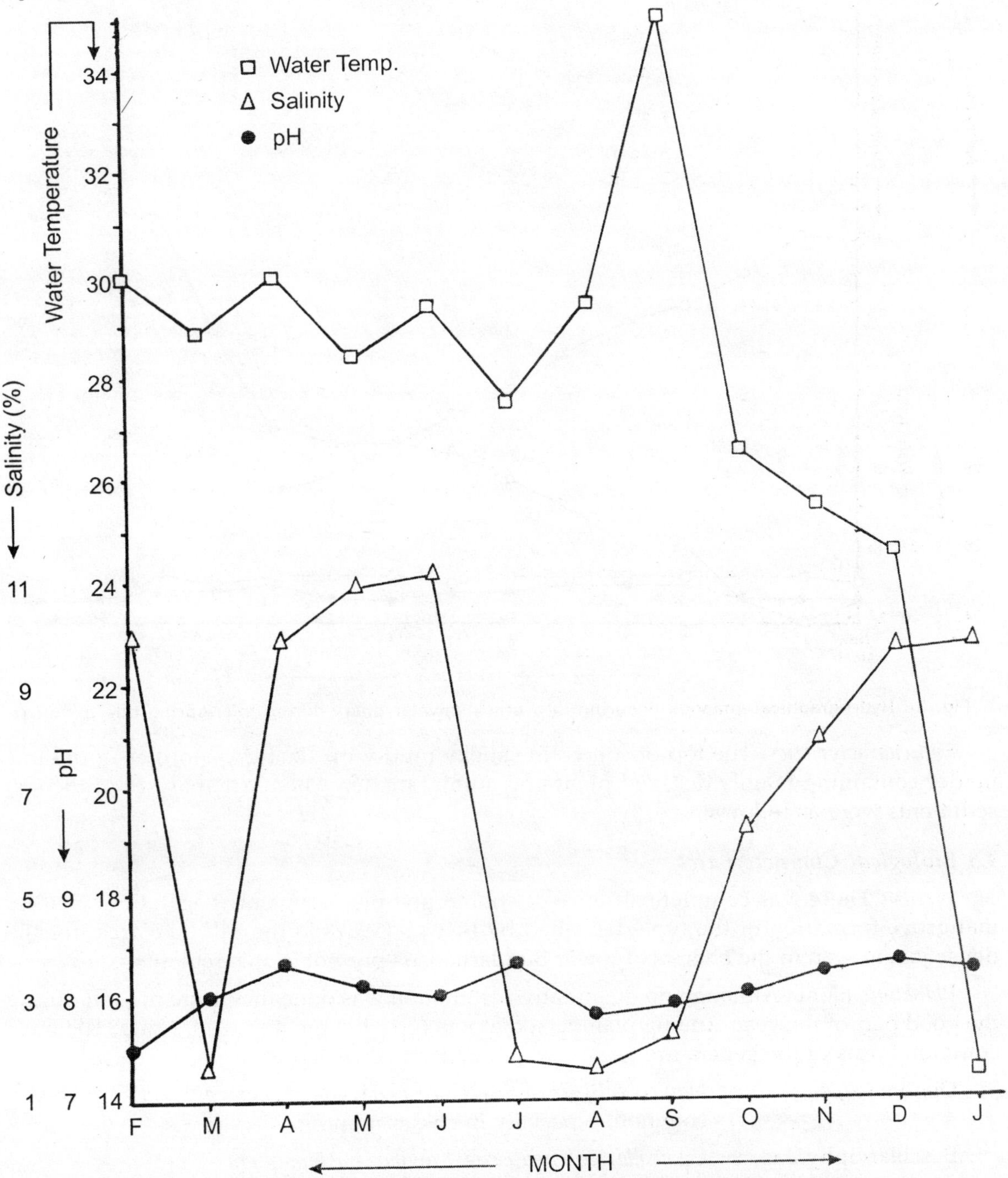

Fig. 2 : Hydrological changes occuring in a brackishwater bhery.

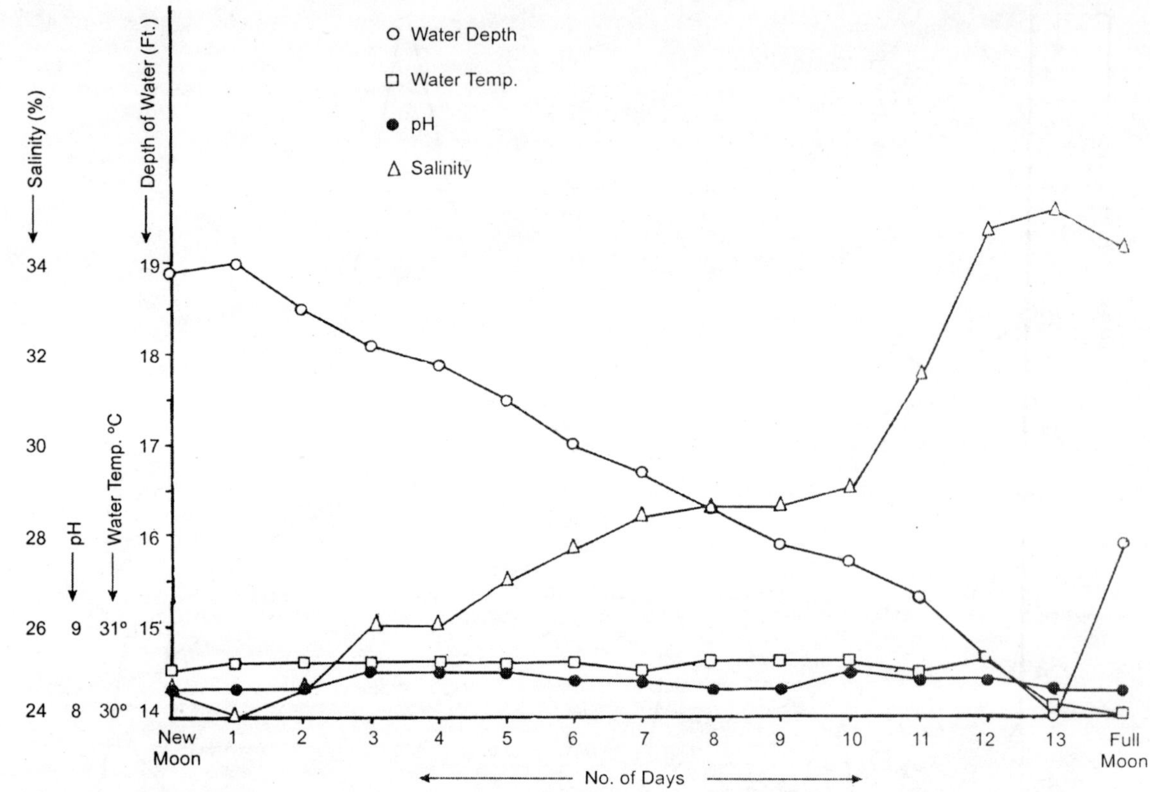

Fig. 3 : Hydrographical changes occuring in a brackishwater bhery during inter-spring tide period.

Soil characteristics: The top of soil is fine sticky mud with sizable quantities of organic matter containing mainly decayed plant and animal matter. Chemical analyses of the soil sediments were as follows:

4.5 Biological Characteristics

Weeds : There was complete absence of submerged phanerogamic vegetation. Neither mangrove formation of the type described by Biswas (1927) in the salt lake area nor the dense jungle seen in the bheries of lower Sundarbans is present in this region.

Plankton : Flora : Although no quantitative estimation was done they were present during the good part of the year. Among plankters, diatoms form the predominant group. The most common forms of the genera are :

Diatomaceae	:	*Navlcula, Pleurostgma, Gyrosigma, Gyrosigma scalproides* are very common, especially in colder months.
Bascillariophyceae	:	*Cymbella, Coscinodiscus, Fragilaria, Amphiptera.*
Myxophyceae	:	*Oscillatoria, Lyngbya, Spirulina, Anabaena, Nodularia* and *Rivularia.*
Chlorophyceae	:	*Protococcus, Pediastrum, Hydrodictyon* and *Ankistrodesmus Cosmarium*
Desmid	:	*Cosmarium.*

Table 7 : Chemical Analysis of Soil Sediments (Pilley, 1954)

Moisture	31 %
Humus	3 %
Chlorides	0.137 %
Silica and other insoluble matter	79.49 %
Iron and Aluminium as oxide	8.88 %
Sulphate as SO_4	0.05 %
Calcium as CaO	3.34 %
Magnesium as MgO	1.12 %
Total Nitrogen	0.42 %
Nitrate-N	0.004 %
Phosphate (P_2O_5)	1.53 %

There were considerable quantities of algal spores in the plankton almost throughout the year. A notable feature is the thick growth of benthic flora on the pond bottom. The different species of Oscillatoria indicated below are the most predominant.

Gloecapsa, Symploca, Protococcus, Enteromorpha, Ghaetomorpha and *Polysiphonia* are also common besides *Lyngbya, Anabaena* and *Microcoleusos* surface plankton.

The presence of many of the benthic algae in the surface plankton is due to the constant stirring of water by the wave and wind action.

Diatoms especially *Gyrosigma scalproides* are found growing on filamentous algae in the benthic niche. The cold season between December and March appears to be the peak period of floral growth, but in the rainy season, the quantity of flora decreases considerably. However, *Oscillatoria* and allied forms are comparatively abundant during rainy season.

Fauna : The quantity of zooplankton is fairly high in the rainy season and is gradually reduced as winter approaches. However, members of different groups encountered are indicated below :

Copepoda :	*Diaptomus, Pseudodiap tomus. Acartla, Cyclopina, Cyclopsis.*	Most pre-dominant item of zooplankton
Cladocera :		Also found in plankton
Zoea, Megalopa larvae		
Mysids young of prawns		
and shirmps Ciliates.		
Flagellates Helminths. Rotifers		Found in small quantities during rainy months and up to the beginning of winters

The dominant elements of macrofauna other than the few polychaete worms and the Gastropod molluscs, *Stenothyra* and *Melanoides*, and the crustaceans and fishes. The common crustaceans are as follows :

Penaeus semisulcatus De Mann,

Penaeus indicus M. Edw.,

Metapenaeus monoceros Fab.,

Leander styliferus M. Edw.,

Palaemon rudis Heller,

Scylla serrata (Forsk)

Varuna littera ta (Feb.)

Metalpax dentipes (Heller)

During the rainy season swarms of megalop larvae of the common crab *Varuna litterata* abound in the pond water. The crabs *Scylla serrata* and *Varuna litterata* are responsible for making holes in the embankments.

5. FISH AND FISHERIES

Since in bheries it is a natural stocking and no transplantation of fry is practised, the overall production is low. Mullets, prawns and *Lates calcarifer* are the dominant species. Cat fish *(Mystus gulio)* and shrimp *(Leander styliferus)* also grow in the ponds in large number. The estimated production from the ponds is about 100-150 lbs (45.4-68.1 kg) only per acre per year. The following fishes have been identified from the pond :

Sub-class Actinopterygii

Family : Clupeidae

Pellona elongata (Bennett)

Family : Engraulidae

Setipinna phasa (Ham)

Family : Bagridae

Mystus gulio (Ham.)

Family : Anguillidae

Anguilla bengalensis (Gray & Hard)

Family : Ophichthyidae

Pisodonophis boro (Ham.)

Family : Belonidae

Xenentodon cancila (Ham.)

Family : Hemirhaphidae

Hemirhamphus limbatus Cuv. & Val.

Family : Cyprinodontidae

Aplocheilus panchax (Ham.)

Thrissocles hamiltoni (Gray)

Family : Cyprinidae

Barbus (Puntius) stigma (Ham.)

Family : Theraponidae

Therapon jarbua Forsk

Family : Lobotidae

Lobotes surinamensis (Bloch)

Family : Leiognathidae

Gerres abbreviatus Blkr.

Gerres setifer (Ham.)

Secutor insidiator (Cuv. & Val.)

Family : Scatophagidae

Scatophagus argus (Bloch)

Family : Eleotridae

Butis butis (Ham.)

Family : Gobiidae

Oryzias melastigma (McClelland)

Family : Mugilidae

Mugil corsula (Ham.)

Mugil parsia (Ham.)

Mugil tade Forsk

Family : Polynemidae

Eleutheronema tetradactylum (Shaw)

Family : Centropomidae

Ambassis baculis (Ham.)

Lates calcarifer (Bl.)

Apocryptes lanceolatus (Bl. Schn.)

Brachygobius nunus (Ham.)

Glossogobius giuris (Ham.)

Gobiopterus chuno (Ham.)

Periophthalmodon schlosseri (Pallas)

Scarleiaos viridis (Ham.)

Stigmatogobius sadanundio (Ham.)

Family : Cynoglossidae

Cynoglossus cynoglossus (Ham.)

It will be seen from the above list that a large majority of the common estuarine fishes of the area is found in the bhery. The above species of fish were grouped by Pillay (*op. cit.*) as follows :

Dominant :

(Animals of outstanding abundance of conspicious influence in the community)	*Mugil parsia*	1"-2" (2.5-5.0 cm) size attain 4"-6" (10-15 cm) in 8 months.
	Mugil tade	3/4"-1½" 1.87-3.75 cm) size Attain 1½"-2" (3.75-5.0 cm) in 8 months.
	Lates calcarifer	1"-2" (2.5-5.0 cm) attain 5" (12.5 cm) in 5 months.
	Penaeus semisculatus	larvae attain 3"-5" (7.5-12.5 cm). From winter to ending of season.

Influent :

(Animals usually effective in modifying the well being or numbers of the dominant group or other influents of the community. They are benefactor competitors or predators.	*Mystus gulio Apocryptes laceolatus Periopthalmodon schlosseri Leander styliferus Metapenaeus monocoros schlosseri Leander styliferus Scyna serrata*

Sub-influent :

(Animals that affect the life of the community but lesser extent than influents	*Aplocheilus panchax Oryzias melastigma Eleutheronema tetradactylum Scatophagus argus Butis butis Gobtopterus chuno Stigmatogobius sadanundio Varuna litterata.*

6. FISH-CUM PADDY CULTURE BHERI–SEASONAL (SEWAGE CONTAMINATED).

Ray *et al.* (1981) made observations on certain aspects of hydrobiological study in a Fish-cum Paddy culture bhery–a moderately productive bhery near the outfall Ghusighata which always received sewage contaminated water directly from the Kulti estuary. Simultaneously they have made a comparative study with a non productive Bhery located away from the previous one. It receives water only after meeting the demand of previous bheries and hence not sufficient and regular.

After an initial drying of the bhery soil, water from the Kulti estuary was drawn sometime in February and allowed to stabilise for 10-15 days to help growing the algal matter. Young prawn (*p. monodan*) and subsequently other brackishwater species like *Lates calcarifer, Mugil tade, Mugil parsia, Mystus gulio* etc. and also *Tilapia* sp. were heavily stocked. Major carp seeds were also stocked during wet months. A constant depth of 0.3-1.0 m water level was always maintained in the bhery. During such tidal phase of full moon and new moon period water was drawn into the bhery and released during low tide for 4-5 days to refresh the ecosystem as well as to maintain proper salinity, which otherwise shoots up during summer.

Harvesting of prawn was mostly completed in 3-4 months time before the onset of monsoon when sudden fall in salinity was expected. At this time the bhery get ready for paddy plantation. Almost simultaneously, when paddy seeds got rooted finerlings of Indian major carps (Rohu, Catla and Mrigal) were stocked. By September, paddy harvesting was complete. Major carps became ready in 6 months while *M. gulio, Tilapia* sp. and other took about 10 months. From November onwards the bhery was dried, ploughed and kept ready for the next stocking. On an average the yield in moderately productive bhery was 900-1000 kg/ha.

During summer months when temperature went up along with salinity occurrence of diseases in prawn and *Tilapia* were common phenomenon.

6.1 Physico-Chemical Conditions

To understand the characteristics of the water quality of a moderately productive and low productive bheries, samples were analysed for different parameters and are presented below in Table 8.

Sampling was done simultaneously in both the bheries under extreme summer condition. In the moderately productive bhery pH, alkalinity and organic matter indicated higher values while in the latter salinity was double.

To understand the productive potentiality of the bheries, 3 hourly diurnal variation of pH and Dissolved Oxygen was studied during the day time. The results are presented below in Table 9.

The high fluctuation in pH and Dissolved Oxygen in the moderately productive Bhery justifies its high productive characteristics. This is also supported by the primary production value, 279 mg $C/M^3/hr$ as gross value against the corresponding values 133 mg $C/M^3/hr$ (GPP) in low productive bhery. Phytoplankton concentration was 146 units and Zooplankton 75 units/litre in medium productive bhery while in low productive bhery the values were 11 and 10 units/lit.

Table 8. Physico-Chemical Conditions

Parameters	*Moderately productive bhery*	*Low Productive bhery*
Turbidity	Clear	Almost clear (Colloidal suspension)
pH	8.9	7.9
Dissolved Oxygen	10.12 ppm	6.0 ppm
Alkalinity	128 ppm	96 ppm
Salinity	10.08%	21.06%
Nitrate (NO_3)	0.84 ppm	0.84 ppm
Phosphate (PO_4)	0.12 ppm	0.11 ppm
O.C.	32 ppm	4 ppm

Table 9 : Diurnal variations of pH and Dissolved Oxygen in a medium and low productive Bhery

		7 a.m	*10 a.m*	*1 p.m*	*4 p.m*	*7 p.m*	
pH	Moderately productive bhery (MPB)	8.3	8.7	8.7	8.8	8.4	Rained heavily.
	Low productive bhery (LPB)	7.9	7.9	8.0	7.9	7.9	
Dissolved Oxygen (ppm)	MPB.	2.84	9.56	14.46	8.28	10.12	
	LPB.	4.36	5.56	7.72	6.64	5.8	

6.2 Bacteriological Conditions

Bacteriolgical counts of the water during summer, the ideal period of microbiol activity, revealed TPC; 1×10^5–4×10^6; Saureus : 1×10^2–2×10^3; Pseudomonas : X-positive growth; *E. coli* : 4×10^2; 9×10^2 while in the low productive bhery the values were, 2×10^6; 7×10^2; 7×10^2 per ml. TPC value was fairly high only once in the productive bhery while others were quite low.

7. LIMNOLOGICAL SURVEY OF BRACKISHWATER IMPOUNDMENTS IN THE DELTAIC SUNDARBANS UNDER DIFFERENT SALINITY

To understand the basic characteristic differences of water, soil biomass etc. of bheries in delitaic Sundarbans under different salinity Saha *et al.* (1986) have enlightened with valuable information in their document from the analysis of random samples for physico-chemical and biological characters from a number of bheries. The results were processed under different salinity – low, medium and high, and are presented in Tables 10 (A and B). Their survey reveals that medium saline zone covers maximum culturable water followed by low saline zone (cf. Fig. 4). *L. parsia, L. tade, L. calcarifer, Mystus,* sp. *E. tetradactylus* and prawn *P. monodon, P. Indicus, M. monoceros* are the important brackish-water fish and prawn cultured. Besides, major carps–*Rohu, Catla, Mrigal* and *T. mosambica* are also cultured.

7.1 *Physico-chemical Characteristics*

Some important physico-chemical characteristics of water and soil of Brackishwater impoundments under different salinity are indicated below.

Table 10. Ranges of some Important Physical and Chemical Parameters of Water and Soil Phases

A. WATER

Nature of Brackish water	*Temp*	*Turb*	*pH*	*E.C m.mhos/ cm*	*T. alka- linity*	*Salin- ity (%)*	*P_2O_5 (ppm)*	*NO_3 (ppm)*	*Ca (ppm)*	*Mg (ppm)*	*D.O. (ppm)*
Low Saline	21.0-32.0	72-360	6.8-7.8	0.9-11.4	84-212	0.1-9.5	Tr-0.32	0.04-0.30	25-230	25-650	4.0-6.8
	(28.1)	(115)	(7.4)	(6.2)	(138)	(3.8)	(0.11)	(0.08)	(65)	(153)	(5.8)
Medium Saline	23.1-32.0	63-350	7.0-8.2	1.1-28.0	90-196	0.27-15.8	Tr-0.24	0.02-0.48	24-220	36-880	4.2-6.2
	(29.1)	(112)	(7.5)	(9.8)	(145)	(6.3)	(0.09)	(0.13)	(94)	(250)	(5.6)
High Saline	20.0 30.6	57-240	7.0-8.3	11.8-51.7	88-155	6.6-36.2	0.03-0.66	0.03-0.32	70-350	280-1640	4.8-6.2
	(28.2)	(135)	(7.6)	(23.0)	(111)	(18.2)	(0.24)	(0.10)	(186)	(843)	(5.0)

B. SOIL

	PH	*EC m.mhos/cm*	*Available nutrients*		*Organic carbon%*
			N	*P_2O_5*	
			(mg/100 g soil)		
Low Saline	6.9-7.5	0.9-7.5	9.8-23.2	5.2-22.0	0.60-0.92
	(7.3)	(3.8)	(16.1)	(10.4)	(0.69)
Medium Saline	6.5-7.4	1.1-12.1	14.1-22.4	6.0-18.0	0.48-0.88
	(7.1)	(5.9)	(17.8)	(8.1)	(0.68)
High Saline	7.0-7.4	9.8-23.2	10.6-22.2	6.0-24.0	0.41-0.62
	(7.1)	(15.4)	(10.4)	(12.4)	(0.53)

The above table reveals that nutrient status of water is optimum. In soil the C/N ratio being low indicates active disintegration of the organics under optimum pH level. By this process the soil organics has a great role in maintenance of nutrient level of bhery water. Besides, the characteristic soil feature silty clay loam has great affinity for adsorbing released nutrients which in turn helps growing of algal matter on the soil surface for use as food by benthic organisms.

7.2 *Biological Characteristics*

Nine genera of algae have been recorded from the survey of deltaic Sundarbans. The principal plankton of the ecosystem are marine, brackish and freshwater forms. Bacillariophyceae are prominent among phytoplankton and copepod, Rotifer, and other Crustacea among Zooplankton are almost equally present, 9 genera of algae and 2 genera each of Rotifera, Copepoda and one of other Crustacean have been recorded in the ecosystem. These are as follows :

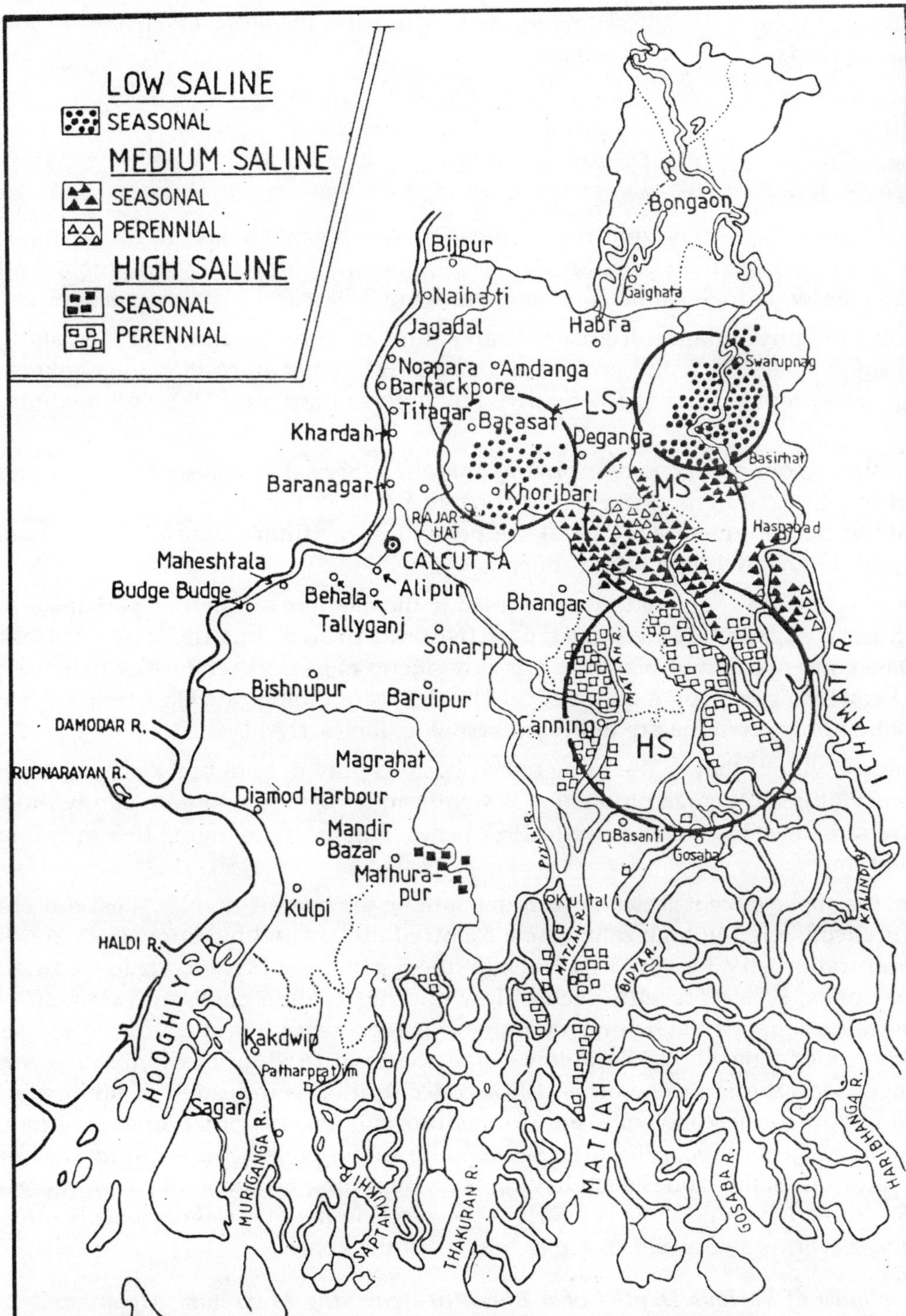

Fig. 4 : Intensity of fish culture under different salinities.

Bacillariophyceae	:	*Melosira, Synedra, Gyrosigma, Bidulphia, Coscinodiscus, Amphora.*
Bluegreen algae	:	*Spirullna*
Green algae	:	*Spirogyra, Oedogonium*
Rotifera	:	*Brachionus plicatilis, Acartiella* sp.
Copepoda	:	*Diaptomus, Cyclops*
Crustacea larvae	:	*Mysid*

Out of 3 zones, low saline zone represented 89.3% zooplankton and 10.7% phytoplankton, in medium saline zone the values are 74.7% (zooplankton) and 25.3% (phytoplankton) and in the same order, in high saline zone the values are 78.9% and 21.1% respectively.

Among the phytoplankton group, high saline zone represented 42.3% followed by medium saline zone 29.6% and low saline zone 28.1%; while in case of zooplankton group low saline zone represented 48.9% followed by high saline zone 32.9% and medium saline zone 18.2%.

Interestingly, out of 9 genera of phytoplankton, 5 (55.5%) are present in low saline zone followed by 3(33.3%) in medium saline zone and 4(44.4%) in high saline zone. In case of zooplankton out of 6 genera 5(83.3%) are present in medium saline zone, followed by 2 (33.3%) each in low and high saline zones.

Various species of zooplankton represented in the medium saline zone perhaps is an ideal medium for the growth of young fry of fish and prawn, but this zone suffers from inadequate concentration of plankton possibly due to mass scale culture of fish and prawn where overstocking is a common practice. The details of biological characteristics, plankton and benthic flora and fauna are indicated below in Tables 11A, B and C.

It is thus evident that in the same bhery, besides prawn, both brackish and freshwater species are cultured. Average production is significantly high. In addition, paddy production which are rare, once in 3 years prior to 1972 is now highly encouraging (8-9 mds/bigha) or 2.22-2448 t/ha.

From the above table it is clearly seen that among the benthic flora, highest concentration (77.6%) is encountered in high saline zone followed by low and medium saline zones in far low concentrations, 19.3% and 3.1% only. Of the 5 genera of benthic algae recorded in all the three zones, 60% represents each in low and high saline zones and only 20% in the medium saline zone. Of the benthic fauna worms represent highest concentration 56.3% in high saline zone followed by 25% in low saline zone and 18.6% in medium saline zone. The four genera representing the worms, 100% concentration is indicated in high saline zone followed by low saline zone (75%) and in medium saline zone (50%). In case of molluscan group, out of 6 genera recorded high saline zone shows 66.7% concentration followed by medium and low saline zones each having 50%. Summing up the above it is interesting to note that although benthic algae is poorly represented in medium saline zone, benthic fauna representing worms and molluscs are richest of all the zones.

7.3 Importance of Various Depths of a Bhery in Achieving Maximum Algae

Pillay (1962) studied the ecological flora in three brackishwater bheries at different depths (5 cm, 5.1-15 cm, 15.1-30 cm and above 30 cm). He has clearly shown in Table 52 the importance of various depths producing benthic flora.

Pillay (1962) opined that the soil nutrients appear to be of greater significance in the production of benthic flora than the nutrients in the water phase. This is undoubtedly a significant finding. The bhery having soil with the highest NPK value (Average value 16.48-4.09-60-94) was found to be the most productive one. The fluctuations in the soil nutrients were found to be correlated with algal production. While construction of bheries the above points (various depth zones) should be given the consideration to achieve desirable plankton concentration in abundance from the ecosystem.

8. THE STOCKING DENSITY

There has been no clearcut indication about the rate of stocking of brackishwater fish and prawn yet. But since prawn is drawing more attention because of its high economic value, some sort of stocking rate is established which is about 30,000/acre, and along with this large number of seeds of brackishwater fishes like *L. parsia, M. cephalus, L. calcarifer, L. tade,* etc. are stocked. Seeds of *P. monodon* are directly stocked in the bheries and are not reared in nurseries at the initial stage. This results in considerable mortality which sometime is as high as 70-75%. Besides, the high rate of stocking of brackishwater fish-fry along with prawn without considering the existing food stock also causes mortality. The repeated stocking or relay stocking is an outcome of such operations. Saha *et al.* (1986), however, have given some stocking density figure for *L. parsia*@1000-1500/ha, Major Carp @ 4000-5000/ha and *Tilapia* fry @ 5000-6000/ha in low saline zone; *P. monodon* (Post larvae) @ 40,000-50,000/ha in medium saline zone. In high saline zone *L. parsia* is stocked @ 700-1000/ha, *L. tade* @7000/ha and *P. monodon* @30,000-40,000/ha.

9. PRODUCTIVITY OF BRACKISHWATER BHERY

Production from brackishwater bheries under different salinity are shown by Saha *et al.* (1986) in Table 12, 13. The production indicates total as well as kg/ha of brackishwater fish and prawn with important species cultured.

Table 11A : Plankton and their abundance (Unit/litre)

Name of Brackishwater	*Phyto plankton*	*Dominant forms*	*Zoo plankton*	*Dominant forms*
Low Saline	18-178 (77)	*Melosira* sp. *Spirulina* sp. *Synedra* sp. *Spiroyra* sp. *Coscinodiscus* sp.	60-2722 (644)	*Brachionus plicatilis* Mysid
Medium Saline	30-198 (81)	*Melosira* Sp. *Spirogyra* sp. Oedogonium sp.	40-1155 (239)	*Diaptomus* sp. *Drechlonus picaulis* Nauplii *Cyclops* sp. Mysid.
High Saline	34-400 (116)	*Gyrosigma* sp. *Bidulphia* sp. *Coscinodiscus* sp. *Amphora* sp.	50-2140 (433)	Mysid *Acartiella* sp.

Dominant Species Composition Under Different Salinity

Phytoplankton	*Low*	*Medium*	*High*	*Zooplankton*	*Low*	*Medium*	*High*
Melosira sp.	+	+		*Brachionus Plicatilis*	+	+	
Spirulina sp.	+			Mysids	+	+	+
Synedra sp.	+			*Diaptomus* sp.		+	
Spirogyra sp.	+	+		*Cyclops* sp.		+	
Coscinodiscus sp.	+		+	*Acartiella* sp.			+
Oedogonium sp.		+		Nauplii		+	
Gyrosigma sp.			+				
Bidulphia sp.			+				
Amphora sp.			+				

Modified from Saha *et al.* (1986)

Table 11B : Abundance of Benthic flora and Fauna

Nature of Bhery	*Benthic Algae (g/m²) (wet weight)*	*Dominant forms*	*Benthic fauna*			
			Worms (n/m²)	*Dominant forms*	*Molluscs n/m²*	*Dominant forms*
Low Salinity	180-1170	*Nostoc* sp.	10-6375	*Gammarus* sp.	675-3601	*Bellamya* sp.
	(675)	*Anabaena* sp.	(1071)	*Tanaids*	(1008)	*Diagoniostoma* sp.
	19.3%	*Oscillatoria* sp.	25.1%	*Polychaetes*	28.1%	*Planobides* sp.
Medium Salinity sp.	89-131 (110)	*Oscillatoria* sp.	135-6400 (2403)	*Gammarus Polychaetes*	550-5945 (1600)	*Bellamya* sp. *Diagoniostoma*
	3.1%		56.3%		44.5%	Thiara sp.
High Salinity	110-4300 (2700) 77.6%	*Lyngbya* sp. *Oscillatoria* sp. *Sptrogyra* sp.	5-6160 (790) 18.6%	*Tanaids Polychaetes Gammarus* sp. Oligochaetes	135-2145 (982) 27.4%	*Bellamya* sp. *Diagoniostoma* sp. *Unio* sp.

Saha et al. (1986)

Dominant Species Composition of Benthic Flora and Fauna Under Different Salinity

Benthic algae	*Low*	*Medium*	*High*	*Benthic fauna*	*Low*	*Medium*	*High*
Nostoc sp.	+			1. (WORDS)			
				Gammarus sp.	+	+	+
Anabaena sp.	+			Tetredes	+		+
Oscillatoria sp.	+	+	+	Polychaetea	+	+	+
Lyngbya sp.			+	Oligochaetes			+
Spriogyra sp.			+	2. (Molluscs)			
				Bellamya sp.	+	+	+
				Diagoniostoma sp.	+	+	+
				Indoplanorbis sp.	+	+	+
				Thiara sp.		+	
				Unio sp.			+
				Lymnaea sp.			+

Modified from Saha *et al.* (1986).

9.1 *Typical Example of Productive Potentiality in a Paddy-Cum-Fish Culture Bhery in Medium Saline Zone*

Information with regard to productive potentiality of a paddy-cum fish culture bhery in a medium saline bhery under Haroa block is indicated below in Table 13. This information is communicated by the secretary, 24-Parganas Fish Producers' Association, Calcutta. Prior to seventies, paddy cultivation was a problem. The monocrop paddy used to grow once in three years and was 1-1½ bags (2-3 mds. 560.7-841-05 kg/ha). Paddy was normally destroyed either due to high saline water or flood. It is now growing 8-9 mds/bigha. (2242.8-2523.2 kg/ha).

Table 11C. Flora in relation to depth of bhery

Zone I	*Zone II*	*Zone III*	*Zone IV*
Depth up to 5 cm (Average salinity 24.3%)	Depth between 5.1-15 cm (Average salinity 18.76%)	Depth between 15.1-30 cm (Average salinity 10.90‰)	Depth above 30 cm (Average salinity 10.57%)
Myxophyceae predominant group	Myxoplyceae and chloro-phyceae almost equally abundant.	Blue green algae are much more abundant than in all other zones.	Maximum growth of green algae
As in Zone IV	*Lyngbya aestuarii* thrive best	As in Zone II	Lyngbya grows fairly well
—	*Oscillatoria tenuis* grows best.	As in Zone II	—
—	Denser growth of *Chaetophora*, *Chaetomorpha* and *Cladophora*	—	Maximum growth of *Chaetophora* *Chaetomorpha* and *Cladophora*.
—	—	*Microcoleus* grow well	As in Zone III
Tribonema more abundant	Same as Zone I	—	—
—	—	*Schizothrix* and *Compsopogon* occurred	—

* Modified from Pillay *et al.* (1962)

It is thus evident that in the same bhery, besides prawn, both brackish and freshwater species are cultured. Average production is significantly high. In addition, paddy production which was rare – once in three years prior to 1972 is now highly encouraging (8-9 mds/bigha) or 2.22-2.448t/ha. Most of the prawn produced are meant for export.

From the above it is evident that the local agriculturists are very much tempted in the integrated system of fish-cum paddy culture practices as it is more paying. Further, from the employment potentialities discussed below in Table 14 also confirms this views.

Table 12. Productive potentiality of Brackishwater Bheries Under Different Salinity (Saha et al. 1986)

Nature of Bhery	*Salinity (%)*	*Total water area (ha)*	*Average Production kg/ha/yr.*	*Estimated total annual production*	*Principal species cultured*	*Area*
Low saline (Above 22° 40'N latitude)	0.1-9.5 (3.8)	9844	878 *70	8641 t *690 t	*P. monodan* *L. parsia* *L. rohita* *C. catla* *C. mrigala* *T. mossambica*	Barasat Deganga Rajarhat Gaighata Baduria
Medium saline	0.27-15.8 (6.3)	15613	749 *214	11692 t *3349 t	As low saline zone	Minakhan Haroa Hasnabad Basirhat Hingalganj
High saline	6.6-36.2 (18.2)	7472	694 *84	5185 t *625 t	*P. indicus* *P. monodon* *P. monoceros* *L. parsia* *L. tade* *L. Calcarifer* *E. tetradactylus* *Mystus* sp.	Canning Mathurapur Jaynagar Parthapra-tima Basanti Kultuli Gosaba Sandeshkhali Kakdwip
Total		32.929	776	25.518 t *4664t (5.7%)		

* indicates prawn (P. monodan) Total production includes prawn.

Table 13 : Production in a 1000 Bigha Medium Saline Bhery (Owner Naresh Das, Haroa) 24-Parganas. W.B.)

Brackishwater fish		
P. monodon	—	45,000 kg. (Average production;) Large percentage exported)
L. calcarifer	—	20,000 kg.
L. parsia	—	9,000 kg.
M. gulio	—	10,000 kg.
Odontomblyopus rubicundas	—	10,000 kg.
Tilapia	—	50,000 kg.
Freshwater fish		
Major carps	—	70,000 kg.
(L. rohita, C. catla, C. mrigala)		
Total production	—	2,14,000 kg/1000 bigha or 1.6 t/ha.

10. COMPARATIVE EMPLOYMENT POTENTIALITIES OF PADDY CULTIVATION, FISH CULTURE AND PADDY-CUM-FISH CULTURE

Various aspects related to employment potentialities in a brackishwater bhery with comparative employment potential ratio between paddy cultivation. Fish culture and paddy cultivation and fish culture together are given below.

Table 14 : Employment Potentialities in a brackishwater bhery

(Perennial bhery : 100 acre)

1.	Dyke construction 50 days × 80 labourers	4000 man days
2.	Watch tower, fish trap making etc. 45 days × 10 labourers	450 man days
3.	Making of bamboo fish carrier 15 days × 5 labourers	75 man days
4.	Dyke repair 300 days × 15 labourers	4500 man days
5.	Monthly paid worker 365 days × 30 labourers	10950 man days
6.	Netting charge (a) Darg netting — from middle of April-middle of December, 215-230 days (average 225 days). Twice daily. Each net requiring 10 labourers for dragging etc. 225 × 10 × 2 = 4500	6,180 man days
	(b) Cast netting Fishing for 4 days/month during each 'Kotal' (Tidal phase) for seven month's season. Total working days — 56 with 30 workers/day, 56 × 30 = 1680	
7.	Night guards Working days (9 months) about 250 days. Number of guards employed 15/day 250 × 15	3,750 man days
8.	Transportation of fish from Bhery to local market (usually centre) Number of additional labourers employed per day-5 Fish Transported for 150 days 150 × 5	750 man days

* Prepared by Sunil Bhattacharya (In *'Sundarbaner Mechogheri'* published by Fish Producers Association, 24 Parganas West Bengal), Calcutta.

From the foregoing account it is clear that in a brackishwater bhery, employment potentialities per 100 acre is 30,655 man days i.e.

Per acre employment potentialities or fish culture = 306.5 man days.

But per acre employment potentialities Paddy cultivation = 70 man days and per acre employment potentialities. Fish culture + Paddy cultivation together = 376 man days.

The comparative employment potential ratio between Paddy cultivation, Fish culture, and Paddy cultivation and fish culture thus comes to –

Paddy cultivation : fish culture : Paddy cultivation + fish culture (together) = 1 : 4.5 : 5.5

On the whole in Sundarbans, for the same plot of land, the employment potential rate in fish culture is more by 4.5 yo 5.5 times when compared to paddy cultivation.

11. PRESENT STATUS OF FISHERIES IN SUNDARBAN DELTA UNDER DIFFERENT SALINITY

A more detailed information on various aspects like (A) production of brackishwater impoundments. (B) contribution of individual groups of fishes in the total production and finally (C) their export, local consumption and supply are now available from Secretary, Fish Producers' Association 24-Parganas, West Bengal. These data speak of the present status of fisheries of Sundarban delta.

Table 15 : Fisheries in Sundarban Delta

Sl. No.	*Ecosystem*	*Nature of Zone*	*Water area (ha)*	*Production Rate*	*Production Total (t)*
1.	Perennial tank fishery (Recorded)	*High Saline Zone* (3200 acre)	1296 ha	1200 kg/ha	1555
		Medium Saline Zone (3300 acre)	1336 ha	1 t/ha	1336
		Low Saline Zone (1500 acre)	607 ha	100 kg/ha	425
		8000 acre	3239		3316
2.	Paddy-cum-fish	*High Saline Zone* (4000 acre) Mainly Canning, Basanti, Kultuli Gosaba Sandeskhali etc.	1619	500 kg/ha	810
		Medium Saline Zone Mainly-Haroa (28000 acre) Minakhan	11331	1 t/ha	11331
		(17000 acre) and	6880		6880
		Hasnabad, Basirhat and Hingalganj (10,000 acre)	4047	800 kg/ha	3238
		Low Saline Zone Mainly Barasat (17000 acre)	6880		
		and Swarupnagar, Deganga, Rajarhat, and Baduria (6000 acre)	2428	300 kg/ha	2792
		82000 acre	33185		25051
3.	Grand Total	90000 acre	36424		28367
	Contribution of Individual Group of fishery				11706 t
4.	*Contribution of freshwater fish* 40-50% (average 45%) of production in medium and low saline zones of 1 and 2, 45% of (1336 + 425 + 11331 + 6880 + 3238 + 2792 t or 26002 from 33509 ha				
5.	Contribution of brackishwater fish and prawn.				28367 t
6.	Contribution of prawn alone (60% of brackishwater fish and prawn)				17020 t

contd...

Table 15 – *contd...*

7.	Contribution of brackishwater fish alone (5-6)	11347 t
8.	Contribution of *P. monodon* (70% of prawn)	11914 t
	Export. Local consumption and supply *P. monodon*	8936 t
	(75% of *P. monodon* after initial size grading) (*Actual quantity of export of P. monodon*) (20% less due to headless shellon)	7149 t
	Local Consumption and Supply	
	Freshwater fish	11706 t
	Brackishwater fish	11347 t
	Prawn other than *P. monodon* (6-8)	5106 t
	P. monodon (Rejected 25% due to unhealthy and undersized)	2979 t
		31138 t

References made for paddy-cum-fish at various places in this document by various workers are actually fish-cum-paddy as the primary produce in the process is fish and not paddy.

12. SEWAGE AS POTENTIAL GROWTH PROMOTER

In the absence of adequate data separately between sewage fed and sewage free bheries of medium saline or high saline zone it is difficult to make any comparative study of their production. Despite such inconvenience some rough conclusion can be derived at by analysing portions of the survey data taken from Pakrasi's survey report (1965) falling in both the zones. The results are as under :

Location	Total water area (acre)	Prawn Production alone	Total fish production	Type of zone	Remarks
Haroa P.S.	1734	2496 (1.4 kg/acre)	6683 kg (3.82 kg/acre)	Medium saline	Sewage fed area
Sandeshkhali P.S.	2301	689 kg (0.299 kg/ acre)	2178 kg (0.947 kg/ acre)	High Saline	Sewage free area

The above data indicates that the common belief of high production either prawn or other brackishwater species in high saline water does not hold true. The production at Haroa whose water is medium saline in character but sewage fed, has shown increased production of over four times in case of prawn and 4 times in case of total production. Similar findings are also reported by Barman (1983).

13. SALINITY NO LONGER A PROBLEM FOR INTEGRATED FISH AND PADDY CULTURE

Although the traditional practice of paddy culture in the Sundarbans was only during monsoon, yet in the years of drought, late monsoon or insufficient rains the production suffered considerably and the land remained abandoned for the rest of the year. The common belief that once saline water has access to the land it will gain salinity which is difficult to remove and make paddy cultivation imposible does not hold good. Such belief is baseless and has been experimentally shown by Natarajan & Ghosh (1982) and subsequently by Chattopadhyay *et al.* (1983a) by comparing the salt content in paddy-cum-brackishwater fish culture areas of upper and mid estuarine regions of West Bengal during paddy cultivation period with those of the nearby paddy cultivation from soils. In the former case salinity (EC value) was found to range between 2.6 and 9.3m mhos/cm (averavge 4.3 m mhos) against the latter cases where salinity ranged between 0.7 and 3.6 m mhos/cm (average 2.0 m mhos) The comparatively higher salinity (only in case of 4 out of 14 analysed) in the former case was due to continuous culture of brackishwater fish alone for over 5 years. Yadav *et al.* (1979) and Das (1983) and other workers are of the opinion that even in the coastal saline zone such salinity after monsoon is a common phenomenon and does not stand in the way of cultivation. Bhumbla and Abrol (1977) cited references to show that rice gives satisfactory yields even when EC value of the surface soils ranged 20-25m mhos/cm. Moreover, under paddy-cum brackishwater aquaculture system generally salt tolerant local varieties are cultivated. Bhattacharya (1976) reported some local varieties of Sundarban area to yield 33 q/ha even under EC values of 11 m mhos.

Chattopadhyay *et al.* (1983b) made observation on the impact of fall in salinity in the soil during rains and the effect of such salinity on paddy crop in fields which were earlier used for culutre of euryhaline species, *M. parsia* and prawn. The experiment was conducted in two 0.015 ha paddy plots of Regional Research Station of Central Soil Salinity Research Insititute, Canning, which is in low lying coastal area of West Bengal.

Brackishwater aquaculture of *P. monodon* and *L. parsia* (1 : 1) at a stocking density of 75000/ha carried out till June 1987 gave an yield of 654 kg/ha fish and prawn in a total period of 86 days. Water salinity ranged between 12.5 and 40.0 m mhos which resulted in salinisation of soil to an extent of 24 m mhos/cm from an initial value 7.8 m mohs/cm.

Due to late and scanty rains the fall in salinity was not appreciable (7 m mochs/cm). By August with further rainfall salinity declined and ranged between 4.8 and 5.0 m mhos/cm when seedlings of three rice varieties were transplanted in each of the plots. Simultaneously fresh water fishes and prawn (*L. rohita, C. catla, C. mrigala, H. molitrix* and *M. rosenbergii*) at a total sotcking rate of 23,500/ha in the ratio of 0.6 : 0.6 : 1.2 : 1.1 : 6.5. Paddy was harvested at the end of November giving an yield of 3062 kg/ha and fish and prawn gave an yield of 511 kg/ha in 83 days.

Thus the paddy plots by integrating pisciculture operation for a total period of 169 days produced on an average 654+511 or 1165 kg/ha fish and prawn in addition to the normal yield of paddy.

14. INTEGRATED FARMING IN SALINE ZONE TO AUGMENT PRODUCTION

Since integrated farming system has shown to be successful its wider application in monoculture paddy field will undoubtedly increase the production of fish and prawn.

In the Sundarban region about 7,50,000 acre (303521 ha) is being used for paddy cultivation only. In case this area is brought under integrated farming an additional yield of 236108 t of fish and prawn can be obtained (36424 ha gives an average yield of 28367 t cf. (table 15.). Thus in the entire Sundarban delta the estimated average total annual production of fish and prawn in perennial, existing paddy-cum-fishery and converted monoculture paddy field into integrated paddy-cum-fishery will be 3316 t + 25021 +236382 t for 264749 t yield of freshwater fish which is mainly from the medium and low saline zones of perennial and paddy-cum-fishery (assuming at the rate of average 45 % of production from brackishwater area) will be 792 t and 10908 t or 11700 t respectively. In the absence of data on water area that will fall under medium and low saline zones of monoculture paddy field average production of freshwater fish cannot be estimated. But assuming a similar production of 11700 t from 303521 ha, the average annual total production of fresh and brackishwater fish and prawn in the Sundarban delta collectively will be 11700 t × 2 + 264749 t or 288149 t.

Jhingran (1978) claims that the recent advances made in brackishwater prawn and fish farming by the Central Inland Fisheries Research Institute, Barrackpore, the technology of mixed farming of different species of prawns and mullets (*P. monodon, P. indicus* and *Metapenaeus monoceros* among prawns and *Liza parisa, L. tade* and *Mugil cephalus* amongst mullets) a net production of 2671 kg/ha/yr was achieved. Earlier, Ghosh (1978) reported that if progressive steps were taken i.e. adding fertilizer, supplementary feed, use of pesticide etc. production of 2.6 t/ha/yr could easily be achieved. Assuming an average yield of 2.5 t/ha/yr is considered for brackishwater fish and prawn from all the three zones-high, medium and low of perennial tanks. Paddy-cum-fishery and monoculture paddy field converted into paddy-cum-fishery of Sundarban delta, the average total annual production will come to 8098 t + 82962 t + 758802 t or 849862 t. This figure (849862 t) when added to the total freshwater yield (23412 t) stated earlier will be 873274 t. Integrated farming can thus help to increase fresh and brackishwater fish and prawn in case improved technology is followed.

15. PRAWN PROSPECTS OF SUNDARBAN DELTA

Prawn being sold at a very high price bhery owner normally introduces prawn larvae following traditional system viz 2-5 lakh/ha without any supplementary feed. Such rate of stocking is abnormally high and affect growth and cause disease resulting in high mortality, 75-90% (Sakthivel, 1980). He further stated that tiger prawn which usually are harvested at the optimum size range 16-20/lb are being harvested at 31-40 size range. Thus pisciculturists are deprived of their maximum benefit.

Contribution of prawn for export from West Bengal as stated by Sakthivel (1980) was 149t in 170 and 2853 t in 1978 which amounts to 0.69 and 3.66% of the total prawn exported from India. The catches are mainly from Sundarban delta. The export figure in recent years has now gone up to 8939 t.

A close look into the analysis of data (cf. Table 15) reveals that paddy-cum-fishery is the major source of prawn supply. Contribution of prawn from Sundarban delta is 17020 t of which *P. monodon* is 11914 t i.e. 70% of total prawn production and is produced from 36424 ha of water area (cf. Table 56). Area of land used for monoculture paddy is 7,50,000 acre (303521 ha). In case this new area (303521 ha) is also converted into integrated paddy-cum-fishery system, the calculated additional crop of prawn that could be achieved at the existing rate (17020 t in 36424 ha) is 141828 t. Thus adding the existing yield of prawn 17020 t to the new production (141828 t) the total prawn production that can be achieved is 158848 t and taking 70% of total prawn as *P. monodon* (existing composition), the figure thus achieved is

111194 t in addition to usual paddy crop. This figure can even go up if the bhery owners pay the attention to the system of nursery, rearing, optimum stocking density (60,000/ha) as suggested by Sakthival (1980), proper food supply and avoidance of predators through intensive extension work. The loss can thus be minimised and the excess of prawn larvae could be well utilised in a national way elsewhere for prawn farming to get maximum benefit.

More recently Kakdwip Research Centre of Central Inland Fisheries Research Institute, Barrackpore (CIFRI News Letter-1985) has achieved a spectacular production rate 1 to 1.5 t/ha of prized tiger shrimp in a short span of 8-12 weeks against their previous rate 200-400 kg/ha in 3-4 months in their experimental ponds. The new technique entails an innovative feed formula, effective stock manipulation and continuous monitoring of the environment. The institute further claims quality seed and feed are essential prerequisite for launching an extensive shrimp production. Such feed can even boost shrimp production to 2t in 2-3 months. 2 to 3 crops can easily be taken in a year depending upon location, tidal regime and favourable salinity. Such findings are not only encouraging but also spectacular. Although economics of feed is not indicated undoubtedly this new feed has helped in achieving a breakthrough in mass scale production of prawn.

In case integrated farming and the suggested technology is followed and assuming an average yield of 5t/ha/year (in 3 crops) in place of 6t is produced the estimated production of prawn alone that can be achieved from the Perennial tanks (16195 t), Paddy-cum-fishery (182120 t) and the converted monoculture paddy field (1517605 t) together will be 1715920 t.

16. PROBLEMS ASSOCIATED WITH BRACKISHWATER FISHCULTURE IN BHERIES

Despite the progress made by the farmers of Sundarbans in utilising the land and water of deltaic West Bengal for paddy and fish culture they have been suffering from certain limitations. This has restricted the production both in case of paddy, fish and prawn. Some of the major problmes are discussed below :

1. Water supply from the Kulti estuary is much less compared to present demand for the bheries. This is due to rise of river bed on account of silt deposition.
2. Sometime, sudden excessive flow of metropolitan waste discharged through DWF canal might be injurious not only for survival and fish migration in the Kulti estuary but also may cause harm to the stocked fry and prawn larvae in the adjoining bheries.
3. Draughts or late monsoon under the present condition may change the hydrobiological condition to a level not congenial for biomass to survive. It becomes difficult to maintain the water level unless udal water reaches sufficiently high.
4. Absence of proper feeder canal supplying water to the bheries.
5. Silting of feeder canal and the bhery causing problem for maintaining proper inflow/outflow and water level.
6. Excessive growth of algae in the bhery.
7. Inadequate information about the technique on integrated farming.

8. There is no information about the extent of damage to fisheries in case pesticides are reasonably used to save the paddy crop.
9. Lack of information on the baseline data in respect of physico-chemical parameters including heavy metals, pesticides and bacteriological load in water, soil and fish in bheries having different salinity both in sewage fed and sewage free water. All these factors can affect healthy growth of fish.
10. In the absence of nursery rearing practice, protection against predators and over stocking directly without due consideration to optimum available food results in high rate of mortality especially in the case of prawn larvae. Epizootics is a common phenomenon in bheries especially in prawn and tilapia.

    Prawn seeds being very costly are collected, along with prawn, seeds of other varieties of fishes estimated to be about 70% of total collection also come. These are simply thrown away and destroyed in the process. This results in considerable loss to fish populatio which could otherwise be cultured and feed the local markets.
11. Lack of cold storage facilities, ice plants, quick transport and communication both in water and land, absence of basic amenities are some of the important factors.
12. Awakening of the villagers with extension work, like training and demonstration with latest scientific, information on bhery management for fish culture - an essential part of their profession through some fishery training school is rather difficult.
13. Poaching in bhery has become a common phenomenon, and according to an estimate between the years 1967-68 to 1971-72 as stated by Gangopadhyay (1980), the loss incurred in 6 bheries was Rs. 1451986 or 75.67 % of the total loss and is followed by flood (18.34%). Further, the poaching in 'Poina fishery' in May 1967 measuring on area 1485 acres (594 ha) caused a loss to the extent of Rs. 584,760. This is also followed by the most recent event on 17th March 1986 of Gobaria bhery having an area of 2500 acre (1000 ha), which earns foreign exchanges worth of 4 crores annually from prawn exprot, is another example. The extent of annual loss estimated to be Rupees 100 crore worth of fish, half of which is from *P. monodon*. The above facts are certainly alarming. As a result of this most of the bhery owners are hesitant in investing any more fund in future. As a last resort they have started culturing tilapia in place of commercially important fishes. Besides, Bankers also become reluctant in releasing any fund in future to the bhery owners of the area. The production not only dwindles but has far reaching effect on the economy of bhery in general.

17. REMEDIAL MEASURES SUGGESTED

In the previous chapter (Part II) some remedial measures are suggested for the conservation of the Kulti estuary. If any such measures are implemented there will be more flow in the estuary and help dilution of the concentrated sewage. At present it has been observed that bheries located away from the river suffer from regular non-supply of river water both in quality and quantity. For example, Ray *et al.* (1981) observed salinity in the unproductive bhery rose as high as double (21%) against 10.08% in the moderately productive bhery. Such high salinity could not be lowered since the bhery was located away from the river and supply of water to such bhery could not be effected.

A range of safe dilution level of metropolitan waste by the river water has to be established then only the water can be drawn in the bheries for the safety of the aquatic biomass. It is possible that bheries nearer to river will have the benefit of enriching them with organic detritus essential for prawn more than the bheries located for away. In that case additional supply of organic manure can be made. But the bhery owner should be careful aganist excessive growth of algal matter in both cases. There should be a proper feeder canal that ensures water supply to all the bheries. Each bhery should have both inlet and outlet points permanently fixed for drawing and releasing water through the feeder canal. This will save the bhery from any adverse situation, if arises, during dry months, draughts and late monsoon. Feeding canal and bheries should be desilted as and when necessary, otherwise maintaining the water level in bhery may cause problem. Special attention is needed to study factors responsible for the mortality, unhealthy growth and disease of fish and prawn fry/larvae. If success is achieved in reducing mortality, seeds can be transported to other areas and deficit of seed can be overcome. Adequate baseline data on physico-chemical parameters for water and soil having different salinity should be collected to help deleniating bheries under low, medium and high saline zones to find out their characteristic differences in productivity. Similar studies for sewage fed and sewage free bheries should also be separately made for the above purpose as difference in production is clearly observed in the survey data of Pakrasi (1965). Further, additional stress need by given for the study of metals, pesticides, microbiol load in water, soil and fish and rate of mineralisation in soil sediments.

Plankton and benthos which are the important secondary producers deserve attention for detailed investigation.

Natural stocking of fish seed from the estuary although may solve a major problem, such stocking because of its varying composition, does not prove to be economically very beneficial. Further, forage fishes are competitors of food. Production of economic variety thus suffers. It is, hence suggested to have selective stocking with compatible species and reared. This is possible by transplantation of seed. The common practice of overstocking fish fry or prawn larvae should be avoided. This will reduce food competition, disease and mortality. Composite culture by integrated farming even in paddy cultivated lands should always be parctised and encouraged to have increased production. Excessive growth of filamentous algae can be controlled by tilapia. Special attention should be given for prawn culture. For this if prawns are reared initially in nursery ponds up to juvenile stage and then released into the stocking pond with optimum stocking density (60,000/ha approximately) after proper selection, adding proper feed and taking adequate protection against predators through intensive extension work, high mortality rate can be avoided and production will increase. Further, the new feed formula developed by CIFRI, which claims to achieve a high yield of tiger prawn, can be put to trial on an experimental basis. Keeping in view of the local demand and export, further research on the following lines may provide valuable information. (i) Whether culture of *P. monodan* more than once in a year is beneficial andup to what percentage of water area should be kept separate for it. (ii) Extent of benefit of culturing integrated farming (paddy-cum prawn) and (iii) Composite culture; fresh, brackishwater fish and prawn especially *P. monodan* in association with paddy cultivation.

The bhery after harvesting (Paddy, prawn and fish) may be allowed to get dry and properly tilted and made ready for next stocking.

Prawn use organic detritus as food and hence organic manure compost prepared out of locally available mangrove leaves can be used. Besides, when there is scarcity of food, manure used in freshwater bheries can also be used. The new feed formula prescribed by C.I.F.R.I., which is claimed to be effective for high yield, may be tried. Bhetki (*Lates catcarifer*) can be cultured separately using trash fishes as their food, otherwise they will prey upon the young fishes and prawns kept in the bheries for culture.

Chorlands/wet areas adjoining to the sides of rivers, especially in the lower Sundarbans are greatly influenced by tidal waters. These lands are not capable of yielding any agricultural crop due to high saline soil. But they can be conveniently converted into shallow reservoirs/bheries by fencing with proper dykes and water can be introduced through sluices during high tides as is normally done in 'Bhasabadha' or recorded fishery. According to Hora, Brackishwater bheries have no water problem if tidal amplitude in the estuaries range between 5 and 12 ft. Bhattacharjee (1986) recoded water in the estuaries of Sundarban to rise as high as 17 ft from sea level when the height of land is only 10/12ft. Seeds of brackishwater fish can be trasplanted from April to July and allowed to be reared until next January before harvesting. The Sundarban chor-lands can thus be profitably utilised. Desalinisation of land can be done appreciably by repeated rain washings and hence is not a problem. Integrated farming even in paddy cultivated lands in saline zone can thus be implemented to increase the production. Simultaneously, Indian major carps, Rohu, Catla and Mrigal can also be cultured as it has been experimentally seen that Indian major carps can grow well up to a salinity of 14,000 ppm (Saha, 1970). Since Sundarban delta consists of network of estuaries, tributaries, creeks, channels infested with reserve forests and fed by regular tidal waters but not easily accessible, communication and transport facilities need be developed. The quick transport of fish seed will not henceforth pose problem of its transport either by water or by road.

From a comparative employment potentialities of paddy cultivation, fish culture and paddy-cum fish culture it is revealed that paddy-cum fish culture system is the most profitable. Futher Table 55 clearly gives that employment potential rate in fish culture is 4.4 to 5.5 times more when compared to paddy cultivation. A similar view is also expressed in the "*Report of the Task force on the best land use options for the saline soil areas in Coastal Districts of West Bengal*", published by Dept of Fisheries, Govt. of W.B. 1987. Hence this integrated farming system even in paddy cultivated lands is recommended. Similarly in perennial brackishwater bheries and chorlands of lower Sundarbans such practice can be implemented to achieve increased production. Special attention has to be given to safeguard the fishery from poaching. Such practice which is gradually increasing, can not only dwindle the production but also change the economy of the bhery fishery of the area.

Finally, for preservation of fish and prawn, installation of cold storage, ice plants etc. at the important fish assembly centres are essential. In case it is not economical due to power problem and other reasons, quick transport system of ice to the centres may be arranged. Facilities for quick transport of preserved fish to the nearby city is of vital importance for their quick disposal or export. The Govt. should extend full cooperation to the pisciculturist in the above matters.

18. DEMAND VIS-A-VIS FISH PRODUCTION

Population of greater Calcutta being 92 lakh (1981) which has increased futher by now, the daily requirement of fish at the rate of 60 g/head/day comes to 552 t.

The existing annual production is from two important sources - Freshwater Bhery (sewage fed) and Brackishwater zone of Sundarbans delta and the details are as follows :

Table 16. Existing Fresh and Brackishwater fish and Prawn Production

1. Fresh water Bhery (Sewage fed)	=	4000t
2. Brackishwater zone		
a Paddy-cum-fish (Integrated)	=	25051 t
b. Perennial tankt of Bhasabadha Fishery (Brackishwater fish and prawn only from a and b)	=	3316 t
c. Freshwater fish from medium and low saline area of a and b.	=	11706 t
Grand Total	=	44073 t

Out of total production 44073 t, 8963 t of *P. monodon* is exported. Thus the production comes down to (44073 t - 8936) or 35137 t. Production per day which becomes available for local consumption is 35137 t +365 or 96 t against the demand of 552 t/day. Thus there is a short fall of 456 t everyday.

19. SCOPE FOR AUGMENTING THE PRODUCTION

Suggestions for further improvement of fish production have been discussed individually in all the three parts I, II and III of this document. All these sources will certainly contribute towards the total production in case suggestions are adequately implemented. It is clear that there is enough of scope for the development of fish and prawn culture in the integrated paddy-cum-fishery following sequential system without affection the usual paddy crop. The following account will clearify the position.

Table 17. Modified annual fish and prawn production

A.	Freshwater Bhery (Sewage fed)*	=	4000 t
B.	Brackishwater area		
	1. Perennial Tanks (Existing) or Bhasabadha Fishery @ 2.5 g/ha (Brackishwater fish and prawn only)	=	8095 t
	2. Freshwater fish (From medium and low saline areas of 1; average 45% of Brackishwater fish and prawn)	=	218
	3. Paddy-cum-fish (Existing) Brackishwater fish and prawn @ 2.5 t/ha	=	82962
	4. Frehswater fish (from medium and low saline areas of 3; average 45% of Brackishwater fish and prawn)	=	37333 t
	5. Monoculture paddy field (303521 ha) converted into integrated Paddy-cum-fishery. (Brackishwater fish and prawn only) @ 2.5 t/ha	=	758303 t
	6. Freshwater fish from medium and low saline area (assuming a similar yield as shown in 4 under existing paddy-cum-fishery.)	=	37333
	Grand Total	=	930712 t

* There is ample scope to raise the yield between 6-10 t/ha/yr in case improved technology as suggested is followed)

Prawn Production and Experts

Contribution of Prawn	=	509917 t
60% of Brackishwater fish and Prawn (849862) t Contribution of *P. monodan* (75% of Prawn)	=	356942 t
Export of *P. monodon* (75% of *P. monodon*)	=	267706 t
Total annual production after export (930712-267706 t).	=	663006 t

Total production per day left for local consumption thus comes to 363006 t + 365 or 1816 against the actual demand 552 t per day 1264 t i.e. excess of the present requirement. The above account clearly indicates that by adopting new technology and system a surplus yield of 1264t per day of fish and prawn can be achieved.

Alternatively if prawn alone is culture in the monoculture paddy field after its conversion into integrated paddy-cum-fishery, by adopting new technology discussed, the production at the rate of 5 t/ha/yr from 3 crops (CIFRI News letter. 1985) comes to 1517605 to of which *P. monodon* (70% of Prawn) will be 1062324 t. If 75% of it which comes to 796743 t is exporterd against the present export of *P. monodon*, 8936 t (8 times more) the total production of prawn that will be left is 720862 t. This figure when added to 134576 t (Production from freshwater sewage fed area and existing paddy-cum-fishery) the yield will be 855438 t. From this per day production that will come is 2344 t against the actual demand of 552 t, thereby an excess of 1792 t will be possible.

Thus utilizing the land and water of Sundarbans delta for increased production of fish and prawn will not only fulfill the demand of Calcutta and the suberbs and to some extent the area of supply but will also put Sundarban in the world map as an important source of tiger prawn supplier.

List of Commission Agents/whole sellers of fish under different P.S. of 24 Parganas and names of Sea food processors and exporters are indicated in appendices II and III.

Addendum A
(Some Tit Bits)

How paddy and fish culture has brought affluance in Haroa P.S.

(Report from 'Bhumilakshmi' - Ananda Bazar Patrika 22.4.1981)

Area of Haroa Block 37760 acre (15292.8 ha)

Formation of salt water Bhery since 1922 28000 acre (11340 ha).

Eighty years before

Number of buildings 2 (Police station and school building)

At present

1. No. of buildings 400 (although many more are coming up).
2. Telephone connections 55
3. Television set during last 3 yrs. 15

4.	Motor cars	12
5.	Motor cycles	30
6.	Fish market committee incurred all the expenditure for construction of brick-laid roads and metalled roads by village Panchayat from the income of bheries	
7.	Cost of land during 1972	Less than Rs. 300/bigha.
8.	At present	Rs. 6000/bigha.
9.	Money deposited with UBI	Rs. 60 lakh.
10.	No. of account holder out of which 8000 are from small savings agricultural labour.	
11.	Money deposited with United Bank of India.	Rs. 36 lakhs
12.	No. of account holder	3300
13.	Inconme as labour charges from salt water bheries, deposited with the Bank by number of labourers.	1300
14.	Every body is owner of a saltwarer Bhery.	
15.	20 percent of the shareholder of the saltwater bhery are landless daily labourer.	
16.	In a single Amon (Paddy) cultivated land the crop produced is	...1 or 1½ bag on ... 9 mds/bigha.
17.	According to agriculture expert paddy used to be cultivated once in 3 years.	

In 1981 Bagda prawn (*P. monodan*) were known as liquid gold of Haroa Salt water bhery to common people and is like liquid gold of Kuwait. The common fishery people expressed their affluence of fish as folk-rhyme like:

Bagda (*P. monodan*) is our son,

Bagda is our son-in-law.

Bagda is just like a tied up

elephant at our door".

Addendum B
Integrated Brackish water Aquaculture Development Project in Meen Dwip.

The Chief Minister, Government of West Bengal has recently inaugurated a Rs. 17 crore "Integrated Brackishwater Aquaculture Development Project" at Naya Chor (Meen Dwip) (Fig. 3 and 4) 3 km. off Haldia in Midnapur east on 20th November 2002 (*Statesman* news service, Bengal plus). The project area spans over 4000 ha and when completed in next February is expected to produce fisheries worth Rs. 30-40 crore annually. Since the distance of Meen Dwip from Haldia is highly industrial belt of Haldia disposing high concentration of mercury, may create problem (due to tidal water) for fish culture as suspected by various organisations. It is suggested to counduct detailed studies to find out whether there is any effect on aquatic life of Meen Dwip Brackishwater fishery. This is also discussed in the Bhagirathi-Hooghly estuarine pollution chapter 2B under capture fishery.

A problem has again appeared about the honourship of the Meen Dwip as indicated by Amitava Das, Haldia (2004) in his interesting article - "whose island it is any way?" indicated below:

The year long dilly daily of the state and land reforms Department over granting lease to 53 entrepreneurs for fish cultivation in Nayachar island is being described as the out come of a tussle between the Haldia Development Authority Chairman and State Fisheries minister.

1. The Nayachor island (recently renamed Meen dwip as it is shaped like a fish) is situated in the middle of the Hooghly river, 2 km. away from Haldia town.
2. The island is an ideal place for eco-tourism and fish cultivation.
3. Ben fish has a guest house here and the fishery Department cultivation prawn in 316 tanks spread over the island.
4. The Haldia Development Authority is also interested in the island which has a total area of 8306.47 acre. They invited officials of the Sahara Group to invest in tourism project in Nayachar in last year.
5. Sahara expressed willingness to invest if a major portion of the island is handed over to them. So HDA authorities, who have already been given 1000 acre on the island started lobby for 3500 acre more.
6. The land and land Revenue Department had already earmarked 24.2 acre of land for each of the 53 entrepreneurs (a total of 1282.6 acre of land). "The land will be utilised for setting up fish farms covering more than 60 percent of the area and the rest will be utilized for plantation of coconut and other fruit bearing trees and allied projects like tourism. They will have to pay lease rent and salami to be fixed by the Land and Land revenue Department." The land was allotted but the lease was not granted. As a result the entrepreneurs have not been able to start fish cultivation and the cash starved Government has been loosing money nearly Rs. 29,000 every year as land rent from them.

The above tussel between the two Departments has led to the uncertainty of any development of this island.

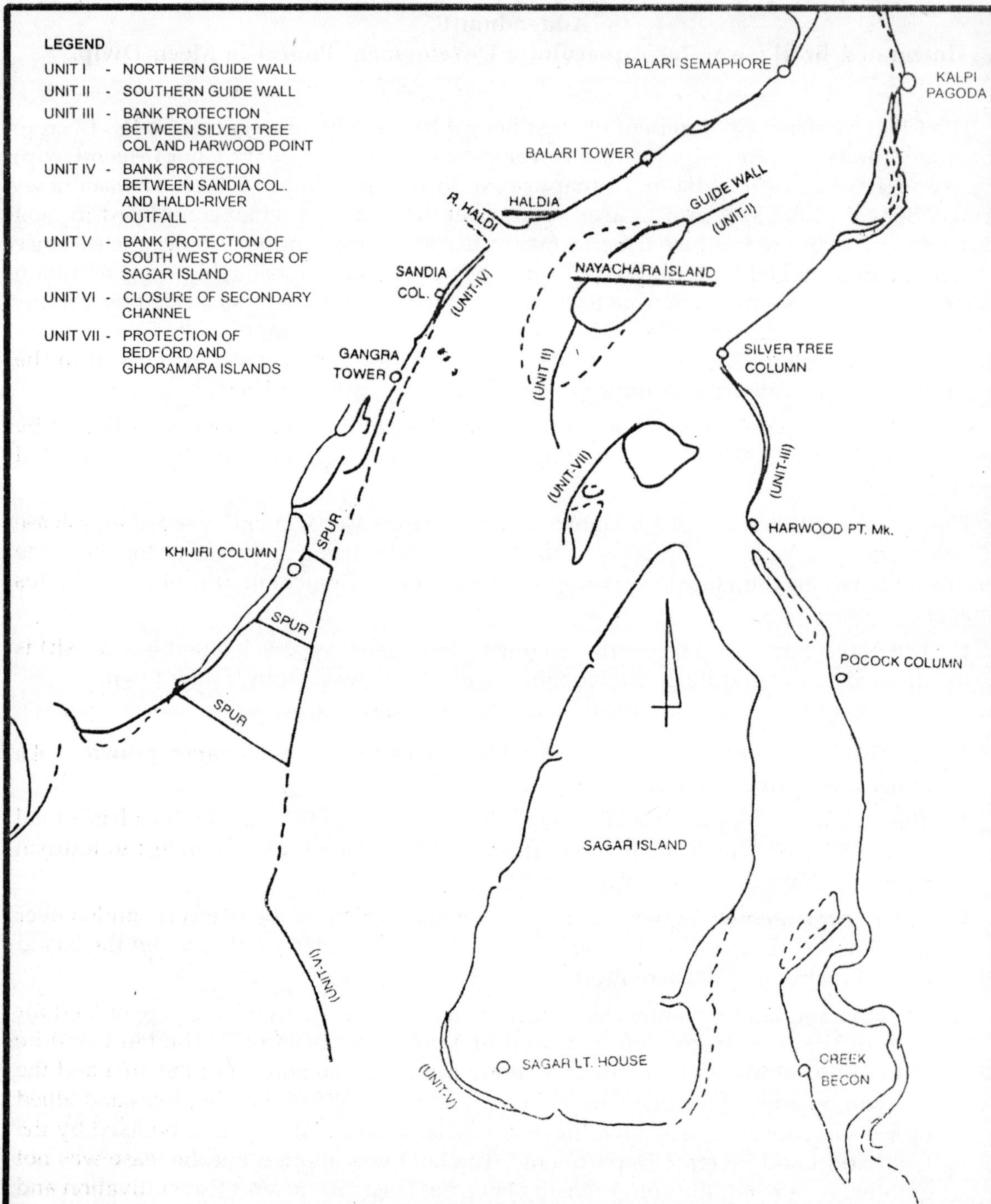

Fig. 3 : Location of Meen dwip (Nayachor island), 3 km. off Haldia in Midnapur East.

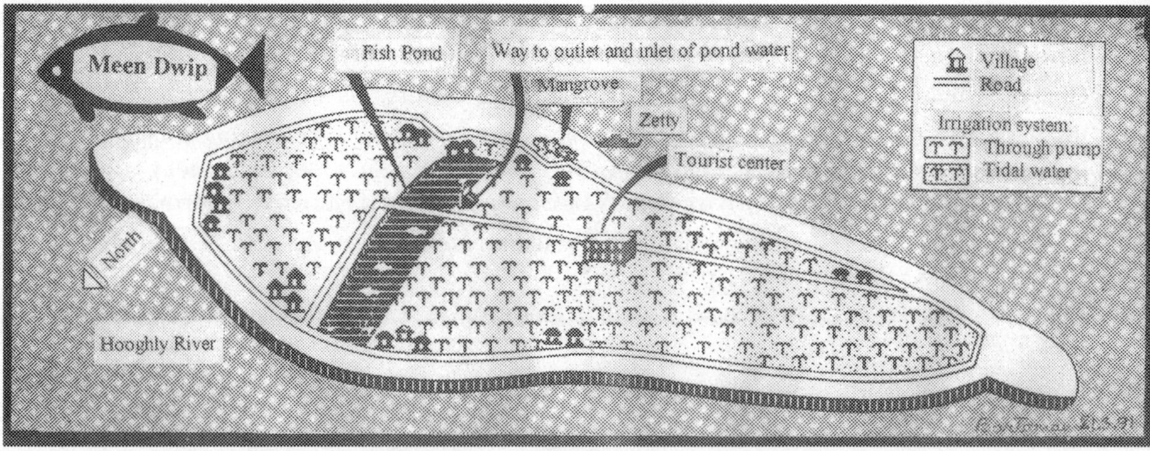

Fig. 4 : Proposed Meen dwip (Nayachor), for fish culture near Haldia, inaugurated on 19.11.2002. Statesman news service. 20.11.2002

REFERENCES

1. Biswas, K.P. (1927). *Flora of the Salt Lakes*. University of Calcutta.
2. Bhattacharya, R.K. (1976). New Salt tolerant rice varieties for Coastal Saline Soils of Sundarbans (West Bengal). *Scl. and Cult.* 42 : 122-123.
3. Bhumbla, D.R. and Abrol, I.P., (1977). Saline and Sodic Soils, *Symp.* on soil and *Rice* IRRI, Philippines.
4. Barman, G. (1983). *Sunderbaner Jamite Dhaner Chash Na Macher Chash Deficiencies in Development Planning in Agriculture of West Bengal.* SOUVENIR. Organised by Institute of Applied Humanities and Socio-economic Development Consortium.
5. Bhattacharjee, Kapil (1986). *Sadhin Bharate Nad-nadi Parikalpana* 'Kalam' 52, Braja Nath Dutta Lane, Calcutta-12.
6. Chattopadhyay, G.N., Ghosh, A. and Saha, P.K. (1983a). Anote on the possibility of Salinisation of Soils under Paddy-cum-Brackishwater Aquaculture in Coastal Saline Soils, *J. Ind. Soc. Coastal Agric.* Res., 1(1) 43-45.
7. Chattopadhyay, G.N., Chakraborty, P.K. Biswas, C.R., and Bandopadhyay, A.K. (1983b). A Pilot study on Paddy-cum Brackishwater Aquaculture in Coastal Saline Soils, *Management of Coastal Saline Soils in Orissa and West Bengal. FAI Seminar,* Bhubaneshwar, 7-8 April.
8. Central Inland Fisheries, Barrackpore (1985). Spectacular Production Rate of Shrimp using a new formula feed, *News Letter,* 8 ; May-August; No. 3 and 4.
9. Department of Fisheries West Bengal (1974). *Master Plan for Fisheries Development,* West Bengal, Part I and III.
10. Das, G.C. (1983). Management Problems of Coastal Saline Soils of Orissa in Rajanagar Block- a micro analysis. *FAI Seminar on Management of Coastal Saline Soils of Orissa and West Bengal, Bhubaneshwar.*
11. Ghosh, H. (1973). *The Fisheries and wet lands in the District of 24 Parganas.* Settlement Office, Alipore

12. Ghosh, A.N. (1978). Better Utilization of Brackishwater Resources through Scientific Aqua-farming, '*Fisheries Journal*' West Bengal Fisheries Extension Officers' Association Vol. I.

13. Gangopadhyay, N.C. (1983). Chabbis Parganar Mechobhery. *Subhodut.* March.

14. Jhingran, V.G. (1978). *Recent Advances in Brackishwater Prawn and Fish Farming in India.* Training in Brackishwater Prawn in Fish Farming. Dec. 26 1977 to March 30, 1978, Held at Kakdwip Research Centre Kakdwip, 24-Parganas W.B. Convened by CIFRI (ICAR), Barrackpore, W.B.

15. Mukherjee, G.D. (1959). Drainage in Greater Calcutta area. *The Journal of Institution of Engineers* (India) Vol. XXXIX, July No. 11 p. 6.1, 1061-1083.

16. Mitra, A. (1954). Census 1951. West Bengal, *District Hand Book - 24 Parganas,* West Bengal Govt. Press.

17. Natarajan, A.V., and Ghosh, Apurba (1982). *Indian Farmers' Digest.,* XV (1) : 32-40.

18. Pillay. T.V.R. (1954). The Ecology of Brackishwater Bhery with special references to the fish culture practices and the biotic interaction. *Proc. nat. Inst. Sci. India.* 20.

19. Pillay. T.V.R., Vijaraghavan, M.R. and Thakur, S.. (1962). An Ecological Study of Benthic Flora in Brackishwatrer Bheries. *Proc. nat. Inst. Sci. India.* 28 (5).

20. Pakrasi, B., (1965). A report on the Priliminary Survey of Brackishwater impoundments in West Bengal. *Survey Report* No. 3, March, CIFRI., Barrackpore, West Bengal, India.

21. Ray P., Saha, S.B. and Banerjee, R.K. (1981). A case study of sue of Calcutta Municipal waste for Fish Culture in the Bidyadhari-Kulti Complex, West Bengal. *International Symposium on Water Resources, Conservation, Pollution and Abatement,* Dec. 11-13, University of Roorkie, Roorkie, India.

22. Sewell, R.B.S. (1934), A study of the Fauna of the Salt Lakes, Calcutta. *Rec. Ind. Mus.,* 36 : 45-49, 58-60.

23. Saha, K.C. (1970). *Fisheries of West Bengal.* Government Printing, West Bengal Government Press. Alipore, West Bengal.

24. Sakthivel, M., (1980). Sundarban - A gold mine for prawn fisheries, *"MATSHACHASI DIWAS PLAN' SEMINAR Convened jointly by State Fishery Department and Bhery-Fish Producers Association,* Calcutta, Dec. 1980.

25. Saha, G.N., Thakur, S.C., Laha G.C., Nandy, A.C., Karmakar, H,C., Naskar, K.R., Das, P.B. and Chatterjee, S.K. (1986). Ecology and Fishery Management of Brackishwater Bheries in West Bengal. *Bulletin No. 46, August. C.I.F.R.I* (ICAR) Barrackpore, West Bengal.

26. Yadav, J.S.P., Bandyopadhyay, A.K., Rao, K.V.G.K., Sinha, T.S. and Biwas, C.R. (1979). *Coastal Saline of India. CSSRI Bull.* No. 5 : 34.

APPENDIX I

Reconnaissance survey of Brackishwater Impoundments in The Sundarban Deltaic Region HOROA (24-Parganas)

Sl. No.	Name of Fish farm	Village were located	Area (acres)	Source of water and fish seed	Approximate annual yield (kg) per acre: Lates Calcarifer	Mullets	Other including catfish	Prawns	Approximate total yield (kg) per acre	Remarks
1	2	3	4	5	6	7	8	9	10	11
1.	Ramchakir Gheri	Ramchakir Gheri	233	Vidya river branch	110	225	110	225	670	Abundant growth of algae
2.	Batagachi Gheri	Batagachi	180	-do-	111	226	102	330	739	-do-
3.	Boalmari Gheri	Munshi Gheri	365	Vidyadhari	123	213	112	336	784	-do
4.	Nebutala Gheri No. 1	Neutala	30	Kurunia	109	258	112	167	646	
5.	Nebutala Gheri No. 2	-do-	30	-do-	102	189	90	257	639	
6.	Nebutala Gheri No. 3	-do-	45	-do-	110	234	103	292	738	
7.	Nebutala Gheri No. 4	-do-	301	-do-	110	258	109	309	786	Excessive growth of algae
8.	Tentuliabad Gheri No. 1	Tentuliabad	400	Vidya river branch	123	258	123	313	817	-do-
9.	Tentuliabad Gheri No. 2	-do-	150	-do-	111	267	144	297	819	-do-
10.	Muchikhola Fishery No. 1	Muchikhola	50	-do-	123	259	110	359	851	-do-
11.	Muchikhola Fishery no. 2	-do-	27	-do-	110	266	145	265	786	
12.	Munshi Gheri	Munshi Gheri	735	Canal of Bidyadhari	—	—	—	—	203	Excessive growth of algae and weeds
13.	Goberia Gheri	Vikarhati	2500	Kurunia river & Bidhadhari	NOT KNOWN					Excessive algae growth
14.	Ramjoy Gheri	Ramjoy Gheri	660	-do-	—	—	—	—	221	Submerged weed

....**Contd.**

Appendix I – *contd...*

1	2	3	4	5	6	7	8	9	10	11
				HASNABAD (24-Parganas)						
1.	Sulkuniabad	Sulkunia	12	Dasta river	60	137	89	130	416	Excessive algae growth
2.	Chakpatli Gheri	Chakpatli	49	-do-	48	118	70	41	277	-do- and silting
3.	Ichhapur Fishery	Ichhapur	80	-do-	70	185	144	230	629	Excessive algae growth
4.	Bhowanipur Fishery	Bhowanipur	70	-do-	41	74	59	100	274	-do-
5.	Ghosalati Fishery	Ghosalati	50	Bidyadhari river	92	133	200	111	537	-do-
6.	Mohanpur Abad Fishery	Mohanpur	27	Kalakhali river	89	185	115	130	519	-do-
7.	Chimta Fishery	Chimta	15	-do-	133	189	148	141	611	
8.	Bholakhali Fishery	Bholakhali	100	-do-	23	21	151	566	769	
9.	Laskarnagar	Lakarnagar	14	-do-	33	156	48	63	300	
10.	Bailari Fishery	Bailari	60	-do-	30	35	12	268 + 120	465	
				BADURIA (24-Parganas)						
1.	Kakrasuti Fishery	Kakrasuti	150	Ichhamati river	6	NOT KNOWN				
2.	Mudiar Gheri	-do-	167	-do-	10	—	—	—	—	
				JAYNAGAR (24-Parganas)						
1.	Nagar Fishery	Shamnagar	25	Thakuran river	149	251	124	76	600	Excessive growth of shrub
2.	Gnhar Fishery	-do-	166	-do-	71	100	37	13	221	-do-
3.	Madhu-sudanpur Fishery	Madhusudan pur	400	Malta and Thakuran	—	—	—	—	447	
4.	Madhabpur Fishery	Madhabpur	165	Thakuran river	—	—	—	—	890	
5.	Katamari Fishery	Katamari	130	Malta and Thakuran	—	—	—	—	679	
				MATHURA PUR (24-Parganas)						
1.	Narayanpur Fishery No. 1	Narayanpur	100	Chhatra River	26	20	—	66	112	Excessive growth of weeds
2.	Narayanpur Fishery No. 2	-do-	86.5	-do-	56	112	27	74	269	-do-

....Contd.

Appendix I – *contd...*

1	2	3	4	5	6	7	8	9	10	11
3.	Narendrapur Fishery	Narendrapur	61	-do-	84	195	70	130	479	
4.	Kuemari Fisher No. 1	Kuemari	65	Raidighi river			NOT KNOWN			
5.	Kuemari Fisher No. 2	-do-	100	-do-			NOT KNOWN			
6.	Dakshin Kashinagar	—	—	—			NOT KNOWN			
7.	Singher Gheri	Dakshin Joy Krishnapur	280	—			NOT KNOWN			
8.	Boser Fishery	Kumrapara	150	Chatra River			NOT KNOWN			
9.	Halder Fishery	Narayanpur	250	-do-			NOT KNOWN			
10.	Surjapur	Surjapur	—	-do-	8	4	24	24	60	—
				SANDESHKHALI (24-Parganas)						
1.	Daudpur Fishery	Daudpur	60	Bidya River	—	—	—	—	17	Excessive growth of submerged weeds
2.	Manmatha Nagar Fishery	Manmatha Nagar	15	-do-					NOT KNOWN	—
3.	Dwari Jangal Fishery No. 2	Dwari Jangal	333	Ghatiara River	48	88	18	88	242	Excessive growth of algae and their decay
4.	Dwari Jangal Fishery No. 4	-do-	400	-do-	44	111	44	122	321	-do-
5.	Dwari Jangal Fishery No. 5	-do-	233	-do-	44	78	11	85	218	-do
6.	Dwari Jangal Fishery No. 6	-do-	170	-do-	56	122	86	118	382	
7.	Bholakhali Fishery	Bholakhali	81.3	Dasha River	33	90	48	48	219	Profuse growth of weeds
8.	Dhamakhali Fishery No. 1	Dhamakhali	70	Kalagachi River	25	48	44	25	142	Silting
9.	Dhamakhali	-do-	67	-do-	19	55	44	50	168	Excessive

....Contd.

Appendix I – *contd...*

1	2	3	4	5	6	7	8	9	10	11
	Fishery No. 2									growth of algae
10.	Atapur Fishery No. 1	Atapur	80	-do-	18	37	58	20	131	Silting
11.	Atapur Fishery No. 2	-do-	67	-do-	20	56	61	22	159	-do-
12.	Atapur Fishery No. 3	-do-	50	-do-	53	85	56	29	223	-do-
13.	Atapur Fishery No. 4	-do-	8	-do-	37	87	73	38	235	
14.	Atapur Fishery No. 4	-do-	16	-do-	39	48	92	53	232	Abundant growth of weeds
15.	Bhangatosh Khali Fishery	Bhangatosh Khali	40	Sandesh khali River	44	74	44	60	222	-do-
16.	Amlamethi Fishery	Amlamethi	800	Bidya River	24	37	22	111	194	
				CANNING (24-Parganas)						
1.	Swarna-Karer Gheri	Parbatipur	366	Malta River			NOT KNOWN			Growth of weeds and shrubs, silting
2.	Bager Gheri	-do-	330	-do-	37	65	56	14	172	-do-
3.	Parbatipur Fishery	-do-	116	-do-	19	46	37	18	120	-do-
4.	Laskarpur Fishery	Laskarpur	243	-do-	56	120	84	110	370	
5.	Nafarganj Fishery	Nafarganj	222	-do-	20	35	20	56	131	Silting
6.	Seafeli Gheri	Laskarpur	274	-do-	50	80	20	85	235	Excessive growth of weed
7.	Kripakhali Fishery No. 1	Kripakhali	90	-do-	—	—	—	—	844	
8.	Kripakhali Fishery No. 2	-do-	750	Karati River off Malta	—	—	—	—	893	
9.	Kripakhali Fishery No. 3	-do-	10	-do-			NOT KNOWN			
10.	Sajnakhali Fishery	-do-	67	-do-	—	—	—	—	940	
11.	Hiranmoypur Fishery	Hiranmaypur	100	-do-	—	—	—	—	890	

....*Contd.*

Appendix I – *contd...*

1	*2*	*3*	*4*	*5*	*6*	*7*	*8*	*9*	*10*	*11*
12.	Budhakhali Fishery	Budhakhali	40	-do-	—	—	—	—	890	
13. a.	Golabari Fishery No. 1	Golabari	83	Karati River off Matla	—	—	—	—	341	
b.	Golabari Fishery No. 2	-do-	50	-do-	7	32	14	47	100	
14.	Rolamin Sahab Fishery	Tamluda	250	Bidhyadhari			NOT KNOWN			
15.	Samad Mia's Fishery	Khagra	500	-do-				-do-		
16.	Prasad Kayal's Fishery	Jibantala	200	-do-						
17.	Adhar Naskar's Fishery	Khagra	40	Karati River	11	9	—	4	24	
18.	Sajad Molla's Fishery	Tamnlda	96	Bidyadhari	19.3	14.7	—	10	44	
19.	Kinu Mondal's Fishery	Gheekhali	200	-do-			NOT KNOWN			
20.	Nibaran Bag's Fishery	Howramari	1000	-do-				-do-		
21.	Akbar Molla's Fishery	Gheekhali	100	-do-				-do-		
22.	Nibaran Bag's Fishery	Payna	500	Karati River				-do-		
23.	Sarengabad Fishery	Sarengabad	330	-do-				-do-		
24.	Biswas's Fishery No. 1	Moukhali	100	-do-	22	15	—	11	48	
25.	Biswas's Fishery No. 2	Moukhali	50	Karati	20	14	—	11	45	
26.	Nibaran Bag's Fishery	Moukhali Kumrakhali	160	-do-	27	16	—	16	59	
27.	Jagesh Pata's Fishery	Baintala	650	Bidyadhari	20	17	—	10	47	
28.	Hozir Fishery	Tamulda	300	-do-			NOT KNOWN			

....Contd.

Appendix I – *contd...*

1	2	3	4	5	6	7	8	9	10	11
29.	Manilker Fishery	Chalikali	150	-do-				-do-		
30.	Bager Fishery	Bibirabad	100	-do-			-do-			
31.	Kayaler Fishery	Homrapalta	400	-do-			-do-			
32.	Main Paler Fishery	Basra	150	-do-			-do-			
33.	Laskarer Fishery	Kulurkhoj	12	-do-			-do-			
34.	Sarderer Fishery	-do-	30	-do-			-do-			
35.	Alam Mollar Fishery	-do-	30	-do-			-do-			
36.	Mollar Fishery	Rajpur	100	-do-			-do-			
37.	Laskarer Fishery	Basra	500	-do-			-do-			
38.	Kayaler Fishery	Netra	350	-do-			-do-			
39.	Sarderer Fishery	Jaldhaka	200	-do-			-do-			
40.	Ebadater Fishery	Jaldhaka	250	-do-			-do-			
41.	Mollar Fishery	Kaparpuri	215	-do-			-do-			
42.	Jagannather Fishery	Chunpuri	350	-do-			-do-			
43.	Mandaler Fishery No. 1	Sarangabad	300	-do-			-do-			
44.	Mandaler Fishery No. 1	-do-	200	-do-			-do-			
45.	Bager Fishery	Bibirabad	100	-do-			-do-			
46.	Boser Fishery	Patikhali	50	-do-			-do-			
	DEGANGA (24-Parganas)									
1.	Chakknlia in paddy fields (Partly paddy culture)	Chakknila	40	Bileghata Khal and Bidyadhari river branch		Annual Production is restricted to 6 months (August to January)		25		Calcutta Sewage cause fish moratlity
2.	Sholakuri (in paddy field)	Gobardhan-pur	65.5	-do-		-do-		21		-do-
3.	Chowhan Molla's Fishery	-do-	60	Beleghata Khal		-do-		17		-do-

....*Contd.*

Appendix I – *contd...*

1	2	3	4	5	6	7	8	9	10	11
				BARASAT (24-Parganas)						
1.	Bamuner Gheri	Khamer Rameswarpur	67	Noai or, Horoa Khal and Paskhali canal (off Bidyadhari)		Annual Production to 6 months August-January		restricted	60	Excessive discharge of Calcutta Sewage silting, flood (1958-59) drought (1954), 6 months fish culture
2.	Baherdhapa	Mudia	27	Paskhali River			NOT AVAILABLE			
3.	Kakrasuti	Noabad Khakar	250	-do-			-do-			
4.	Kazir Bheri	Khamar Noabad	366	Paskhali			-do-			Source of seed affected by sewage silting
5.	Bhulanpur Fishery	Sankeria Abad	250	Noai Khal	—	—	—	—	30	-do-
6.	Bardesia Kaliprasanna Janta Sambaye Krishi Samiti L.T.D.	Bardesia	133	Paskhali River			NOT AVAILABLE			Fish seed affected by sewage, silting. Mostly Paddy cultivation.
7.	Biswaser Fishery	-do-	67	Paskhali Canal					188	Excessive growth of algae, silting seed affected by sewage
8.	Koarir Dhapa	Par Khariberia	72	-do-			NOT AVAILABLE			Excessive growth of gastropod, silting
9.	Pulian Beel	-do-		Converted into Paddy Field						
10.	Sarkarer Math	Kamdhenu	66	Surti River		Converted to freshwater fishery				Flow of feeder channel stopped due to silting
11.	Bilbauchandi Fishery	Bilbauchandi	300	-do-			Partly freshwater and partly brackishwater fishery			-do-

....Contd.

Appendix I – *contd...*

1	*2*	*3*	*4*	*5*	*6*	*7*	*8*	*9*	*10*	*11*
12.	Taplir Bheri	Nabapur	30	-do-			-do-			-do-
13.	Muder Bheri	Mudia	150	Paskhali			NOT AVAILABLE			
14.	Orakure	Mal Ram-eswarpur	—				Recently abandoned			
15.	Baisbigha Fishery	Baisbigha	38	Paskhali			NOT AVAILABLE			—

* Modified from Pakrasi, B.B. (1965)

APPENDIX II

Commission Agents/Wholesalers of Fish in 24 Parganas

BARASAT P.S.

Sl. No.	Name of the Trader	Name of the Market	Name of the P.S.	Type of Traders	Remarks
1.	Md. Omar Ali Sahaji	Kharibari Market	Barasat.	C.A.	(Commission agents.)
2.	Sh. Krishna Kanta Pal & others	Barasat	"	"	
3.	Md. Abdur Razzaque	"	"	"	
4.	Sree Susil Kr. Some	"	"	"	
5.	Sree Dukhiram Mondal	"	"	"	
6.	Sree Haripada Mondal	"	"	"	
7.	Sree Mihir Kumar Roy	"	"	"	
8.	Smt. Mochhan Bibi & Smt. Jabarani Mukherjee	"	"	"	
9.	Sk. Panjab Ali	Khanfura Fish Market	Barasat	C.A.	
10.	Nayrul Islam & Laxman Ch. Ghosh	Chanpadali Fish Market	"	"	
	P.S. Habra				
1.	Bejoy Krishna Biswas	Maslanda Pore	Habra	C.A.	
2.	Chittaranjan Biswas	"	"	"	
3.	Asutosh Ray	"	"	"	
4.	Bhasan Chandra Mazumdar	"	"	"	

Source : For Appendices I & II

The Report of the Fact Finding Committee set up by the West Bengal State Land use Board, March, 1984, Director of Land Records and Surveys, West Bengal.

BASIRHAT P.S.

Sl. No.	Name of the Trader	Name of the Market	Name of the P.S.	Type of Traders
1.	Md. Afsuddin Mandal	Andulpati	Basirhat	C.A.
2.	Biswanath Kundu	Trimohini	"	"
3.	Md. Abdul Kedangagi	"	"	"
4.	Nitya Chatterji	"	"	"
5.	Md. Tajmohumad	"	"	"
6.	Md. Monohar Gaji	"	"	"
7.	Nirmal Barman	"	"	"
8.	Nemai Ch. Das	"	"	"
9.	Dinabandhu Sarkar	"	"	"
10.	Md. Abdur Rahim Gaji	"	"	"
11.	Monoranjan Sarkar	"	"	"

....Contd.

Appendix II – *contd...*

Sl. No.	*Name of the Trader*	*Name of the Market*	*Name of the P.S.*	*Type of Traders*
12.	Md. Abu Ibrahim Gaji	"	"	"
13.	B. Chatterjee	"	"	"
14.	Gouranga Majumdar	"	"	"
15.	Suvash Pal	"	"	"
16.	Hamidul Haque Mandal	"	"	"
17.	Md. Unus Ali Ganji	Andulpatti	"	"
18.	Abdus Sattar	"	"	"
19.	Abu Siddique	"	"	"
20.	Jobed Alimmal	"	"	"
21.	Md. Afsar Ali	"	"	"
22.	Md. Johurul Islam	"	"	"
23.	Ratan Kr. Saha	"	"	"
24.	Sanatan Mandal	"	"	"
25.	Md. Irad Ali Mandal	"	"	"
26.	Netai Ch. Halder	Trimohini Market	"	"
27.	Md. Abubakkar Mandal	Andulpati	"	"
28.	Abu Semar Gaji	Trimohini	"	"
29.	Azizul Haque	Andulpati	"	"
30.	Mustakin Gaji	"	"	"
31.	Swakat Ali & others	"	"	"
32.	Mihir Samaddar	Asfelia Market	"	"
33.	Gobinda Prosad Maity	"	"	"
34.	Salauddin Akaji	"	"	"
35.	Obidul Haque	"	"	"
36.	Md. Rafikul Rahamat	"	"	"
37.	Sirajul Amuji	"	"	"
38.	Md. Rahul Amed Mahick	"	"	"
	RAJARHAT P.S.			
1.	Chandicharan Pramanick	Krishnapur mission fish market.	Rajarhat	Commission Agent.
2.	Sudhir Kr. Sander	"	"	"
3.	Jotish Ch. Pramanick	"	"	"
4.	Phanindra Nath Naskar	"	"	"
5.	Nripendra N. Pramanick	"	"	"
6.	Joydev Mondal	"	"	"
7.	Sekhar Pramanick	"	"	"
8.	Gangadhar Sunder	"	"	"
9.	Samir Kr. Mondal	"	"	"

....Contd.

Appendix II – *contd...*

Sl. No.	*Name of the Trader*	*Name of the Market*	*Name of the P.S.*	*Type of Traders*
10.	Lalit Modal	"	"	"
11.	Dharma Das Pramanick	"	"	"
12.	Hazari Lal Barui	"	"	"
13.	Rabindra Nath Mandal	"	"	"
14.	Provash Ch. Mandal	"	"	"
15.	Dhirendra Nath Mandal	"	"	"
16.	Prosad Mandal	"	"	"
17.	Sailen Naskar	"	"	"
18.	Bankim Naskar	"	"	"
19.	Sankar Pramanick	"	"	"
20.	Bejoy Kr. Mondal	"	"	"
21.	Mihir Mondal	"	"	"
22.	Ajit Kr. Naskar	"	"	"
23.	Sudhanya Kr. Majhi	"	"	"
24.	Chandi Kr. Prosad Halder	"	"	"
25.	Anil Kr. Pramanick	"	"	"
26.	Subinoy Kr. Mondal	"	"	"
27.	Sanjay Kr. Mondal	"	"	"
28.	Sudhir Kr. Pramanick	"	"	"
29.	Md. Rafiqul Middaya	Lauhati Fish Market	"	"
30.	Sk. Safiulla	"	"	"
31.	Md. Rafiqul Islam	"	"	"
32.	Asutosh Mondal	Krishnapur Mission Market	"	"
33.	Johurul Islam	"	"	"
34.	Md. Idrish Ali & Hossain Ali Sarkar	" "	" "	" "
35.	S.K. Hamidul	Lauhati fish market	"	"
36.	S.K. Matobbar	"	"	"
37.	Md. Rajjak Ali	"	"	"
	HAROA P.S.			
1.	Md. Ayub Ali Mondal	Gopalpore Trimohini Fish Market	Haroa	Commission Agent
2.	Md. Laysal Haque	Haroa	"	"
3.	Sri Palan Ch. Gain	Ghusighata	"	"

....Contd.

Appendix II – *contd...*

Sl. No.	*Name of the Trader*	*Name of the Market*	*Name of the P.S.*	*Type of Traders*
4.	Sri Samir Banerjee & Sahdul Hossain	Haroa	"	"
5.	Md. Kashem Ali Mondal	Gopalpore Trimohini	"	"
6.	Sri Shyamal Mukherjee & Rakhal Ghosh	Haroa	"	"
7.	Sri Sunil Kumar Sardar	Ghusighata	"	"
8.	Md. Henat Ali Molla	"	"	"
9.	Sri Bhaskar Kr. Mondal	"	"	"
10.	Sri Ramchandra Halder	"	"	"
11.	Md. Abdul Malak	"	"	"
12.	Md. Saimuddin Gazi	Haroa	"	"
13.	Tri Tapan Kr. Saha	Gopalpore Trimohini		
14.	Suresh Ch. Das	Gopalpore Trimohini Fish Market	Haora	Commission Agent
15.	Mr. Pitter & Duke	"	"	"
16.	Md. Abdul Jalil Mondal	"	"	"
17.	Sri Ajit Kr. Paul	"	"	"
18.	Md. Ajibar Rahaman	"	"	"
19.	Sri Budhiswar Pramanick	"	"	"
20.	Sri Jaydev Kr. Biswas	"	"	"
21.	Md. Samserali Mondal	"	"	"
22.	Md. Mofizul Haque	"	"	"
23.	Md. P. Raj	"	"	"
24.	Sri Avijit Saha	"	"	"
25.	Sri Basudev Saha	"	"	"
26.	Sri Sanat Kr. Saha	"	"	"
27.	Md. Fozerali Malick	"	"	"
28.	Sri Jaydev Saha	"	"	"
29.	Sri Laxmi Kanto Das & Pradip Kr. Santra	" "	" "	" "
30.	Md. Babureli Mondal	"	"	"
31.	Md. Asif Ickbal	"	"	"
32.	Md. Ayub Ali Mondal	"	"	"
33.	Sri Tarak Nath Das	"	"	"
34.	Sri Gautam Sandel	"	"	"
35.	Md. Farukuzzaman	"	"	"
36.	Md. Mosaraf Hossain	"	"	"
37.	Sri Anil Das & Usubali	Haora Fish Market	Haroa	
38.	Sri Sudhin Dutta	"	"	"

....Contd.

Appendix II – *contd...*

Sl. No.	*Name of the Trader*	*Name of the Market*	*Name of the P.S.*	*Type of Traders*
39.	Md. Akram Ahamed	"	"	"
40.	Sri Ranjan Biswas	"	"	"
41.	Sri Direndra Nath Mondal	"	"	"
42.	Sri Phonindra Nath Mondal	"	"	"
43.	Sri Suhodh Ch. Sarkar	"	"	"
44.	Haripada Paul & Prabir Paul	"	"	"
45.	Gobinda Lal Paul	"	"	"
46.	Sri Gasto Behari Mondal	"	"	"
47.	Sri Tari Kul Alam Haroa	Haroa Fish Market	Haroa	Commission Agent
48.	Md. Jalil Molla	"	"	"
49.	Smt. Aparna Deb	"	"	"
50.	Sk. Ashanar Ahamed	"	"	"
51.	Md. Kawsar Ali	"	"	"
52.	Sri Horendra Nath Ghosh	"	"	"
53.	Sri Tarak Sawdagar	"	"	"
54.	Md. Madar Mondal	Madartala	Haroa	"
55.	Sasadhar Maity	Ghusighata	"	"
56.	Sri Nagendra Nath Mondal	Haroa	Haroa	"
57.	Sri Jyotirmoy Mondal	"	"	"
58.	Sri Gobinda Sarkar	"	"	"
59.	Md. Akram Ahamed	"	"	"
60.	Md. Abul Khalan	Haroa	Haroa	Commission Agents
61.	Sri Nirmal Kr. Nandi	"	"	"
62.	Md. Monirul Islam	"	"	"
63.	Md. Nurul Islam	"	"	"
64.	Jaydeve Saha	1 No. Goberia Kata Khali (Whole Seller)	Goberia	"
65.	"	Ran Chakri (Eash) (Ghery Whole seller)	Ran Chaki	"
	HASNABAD P.S.			
1.	Chintaharan Battacharjee	Hasnabad Fish Market	Hasnabad	Commission Agent
2.	Md. Abdul Zobbar & Bors.	"	"	"
3.	Md. Rafiqul Islam	Murari Saha	Shownally	"
4.	Md. Golam Rabbani	Chaupati	"	"
5.	Md. Abdul Karion Mondal	Hasnabad	Hasnabad	"

....Contd.

Appendix II – *contd...*

Sl. No.	*Name of the Trader*	*Name of the Market*	*Name of the P.S.*	*Type of Traders*
6.	Md. Abdul Hossain gazi	"	Hasnabad	"
7.	Md. Sobrat Ali Gazi	"	"	"
8.	Md. Harunal Rasid Gazi	"	"	"
9.	Md. Mucharali Mea	"	"	"
10.	Md. Nurul Islam Gazi	"	"	"
11.	Sri Mohadev Ghosh	"	"	"
12.	Md. Ismail Mallick	"	"	"
13.	Md. Motiar Rahaman Gazi	"	"	"
14.	Md. Nurul Basar Gazi	"	"	"
15.	Sri Sudhir Kazi Mondal	Garakupi	"	"
16.	Md. Saidev Dafader	"	"	"
17.	Sri Amrita Kr. Ray	"	"	"
18.	Md. Razzak Dafader	"	"	"
19.	Md. Khokon Mondal	"	"	"
20.	Md. A. Hossain Mondal	"	"	"
21.	Sri MadhushudanBachhar	"	"	"
22.	Md. Mozuman Molla	"	"	"
23.	Sri Radhesham Bachhar	"	"	"
24.	Sri Paritosh Bachhar	Gurakaji	Hasnabad	"
25.	Md. Razzakali Molla	"	"	"
26.	Md. Abdul Chhalan Peada	"	"	"
27.	Sri Radha Krishna Sarkar	"	"	"
28.	Md. Soharab Hussain	"	"	"
29.	Sri Tarapada Mondal	"	"	"
30.	Sri Avilash Bachhar	"	"	"
31.	Sri Karuna Bachhar	Gurakaji	Hasnabad	Commission Agent
32.	Md. Abdul Kasham Molla	"	"	"
33.	Md. Jobed Ali Mondal	"	"	"
34.	Md. Rafiqul Mondal	"	"	"
35.	Md. Dinmohamad Mondal	"	"	"
36.	Md. Alier Mondal	"	"	"
37.	Sri Chandra Shekhar Roy	"	"	"
38.	Md. Abudakkar Baidya	Murari Saha	"	"
39.	Sri Provash Kr. Mondal	"	"	"
40.	Md. Answar Ali Gazi & Md. Abdul Kalam	Bhetia Chumatha	Hasnabad "	C.A. "
41.	Md. Biru Gazi	"	"	"
42.	Sita Prasad Rayal	"	"	"

....Contd.

Appendix II – *contd...*

Sl. No.	Name of the Trader	Name of the Market	Name of the P.S.	Type of Traders
43.	Md. Abu Bakkar Molla	"	"	"
44.	Md. Abul Basar Gazi	"	"	"
45.	Md. Majit Gazi (Chostu)	Hasnabad	"	"
46.	Md. Nur Ali Gazi	"	"	"
47.	Bhanta Das Mondal	"	"	"
48.	Abdul Majid mia (Daro)	"	"	"
49.	Md. Zammir Sardar	"	"	"
50.	Md. Dond Ali & Bus.	"	"	"
51.	Md. Abuhar Rahaman	"	"	"
52.	Md. Abu Mia	"	"	"
53.	Gopal Ch. Markell	"	"	"
54.	Phani Bhusan Banerji	"	"	"
55.	Md. Mazahar Ali Gazi	"	"	"
	MINAKHAN P.S.			
1.	Md. abdul Hamik Molla	Chaitalghat	Minakhan	Commision Agent
2.	Jamirali Molla & Monsurali	Malancha	"	"
3.	Bhudev Ch. Das	"	"	"
4.	Gour Mohan Mondal	"	"	"
5.	Nasiruddin Sahaji	"	"	"
6.	Raghunath Pramanick	"	"	"
7.	Chittaranjan Mondal	"	"	"
8.	Md. Walliulla	"	"	"
9.	Sudhir Ch. Das	"	"	"
10.	Nirmal Kr. Narpati	"	"	"
11.	Provanjan Kr. Mondal	"	"	"
12.	Md. Abdul Ali	"	"	"
13.	Md. Kancher Ali	"	"	"
14.	Md. Sukur Ali Molla	"	"	"
15.	Md. Jamat Ali Molla	"	"	"
16.	Netai Ch. Mondal	"	"	"
17.	Sudhir Kr. Mondal	"	"	"
18.	Atiar Rahaman Mondal	"	"	"
19.	Md. Hafifur Rahaman	"	"	"
20.	Md. Samsul Haque Molla	"	"	"
21.	Narim Box Molla	"	"	"
22.	Mohiuddin Goldai	"	"	"
23.	Rowsan Ali Molla	"	"	"

....*Contd.*

Appendix II – *contd...*

Sl. No.	*Name of the Trader*	*Name of the Market*	*Name of the P.S.*	*Type of Traders*
24.	Nitya Nanda Das	"	"	"
25.	Md. Jiad Ali Molla	"	"	"
26.	Md. Wachelali Molla	"	"	"
27.	Md. Khali Ahamed	"	"	"
28.	Dulal Ch. Patra	Malancha	Minakhan	Commission Agent
29.	Answarali Molla	"	"	"
30.	Nabab Ali Molla	"	"	"
31.	Sambhu Nath Biswas	"	"	"
32.	Md. Yousufali Molla	"	"	"
33.	Haripada Naskar	"	"	"
34.	Samsul Haque	"	"	"
35.	Ismail Molla	"	"	"
36.	Safique Ahamed	"	"	"
37.	Sukumar Pal	"	"	"
38.	Israil Molla	"	"	"
39.	Manusr Ali	"	"	"
40.	Abdul Latif Molla	"	"	"
41.	Sibapada Sarkar	"	"	"
42.	Md. Anwarali Molla	"	"	"
43.	Md. Tarikul Islam	Bolichara	"	"
44.	Narayan Ch. Mondal	"	"	"
45.	Md. Majid Ali Mondal	"	"	"
46.	Sk. Fazlur Rahaman	Malancha	"	"
47.	Md. Nazrul Islam	"	"	"
48.	Md Abul Kashem Mallick	"	"	"
49.	Rafinul Islam Mondal	Bollchara	"	"
50.	Md. Sahajahan Molla	"	"	"
51.	Fachin Molla	"	"	"
52.	Md. Resonu Ali Molla	"	"	"
53.	Md. Naimuddin Gazi	"	"	"
54.	Md. Ziad Ali Molla	"	"	"
55.	Rabindra Nath Bari	"	"	"
56.	Hem Chand Roy	"	"	"
57.	Gopal Ch. Ghosh	"	"	"
58.	Moni Mohan Sarkar	"	"	"
59.	Md. Abul Hossain Tarafdar	Bollchara	Minakhan	Commission Agent
60.	Md. Sattar Sardar	"	"	"
61.	Abul Kasim Gazi	"	"	"

....Contd.

Appendix II – *contd...*

Sl. No.	*Name of the Trader*	*Name of the Market*	*Name of the P.S.*	*Type of Traders*
62.	Md. Ayub Ali Biswas	"	"	"
63.	Md. Alimuddin Mondal	"	"	"
64.	Raban Ch. Sardar	"	"	"
65.	Md. Salim Gazi	"	"	"
66.	Md. Rakhiful Islam Molla			
67.	Rasiruddin Mondal	"	"	"
68.	Balai Ch. Mondal	"	"	"
69.	Atiar Rahaman	"	"	"
70.	Abdul Jalil Molla & others	Malancha	"	"
71.	Md. Jamshed Ali Molla	"	"	"
72.	Md. Mobarak Molla	Ghasighati	"	"
73.	Md. Abdul Wahab Molla	Malancha	"	"
74.	Md. Niyamali Ali Tarafdar	"	"	"
75.	Dhruba Gopal Das	"	"	"
76.	Harekrishna Das	"	"	"
77.	Md. Sha Ali Molla	"	"	"
78.	Samar Ketan Das	"	"	"
79.	Surja Kanta Das	"	"	"
80.	Ram Krishna Roy	"	"	"
81.	Harendra Nath Mondal	"	"	"
82.	Alim Box Molla	"	"	"
83.	Md. Hazur Ali Molla	"	"	"
84.	Badal Mahato	"	"	"
85.	Niranjan Mondal	"	"	"
86.	Mihir Ch. Patra	"	"	"
87.	Bhim Sen Halder	"	"	"
88.	Md. Jamshed Ali Tarafdar	Bokchara Fish Market	Minakhan	C.A.
89.	Monoranjan Mondal	Malancha Fish Market	"	C.A.
90.	Md. Irnnat Ali Molla	Chalk Ahmedpur	"	C.A.
91.	Amal Kr. Mondal	Chak Ahmedpur Malancha Fish Market	" "	C.A. C.A.
92.	Md. Abul Hossain Molla	Chak Ahmedpur	"	C.A.
93.	Md. Rapajuddin Molla	Miron Mitra Fish Market	"	C.A.
94.	Md. Nousad Ali Molla	Chak Ahmedpur	"	C.A.
95.	Md. Abdul Kalam Molla	"	"	C.A.

....*Contd.*

Appendix II – *contd...*

Sl. No.	*Name of the Trader*	*Name of the Market*	*Name of the P.S.*	*Type of Traders*
96.	Md. Wajed Ali Gari	Bokchara Fish Market	"	C.A.
97.	Nasiruddin Molla	Chak Ahmedpur	"	C.A.
98.	Rosul Amir Molla	Bokchara Fish Market	"	C.A.
99.	Sri Bijoli Kr. Gola	Chak Ahmedpur	"	C.A.
100.	Himanshu Ray	Malancha Fish Market	"	C.A.
101.	Haripada Ghosh	Chak Ahmedpur	"	C.A.
		SANDESHKHALI P.S.		
1.	Md. Kabir Ali Gazi	Nazat Fish Market	Sandesh Khali	Commission Agent
2.	Sri Adhir Kr. Mali	"	"	"
3.	Sri Subodh Kr. Mondal	Sarberia Fish Market	"	"
4.	Sri Sachindranath Bhuniya	"	"	"
5.	Miss Chhamida Khatun	"	"	"
6.	Miss Protima Basu	"	"	"
7.	Sri Sandip Kr. Gupta	"	"	"
8.	Sri Rajendra Kr. Sardar	"	"	"
9.	Miss Tapati Baidya	"	"	"
10.	Md. Momin Khan	"	"	"
11.	Md. Isad Ali Molla	"	"	"
12.	Md. Maola Bax Molla	"	"	"
13.	Md. S.K. Mostan Ali	"	"	"
14.	Md. Din Islam Akunjee	"	"	"
15.	Md. Gaffar Ali Molla	"	"	"
16.	Md. Sajed Ali Mallick	"	"	"
17.	Md. Ahed Ali Mallick	"	"	"
18.	Md. Mojibar Rahaman	Nazat	"	"
19.	Sri Amra Kr. Mahato	"	"	"
20.	Md. Mosiar Rahaman	Nazat	"	"
21.	Sri Nikhil Kr. Patra	"	"	"
22.	Sri Brisokatu Mondal & Nirmal Mondal	"	"	"
23.	Sri Dulal Ch. Ghosh	Kanmari Fish Market	"	"
24.	Sri Subil Ch. Das	Kanmari Fish Market	Sandesh Khali	Commission Agent
25.	Sri Bhebesh Ch. Bag	"	"	"
26.	Sri Bhabosindhu Gok	"	"	"
27.	Sri Saroj Kr. Pramanik	"	"	"
28.	Md. Kashem Ali Molla	"	"	"
29.	Sri Sita Nath Pramanik	"	"	"
30.	Sri Dulal Ch. Mondal	"	"	"

....Contd.

Appendix II – *contd...*

Sl. No.	*Name of the Trader*	*Name of the Market*	*Name of the P.S.*	*Type of Traders*
31.	Sri Dulal Ch. Mondal	Nazat	"	"
32.	Sri Ratan Ch. Das	"	"	"
33.	Sri Horendra Nath Das	"	"	"
34.	Sri Amal Krishna Das	"	"	"
35.	Sri Bimal Ch. Halder	"	"	"
36.	Sri Nirapada Das	"	"	"
37.	Sri Jatindra Nath Sardar	"	"	"
38.	Sri Satyendranath Sardar	"	"	"
39.	Sri Sudhanya Bhuniya	"	"	"
40.	Sri Badal Ch. Das	"	"	"
41.	Sri Manik Ch. Prodhan	"	"	"
42.	Sri Satish Ch. Das	"	"	"
43.	Md. Abdul Hamid Sardar	"	"	"
44.	Sri Tarak Mallick	"	"	"
45.	Sri Madhai Sardar	"	"	"
46.	Sri Pampada Das	"	"	"
47.	Sri Surendra Nath Halt	"	"	"
48.	Sri Narendranath Patra	"	"	"
49.	Md. Estam Ali Moll and others	"	"	"
50.	Sri Puranjan Das	Nazat Fish Market	Sandesh Khali	Commission Agent
51.	Sri Debasish Mukherjee	"	"	"
52.	Sri Basanta Kr. Jana	"	"	"
53.	Sri Arabindu Maity	Kanmari Fish Market		
54.	Majibar Molla	"	"	"
55.	Mojibar Rahaman Molla	"	"	"
56.	Sri Gobinda Das	"	"	"

APPENDIX III

Name of the Seafood Processors & Exporters

1. M/s. A.N. ASH & Co. Pvt. Ltd., 152. New Alipore, Block-B, Calcutta - 700053
2. M/s. Agents (India) Pvt. Ltd. Tapuriah House 42/1, Strand Road, Calcutta-7000001 Tele: 33-6843, 33-2754 & 33-0886 Telex: 541 CA; Cable: Simplicity
3. M/s. Apec Seadfood Pvt. Ltd. 3/12, Poddar Nagar Calcutta-700068 Tele: 42-6298
4. M/s. Aqua Fauna India Pvt. Ltd. 67, B.R.B. Basu road Calcutta-700001, Tele: 23-7236
5. M/s. Associated Commercial Co. of India Pvt. Ltd. 7, Hastings streets Calcutta-7000001, Tele: 23-6650/8686
6. M/s. Associated International Corpn. 11, Pollock St., Calcutta-7000001 Tele: 22-1005/22-8713 Cable: Assintcorp
7. M/s. Atlas Distributors 62/13, Baburam Ghosh Rd. Calcutta-700040, Tele: 46-6813 Cable: Adibutors
8. M/s. Avanti Marine Agencies 82 E, Dr. Sudhir Bose Road Kidderpore, Calcutta-700023
9. M/s. Beejay Enterprises 2/1C, Nawab Ali Lane, Calcutta-700023, Tele: 45-5307 Cable: Jonesent Calcutta
10. M/s. Bengal Export House 11, Amartolla Lane, Calcutta-7000001
11. M/s. Bengal Seafoods (P) Ltd 4, Achambit Shaw Rd. Calcutta-700022 Tele: 45-3169, Telex: 021-2796
12. M/s. Blue Sea Products 69, Diamond Harbour Road, Calcutta-700023, Tele: 45-7610
13. M/s. Bonndr Industrial Co. Marine Enterprises, 1st Floor, Room-23, K.S. Roy Rd., Calcutta-7000001, Tele: 231075
14. M/s. Britania Seafoods 3/A, gas Street, Calcutta-7000009 Tele: 35-4105
15. M/s. Calcutta Seafoods 10, Crocked Lane, Calcutta-700069 Tele: 23-0097/4017/7236 Telex: CA-3488
16. M/s. Calex Trading Corpn. Peruez Shahidi Range, (Circus Avenue), Calcutta-700019, Tele: 44-4658
17. M/s. Chimanlal Desai & Co. 54, Bentinck Street, Calcutta-1 Tele 23-3535, 23-356 Resi: 45-1011
18. M/s. Consumers Service Corpn. A/8, 35. Cossipore Road, Calcutta 7000002.
19. M/s. Crescent Enterprises 5, Bolai Dutta Street, Calcutta-7000001. Tele: 34-3854
20. M/s. East Coast Fisheries Pvt. Ltd. 1-G, Sarovar, 102, Southern Avenue, Calcutta-700029, Tele: 72-2297 Gram: Coldbero
21. M/s. Eastern Cold Storage Food Products, 43, Hindustan Park, Calcutta-700029
22. M/s. Forestees Expots (I) Pvt. Ltd. 52/3a, Harish Mukherjee Road, Calcutta-700025, Tele: 47-1195
23. M/s. Gemex International Pvt. Ltd. 2A & B, Park Centre, 24, Park Street, Calcutta-700016. Ph: 23-8976 & 44-5514 Cables: Corundum
24. M/s. G.A. Randarian Ltd. Commerce House, Gulabi Block 2, Ganesh Ch. Avenue, Calcutta-700013 Tele: 23-1073/4 Telex: Gurabi 7012

25. M/s. Indian Expro Aquarium (P) Ltd. 24 - R.N. Mukerjee, Calcutta-700071 Tele: 23-5162 & 23-4488 Telex: (021) 2387
26. M/s. India's Hobby Centre P. Ltd. 15, Sakespeare Sarani, Calcutta-700016 Ph: 44-9957
27. M/s. I.T.C Limited., 37, Chowringhee Road, Calcutta-700071 Tele: 24-8141 (10 lines) Telex: 021-2515
28. M/s. India Exports Pvt. Ltd. 44B, Samsul Huda Rd, Calcutta-700017.
29. M/s. Kalaman Cold Storage, 18, Free School Street, Calcutta-700001. Ph: 24-6707
30. M/s. Keeisen Cold Storage 6, Palm Place, Calcutta-700019, Tele: 44-2557.
31. M/s. Kundu's Fancy Fish Suppliers & Porltry Firm Shanpur (Shibtala), Dasnagar, Howrah.
32. M/s. Marda agents, 102, C.I.T. Road, Calcutta-700054. Tele: 35-6283 & 35-6261 Telex: Niryat 021-3555 Telegram: Hardworker
33. M/s. Marico Engineering (P) Ltd. 118/1, Brindavan Mullick lane Howrah-9
34. M/s. The Metal Box Co. Op India Ltd. Barlow House 59C, Chowringhee Road, Calcutta-700020. Ph: 44-235/18, Telex: 021-7278
35. M/s. Minbk Agencies 71, Ganesh Ch. Avenue, Calcutta-700013. Ph: 24-5308
36. M/s. Momin Baksh 86, Collin Street, Calcutta-700016. Cable: Buxfeather, Calcutta.
37. M/s. Monifreeze Pvt. Ltd., 8/1-H, D.H. Road, Calcutta-700027, Ph: 45-4012, Telex: Moni-CA-2889, Cable: Monifreeze
38. M/s. P.M. Packi & Co. 1, Tiretta Bazar Street, Calcutta-700001, Tele: 31-1892
39. M/s. Poly Plates India, Place Court 1, Kyd Street, Calcutta-700016. Tele: off: 21-3192, 24-8910 Fact. 44-8853
40. M/s. Puri Seafoods (Pvt Ltd Indian Fisheries) 4, Achambit Shaw Rd, Calcutta-700022. Ph: 45-3160, Telex: (021) 2796
41. M/s. Quality Seafoods (P) Ltd., 3/12, Poddar Nagar, Calcutta-700068. Ph: 42-6298
42. M/s. R.Akoojee Jadiwet & Co. 10, Clootola Street, Calcutta-700073 Tele: 34-2690, Telex: CA-7039
43. M/s. R.K. Enterprises, 24, R.N. Mukherjee Rd, 5th Floor, Calcutta-700001. Ph: 23-0232. Cable: Okirag
44. M/s. Rex Corporation (India), 9H/14, Topsia Lane, Calcutta-700046 Ph: 44-8783
45. M/s. Samar Singh Jayaswal (P) Ltd, 27-B, Camac Street, Calcutta-700016 Ph: 44-7234, 44-7235, 44-8209
46. M/s. Sanzana Sea Foods (P) Ltd. P-294/1, Dargah Rd., Calcutta-700017. Ph: 44-7091
47. M/s. Sanvasco 11/2, Rani Sankari Lane, Calcutta-700026. Ph: 47-1667, 47-7350
48. M/s. Sourab 84, Kilkhusa Street, Calcutta-700017. Ph: 44-5383
49. M/s. Singhvi International D-5, Gillandar House 8, Netaji Subhas Rd, Calcutta-700001
50. M/s. Shagarika Sea Food (P) Ltd. 76, Park Street, (1st Floor), Calcutta-700017. Ph: 44-3700; Fax: 35-2380 off: 43-1490, Cable: Forfish

51. M/s. Spencers & Co. Ltd. 70 Diamond Harbour Road, Calcutta-700023, Tele: 45-5621 (3 lines)
52. M/s. Southern Cold Storage & Industries, 128, Rash Behari Avenue, Calcutta-700029, Ph: 46-0297 & 46-3223
53. M/s. Southern Cold Storage P. Ltd. 128, Rash Behari Avenue, Calcutta-700029. Ph: 46-0297, Resi: 46-3223, Cable: Marine Food
54. M/s. South India Sea Foods, 9 Cashmene Mansion, 25, Welleslay Street, Calcutta-700016, Ph: 24-7012 Resi: 44-7410
55. M/s. Neogi & Company (Food Grains) 14, Subol Ch. Lane, Calcutta-700009. Ph: 35-3907, 35-8949, 610-409. Telegram: Esnogi
56. M/s. State Fisheries Devp. Corpn. Ltd. 43, Shakespeare Sarani, Calcutta-700017. Ph: 44-2688, 44-9302
57. M/s. Supreme Food Products (I). 3/1, Gas Street, Calcutta-700009.
58. M/s. Tea Sales & Allied Industries (I) Pvt. Ltd. 9, R.N. Mukherjee Rd, Calcutta-700001. Ph: 23-8921/26, 23-7514, Gram: Teasour-Calcutta Telex: 7480-Cerritt CA
59. M/s. Temple Engineering Works, 64/2, Belilous Road, Howrah-711101. Tele: 22-3727
60. M/s. Texmaco Limited International Division. 9/A. R.N. Mukherjee Rd, Calcutta-700059. Ph. 22-2316 & 23-0138
61. M/s. Tristar Corporation, 17-A, Everest House, 46-C, Chowringhee Road, Calcutta-700071, Tele: 14-2990
62. M/s. United Commercial Corpn. 29. Ganesh Ch. Avenue, Suite No. 402, Calcutta-700013, Tele: 23-4909
63. M/s. Usha Martin Black (Wire-Rope) Ltd., 14, Princep Street, Calcutta-700071. Tele: 23-9516 (5 lines), Telex: Usharop CA - 7483, Gram: Ushapore.
64. M/s. V.M. Dhawan & Co. (Sundarban Sea Food) 9, Cashmene Mansion, 25, Welleslay Street, Calcutta-700016, Tele: Res: 44-7410, Fax : 24-7012, Cable: Redsaders
65. M/s. V.S. P. Expo (P) Ltd. 24-R.N. Mukherjee Rd, Calcutta-700001 Ph: 23-5162/9488, Telex: CA-2387 West Bengal
66. M/s. Veejoy Impex 4, Achambit Shaw Rd, Calcutta-700022.
67. M/s. Triveni Food Products, D-1, Hide Road, Calcutta-700043.
68. M/s. Bay Sea Food Explores & Associates (P) Ltd, 33, Canal Road, Calcutta-700009.
69. M/s. Bengal Marine (P) Ltd., Ganga Nagar, 24 Parganas
70. M/s. Rallis India Ltd. 3/1 Gas Street, Calcutta-700009
71. M/s. Monali Sea Food (P) Ltd. Sukanta Nagar, 24 Parganas.
72. M/s. A.K. Corporation (P) Ltd., 76 A.J.C. Bose Road, Calcutta-700014.
73. M/s. Devi Marine Food, 76. A.J.C. Bose Road, Calcutta-700014.
74. M/s. Blue Angel Sea Food. 63/2B, Beliaghata Road, Calcutta-700037
75. M/s. G.F. Kellner & Co. 70, D.H. Road, Calcutta-700023

In operative at present as it could be ascertained from Marine Products Export Dev. Authority.

II
Capture Fishery

3
The Kulty Estuary

1. INTRODUCTION

Sundarban region is a part of the 24-Parganas district located between 21°32′ and 22°40′ North latitude and between 88°05′ and 89°0′ East latitude comprising Sagar, Namkhana, Jaynagar (Part), Canning, Barasat, Minakhan, Haroa, Hasnabad (Part), Hingalganj, Sandeshkhali, Gosaba P.S. and constituting the western most part of the Ganga delta.

The rivers like Saptamukhi, Thakuran, Matla and Gosaba although originally connected with the main channel or its tributaries are long disconnected from the later due to silting up at the upper region and have changed their course of water flow. Since they do not receive any spill directly from any major river, they may be considered as estuarine inlets of the sea.

Average annual rainfall is about 1800 mm with about 90% falling between May and October. The region experiences frequent cyclonic storms, sometimes with devastating intensity, the last one of such nature occuring in December, 1981.

Most of the area was under mangrove forests until about 1784 when systematic colonisation in the form of construction of protective embankments, clearing the forests, cultivation, habitation etc. began. The settled areas now consists of 54 islands most of which are seperated from each other by tidal rivers and protected by about 3500 km of earthen embankments.

The fact that supply of upland water essential for preservation of tidal channels permanently has been clearly discussed by Majumder (1942) in his valuable document "Rivers of Bengal Delta".

Provision of spill areas by removal of marginal embankment seems to be an urgent necessity for prolonging the life of these tidal channles. But this is not enough. If these channels have to be given a permanent lease of life, the supply of upland water seems to be an essential requirement. In this context physical characteristics of tidal rivers requires to be understood properly.

The energy which creates the tidal flow viz. attractions of the sun and the moon in the deep sea is very limited and is manifested partly as potential energy i.e. velocity of flow. The former causes the tidal flow up the rivers and the later determines its power of transporting silt. As the velocity of tide when it enters the mouth of these tidal rivers is high and as there is a vast reservoir of unconsolidated silt in suspension along the delta face, the tide as it flows up these rivers is highly charged with silt. It is generally observed. The duration of ebbtide in a tidal river is much longer than that of the flow tide. As the same quantity of water must ebb/out as flowed in, it therefore follows that the average velocity of ebb is less than that of the flow. Now the capacity of water to transport silt depends on its velocity. It therefore follows that the ebbtide is generally unable to transport back fully the silt that has been carried up these tidal rivers by the flow tide. Even a slight deposition of silt will go on accumulating as the tides function twice daily throughout the year and the channel will begin to deteriorate. The deterioration would impede the propagation of tidal wave which would cause further deterioration and the vicious circle would continue till the channel is completely dead.

It would appear from the above that to maintain the life of a tidal river what is necessary is an additional supply of water not saturated with silt i.e. which has reserve capacity to pick up more to suppliment the tidal flow during ebb so as to scour out fully the silt that has been admitted into the river by the flood tide. This could be affected either by the supply of upland water or by local drainage or by throwing the existing spill area open to the daily ebb and flow or by adding new spill areas if possible. The spill area is of course an important flushing agent that it would function only for a limited period at a gradually diminishing rate till the whole area is raised to high tide level and its storage capacity is reduced to nil.

The third agent viz. the unplanted water is certainly the most important and if it were possible to arrange for its supply perinially so as to flush the river even during the dry season, there is no reason why the tidal river should not live and continue her beneficient activities forever.

The above findings clearly indicate how tidal rivers large or small lying Criss Cross the Sundarban delta are heading fast towards silting. River Kulti is an example in which gradual high value in shape factor an indication of deterioration is very clearly shown. Annual average value of shape factor of Kulti estuary ranged from 15.16 to 21.84 (in 1966) which further down at Haroa has ranged between 23.25 and 22.00. At Malancha the values ranged from 25.14 to 30.07 and finally at Bermazone the values were as high as 81.92-96.40. At Malancha the average values of cross sectional area decreased progressively during 1952-59.

In the absence of adequate perennial head water supply, such deterioration has become possible. Such is the case with all the estuaries of the delta. Incidents are quite common that during high tidal bores (August and September during monsoon) high erosion at the marginal areas takes place. This process is also greatly helped by mangrove forest which helps settling the silt and by this process there is a gruadual rise or increase the area of delta bed and slowly encroaches the surroundings. The area of delta thus gradually increases (in 40 years the increase is by 33.7%). Depth of network of channels criss crossed the delta accumulate silt in the spill area to raise their bed. Increase of shape factor indicate considerable deterioration of the tidal channels lying around 52 islands of the delta. Thus during the period of high tidal bores, the high silt load brought in along with the water round the year are deposited in the channel bed along with their spill areas in the absence of adequate head water flow. Similar picture holds good in case of Bidyadhari-Hooghly.

Physiography and hydrology of the two estuaries, Kulti – small estuary flowing in the east and Bhagirathi-Hooghly a mighty tidal estuary flowing on the west of Sundarbans presented, gives the characteristics of individual. Kulti has almost lost its identity while the Bhagirathi-Hooghly is heading fast towards its deterioration. Both of these estuaries in the past played major roles in the development of Sundarban forest which at present has almost ceased. Ray (1993 and 1998) has given detailed account of the estuaries through the valuable documents on (1) Aquaculture in the Sundarban delta – its perspective and (2) Ecological imbalance of the Ganga river system – its impact on aquaculture.

2. THE KULTY ESTUARY

As described in Part 1 (Culture Fishery) of this work the river Bidyadhari used to receive major discharges from the river Ganga prior to 12th Century through the Jamuna, a tributory of the Hooghly river (cf. Fig. 1 a&b). Deterioration of the Jamuna probably started soon after the diversion of the main flow of Ganga due to high floods towards the end of the 15th or early 16th century. But the river continued to receive, at least during rains, a share of the Damodar flood which had one of its outfalls into the Hooghly near Kalna and near Neasarai till about the middle of the 18th century, both places being above the offtake of the Jamuna at Tribeni. Later diversion of the Damodar, down its present channel with its outfall below Uluberia, finally sealed the fate of the Jamuna together with its spill channels (Fig. 3.1a & b).

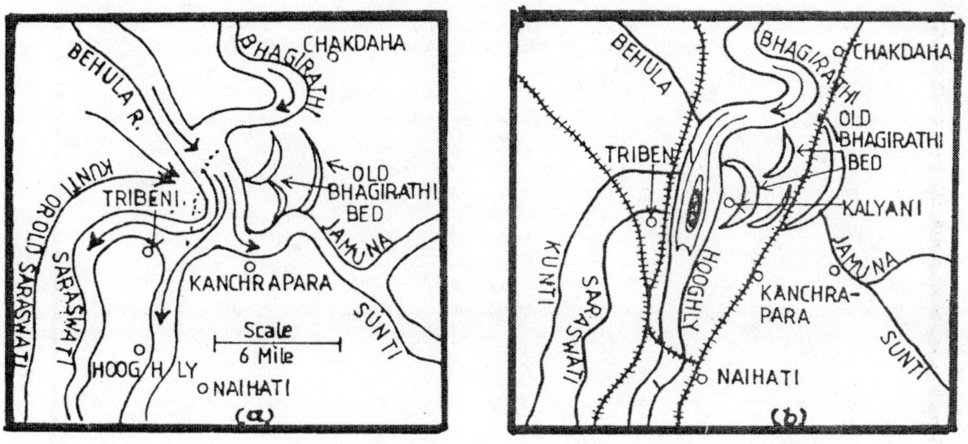

Fig. 1 : (a) Active Ganga, jamuna and Saraswati during seventeenth century. (b) Conditions of these rivers after the Ganga and the Damodar (Bchula) shifted their courses.

Deprived of the upland water supply, the Bidyadhari remained dependent entirely on her other resources, *viz.*, local drainage and spill area, to prolong her life as long as she could. At this stage another characteristics peculiar to tidal rivers carrying a large portion of silt, hastened her decay. The Kulti Gong, the next parallel tidal channel to the east, had already opened up and by means of connecting channel, now known as Haroagong, provided a meeting ground for tides coming in both Kulti and the Bidyadhari rivers. Owing to this check there was always heavy deposition of silt at the tidal meeting ground; but so long as these rivers received a plentiful supply of upland floods, the silt could be cleared by flushing during floods. This check resulted in a grave menace to the preservation of both the channels. In the above context the Kultigong. Which was originally discharging water into the Matla through Kumerjolgong, has her present outfall into the Raimangal estuary the Raimangal being the deepest estuary in the delta and the conditions for the propagation of tidal water being more favourable, the stronger tide, up the Kultigong, could travel higher and higher up the connecting channel (Haroagong) thus shifting the tidal meeting more and more down the Bidyadhari. The Nonagong, the Sunti and the Nowi, which originally used to feed the Bidyadhari, were thus gradually absorbed by the Kultigong for her own sustenance. By robbing the Bidyadhari of the drainage, the spill areas of these feeders, the Kultigong began to develop at the expense of the Bidyadhari and the latter deteriorated rapidly (Mazumdar, 1942) (Fig. 2).

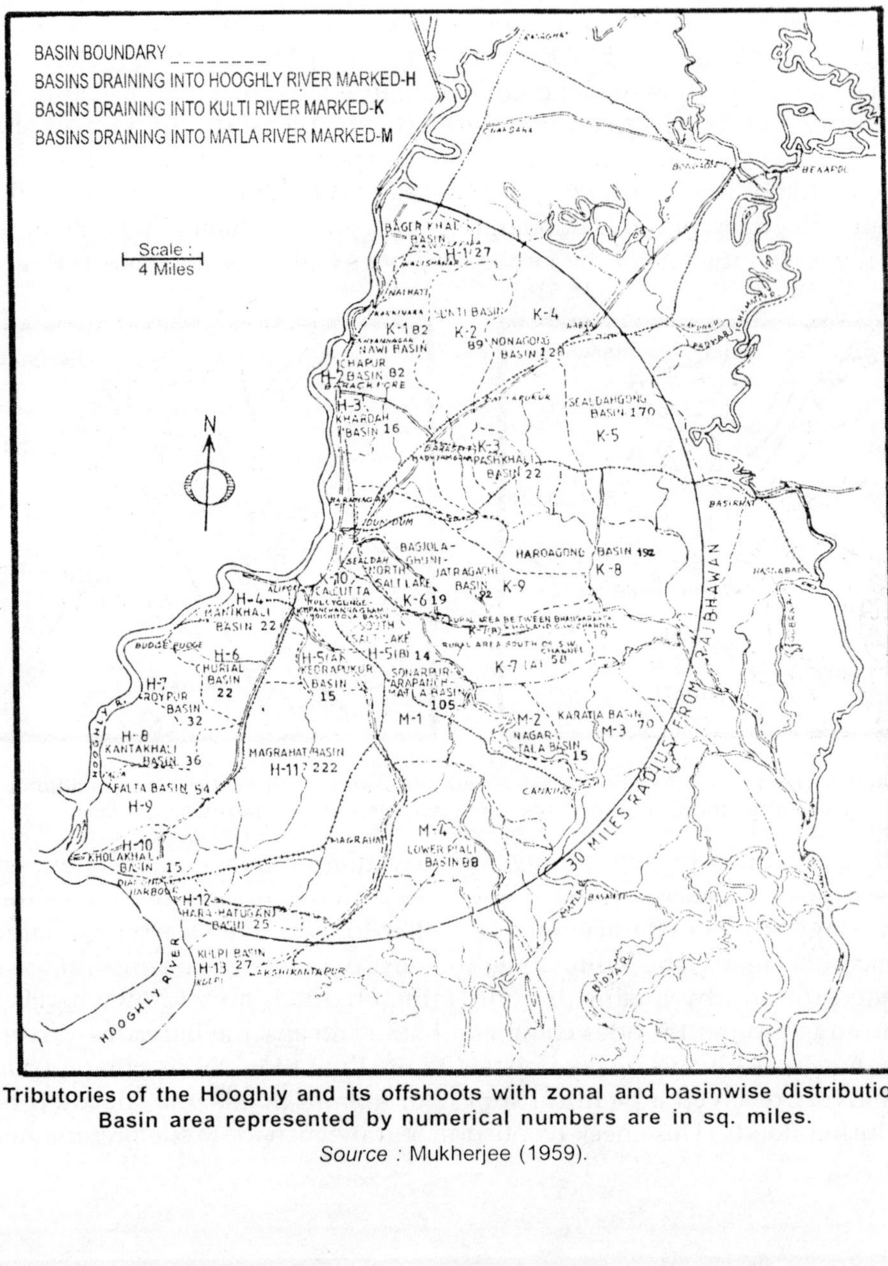

Fig. 2 : Tributories of the Hooghly and its offshoots with zonal and basinwise distribution of area. Basin area represented by numerical numbers are in sq. miles.

Source : Mukherjee (1959).

Mainly on account of deterioration of the Nadia rivers and the head-waters of the Hooghly river, the embankments put up all over 24 Parganas for cultivation and consequent interference with the process of land formation in the east and south of the district the enforced salinity of the large tracts in the south, the drainage problems of the swampy areas of the district loom large. There are no separate problems of irrigation apart from those of drainage and the task of keeping waterways open is also dependent on the thorough draining of the district. Accordingly, the district was divided into 4 blocks under the following main categories (Mitra, 1954).

1. Blocks draining into the Hooghly river.
2. Blocks draining into the rivers of Kulti series.
3. Blocks draining into Matla-Piali-Bidyadhari series.
4. Blocks draining into the Ichhamati series.

Of the 4 block 'Blocks draining into the rivers of Kulti Series' is important in the context of drainage and water ways and hence are discussed below.

a. The Burice Beel in Shyamnagar, Ichhapur and Palta areas partly drain into the Kultigong via the Nowii river and Haroagong and partly into the Hooghly.

b. A large block served mainly by the Sunti river covering an area of about 200 sq miles (51780 ha).

c. A block of 80 sq miles (20712 ha) served by the Nonagong.

d. A block, technically called block V, covering an area of 190 sq miles (4919 ha).

e. A block, technically called block VI, drains into the Kultigong extending over 85 sq miles.

If the above 5 blocks are well-drained they will supply additional headwater to the Kultigong during monsoon. At the present moment owing to the absence of any drainage channel a large portion of the rain water evaporates and does not benefit the Kultigong.

f. and (g) Blocks VI and VIB having areas of 10 and 8 sq miles (2589 and 2071.2 ha) respectively, have no drainage difficulty. They drain into the Bhangar canal from the South-wet canal.

g. A country block covering the northern Salt lake with an area of 18 sq miles (4660.2 ha).

h. The block covering Kasba and Panchanangram area with a spread of 6 sq miles (1553.4 ha). This area is partially drained into the storm water canal of Calcutta Corporation through the Chowbhaga sluice.

i. An area South of storm water canal of Calcutta Corporation extending over 25 sq miles (6472.5 ha). This area drains through the storm water canal of the Calcutta Corporation into the Kultigong. There is some drainage difficulty here.

In the absence of proper drainage canal the river Kulti is deprived of a major volume of water for her sustenance although she has a vast drainage area like Blocks (a) to (e).

2.1 Physical Features

The Kulti estuary is a part of an off-shoot from the Rayamangal estuary and is variously known as Barackalaghi, Kalagachi, Hatgachiagong, Kultigong and Haroagong to its termination into three narrow arterial channels, the Nonagong, the Sunthe and the Nowii. The latter three are tidal rivulets spilling into the brackish-water culture fisheries around Kharibaria and swamps in the North Eastern Calcutta (Fig. 3).

There is no perennial source of flush water to the Kulti above the point of sewage outfall, but the Sealdahgong, about 8 km below, brings a fair amoung of tidal water from the Ichhamati through Hinsee Khal. Some fresh-water is also brought by it during monsoon months from the Ichhamati which receives considerable amount of spill water at that time from the flooded Padma river. Run-off water of a catchment of about 250 sq miles (64725 ha) through the arterial channel of the Nowii, the Sunthe and the Nonagong as well as about 150 sq miles (38835 ha) of Catchment of Calcutta and adjacent areas render the Kulti fresh for two months, August and September (David, 1959).

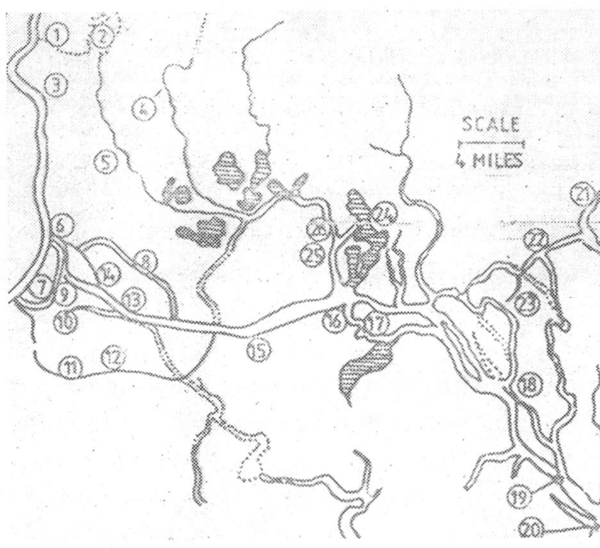

Fig. 3 : Kulti sewage outfall in relation to the Cultivated Fisheries. (1) Ichapur; (2) Barti Beel; (3) Barrackpore; (4) Barasat; (5) Nowee nadi; (6) Chitpur Lock; (7) Calcutta; (8) Krishtopore Canal (Krishnapore); (9) Sewage outlet; (10) Stormwater outlet; (11) Tollygung; (12) Southern Saltlake; (13) Bantala Sewage Pumping Station; (14) Northern Salt lake; (15) Hanakhali; (16) Kulti outall (Ghusighata); (17) Minakhan; (18) Kalinagar; (19) Barakalagachi; (20) Raimangal estuary; (21) Ichamati river; (22) Hasnabad; (23) Hinsee Khal.

Maitra (1958) recorded monthly average of river water level (highest high water and lowest low water level) for 27 years from 1931 to 1957 at Haroahat, Ghusighata and Malancha (Fig. 3.4a, b & c) H.H.W. level at Haroahat ranged between 10.0 and 10.9 ft (3.06-3.3 m). The corresponding L.L.W. level generally ranged between 2.1 and 3.4 ft (0.64 and 1.0 m). At Malancha such levels ranged mostly between 9.2 and 11.9 ft (2.79 and 3.61 m) and 4.1 and 6.8 (1.24 and 2.06 m) while at Ghusighata they were 10.3-10.9 ft (3.12-3.3 m) and 4.1-

5.5 ft (1.24-1.67 m) respectively for the years 1937-53. Monthwise maximum discharge figures at Ghusighata during 1966 indicated higher values (30-52 cu m/sec.) during March, September and October (Figs. 4(a-c) & 5). around 1940, prior to discharge of Metropolitan wastes, the following observations near the outfall point at Ghusighata were recorded (David, 1959).

The author further recorded that maximum flood discharge was 6.14 m m^3 and ebb discharge was 5.46 m m^3 at the outfall region during high tide and was 1.82 m m^3 - 2.05 m m^3 at low tides (the period of maximum sewage discharge) and that the estuary gained 5.2 m in water level H.H.T and 2.4-3.3 m during monsoon which during dry weather came to 95:1 while David (1959) estimated the ratio between 1:12 and 1:43 within 3 hours at the same place. Presently, velocity between 1.45 and 2.44 m/sec was recorded at flow and ebb currents respectively near the outfall. The cross sectional studies made at 4 centres by Pramanik and Roy (1972) on the estuary revealed that the discharge figure at Haroa was less than half of that at the outfall, double at Minakhan and 5-6 times at Malancha.

These workers also studied the rate of silting from the cross sectional area and shape factor data (Table 2 A & B) of the estuary at 5 centres *viz.* Nowii-Suntee junction. Horoahat, below outfall, Malancha and Bermajore the major stretch of the Kulti estuary from 1952-66. Monthwise silting was also studied for the year 1966. From Table 2A it is evident that at Nowii-Suntee junction, the averages sectional area decreased progressively during 1952-58 while the averages values of shape factor showed a rising trend during those years (cf. Table 2B). This indicated siltation. From cross-section and shape factor curves it appeared that during 1959-60 the deterioration was checked, during 1961-65 there had been slight improvement, but in 1966 rapid siltation was noticed.

Table 1 : Some Hydrological Features of Kulti at Ghushinghata Prior to Discharge of Metropolitan Waste

Depth during high tide	-	upto 42 ft (12.7 m)
Margin between L.L.T. and H.H.T.	-	17 ft (5.15 m)
Margin during Neap tide period	-	8-11 ft (2.42-3.3 m)
Average high water level	-	14.50 ft (4.39) Above
Average low water lavel	-	3.33 ft (1.0 m) datum
Maximum flood discharge	-	27.000 cusecs
Maximum ebb discharge	-	24.000 cusecs
At low tides	-	8000-9000 cusecs
Average discharge at 3 ft (0.9 m)/sec	-	17.000 cusecs

At Malancha, the average values of cross sectional area decreased progressively during 1952-59, while those of shape factor showed a rising trend (Table 2 A & B). These were the indications of deterioration. From 1960 to 1963, the average values of sectional areas increased progressively while those of shape factor decreased more or less in the same sequence, indicating progressive improvement. But from 1964 onwards upto 1966 clearly indicated deterioration.

Table 2A : Annual Average Values of Cross Sectional Area Below 1.14 m (3.75 ft) P.W.D. at Different Centres of the Kulti Estuary from 1952-66

Year	*Nowii-Sunti Junction*	*Haroahat*	*244 m. down Kulti O.F.*	*Malancha*	*Bermajore*
1952	33	305	685	2.123	4.323
1953	34	304	719	2.076	5.025
1954	31	302	690	1.893	4.835
1955	30	311	718	1.793	4.872
1956	29	343	734	1.843	4.968
1957	27	336	721	1.787	5.612
1958	24	330	740	1.690	4.886
1959	24	301	665	1.474	4.692
1960	23	321	671	1.531	4.758
1961	30	335	648	1.893	4.171
1962	22	321	725	2.001	4.919
1963	26	378	762	2.246	4.576
1964	26	352	783	2.037	4.281
1965	26	340	875	1.897	4.325
1966	19	371	929	2.002	4.621

* Pramanik. H.R. and Roy, P.B. (1972).

At Malancha, the average values of cross-sectional area decreased progressively during 1952-59, while those of shape factor showed a rising trend (Table 2 A & B). These were the indications of deterioration. From 1960 to 1963, the average values of sectional areas increased progressively while those of shape factor decreased more or less in the same sequence, indicating progressive improvement. But from 1964 onwards upto 1966 clearly indicated deterioration.

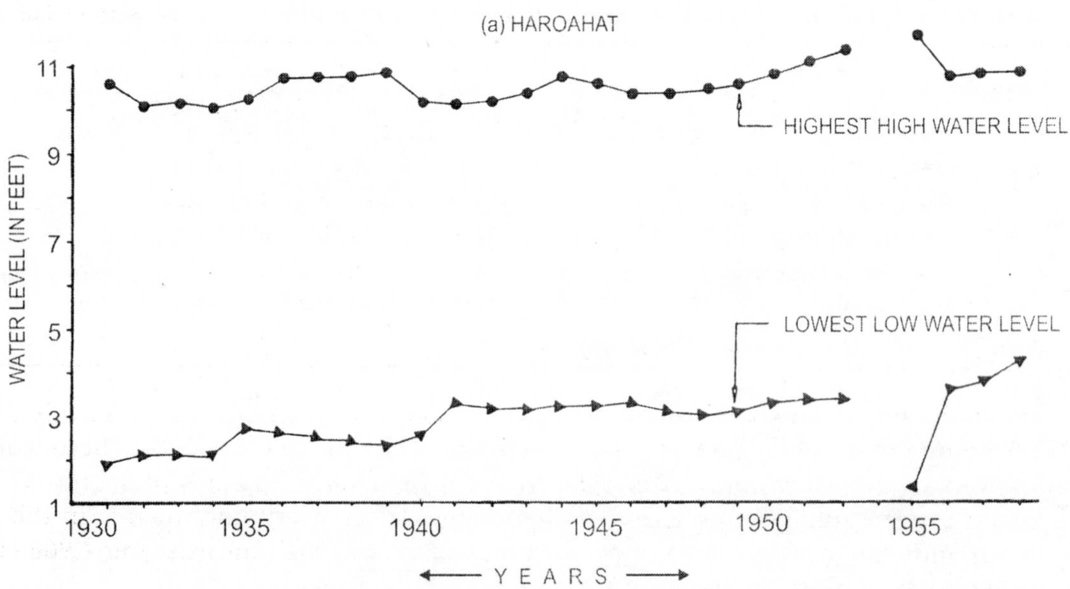

Fig. 4a : Monthly average of water level of the Kulti at Haroahat (1931-1957)

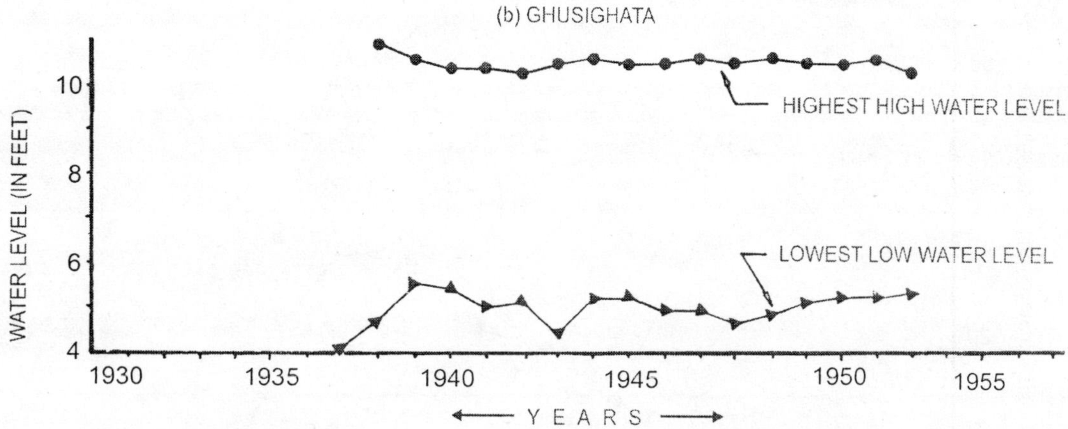

Fig. 4b : Monthly average of water level of the Kulti at Ghusighata (1937-1953)

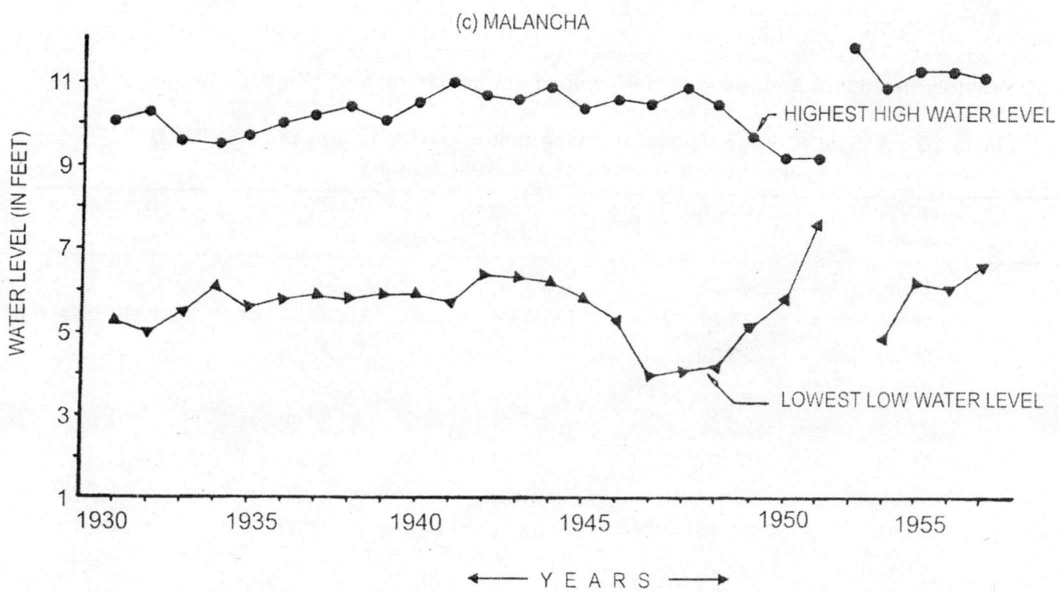

Fig. 4c : Monthly average of water level of the Kulti at Malancha (1931-1957)

At the Kulti outfall (224 m below outfall during 1952 to 1958 there were fluctuations in the average values of sectional areas, and slight deterioration in 1959-61. But from 1962 till 1966 the average values of both the sectional areas and shape factors indicated definite improvement of the river regime (Table 2A & B).

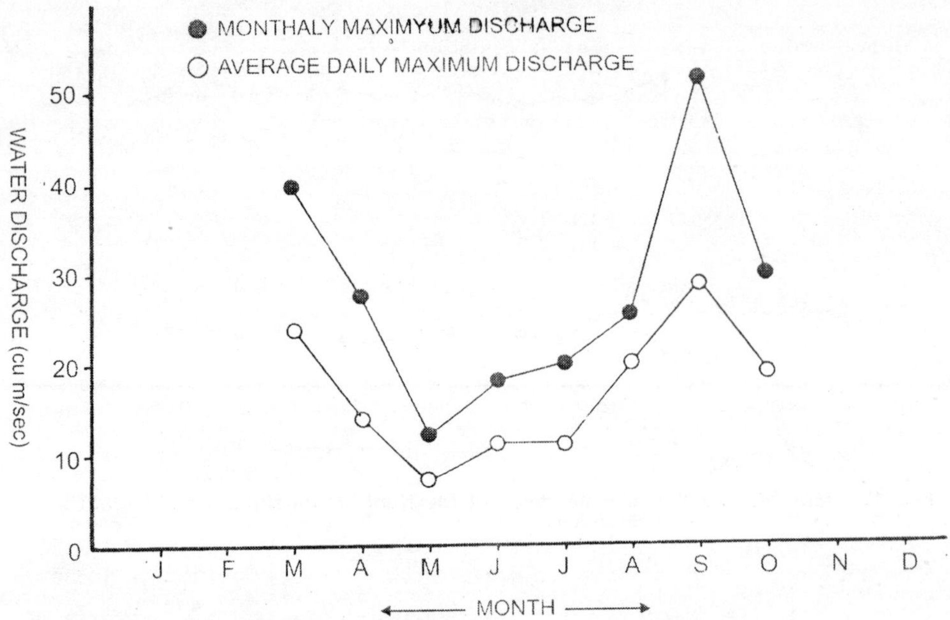

Fig. 5 : Monthly maximum and average daily maximum descharges at Ghusighata during 1966.

Table 2B : Annual Average Values of Shape factor. Below 1.14 m (3.75 ft) P.W.D. at Different Centres of the Kulti Estuary*

Year	*Nowii-Sunti Junction*	*Haroahat*	*244 m. down Kulti O.F.*	*Malancha*	*Bermajore*
1952	15.16	23.25	22.43	25.14	81.92
1953	15.44	23.21	21.88	22.04	80.07
1954	15.50	22.29	22.92	28.45	83.74
1955	16.48	23.89	21.78	28.34	83.13
1956	17.53	20.99	21.15	27.98	81.90
1957	16.29	19.97	21.34	29.76	80.84
1958	16.22	20.40	21.70	32.00	80.50
1959	15.80	23.70	23.75	37.80	82.50
1960	15.40	22.26	23.80	36.80	80.80
1961	15.90	23.10	25.60	32.00	96.40
1962	15.80	22.00	23.80	28.50	76.20
1963	13.30	19.50	22.00	26.20	77.20
1964	12.90	20.00	21.60	27.90	81.20
1965	15.19	20.00	22.43	30.05	83.10
1966	21.84	19.12	21.18	30.07	82.10

* Pramanik. H.R. and Roy, P.B. (1972).

At Haroahat and Bermajore, fluctuations in the average values or both the sectional areas and shape factors from 1952 to 1966 were very much pronounced indicating no definite trend of improvement of deterioration.

The monthwise shape factor studied during 1966 at the above centres indicated silting (Table 2C). At Nowii-Suntee junction higher values were indicated during May to November; Haroahat during February, March, June, August and December; at the outfall, during January, February, April, May and October-December; at Malancha, during January-March, June-September and November and December; and finally at Bermazore, during January-April, August, September and November and December. All such values were indicators of degradation due to silting.

2.2 *Sewage Disposal and Mode of Dissipation*

The outfall canal is provided with 16 vents with a total capacity of 220 cusecs discharge. But in actual practice, as observed by David (1959), 5 vents were being kept open, except during monsoon bursts or near drought condition, discharging about 668 cusecs or 4,286 gallons per second. Duing monsoon months additional gates were opened to prevent spilling of septic sewage on the areas and to drain storm water from the city and around the canal itself. Between February and May all the vents, except one, remained closed for short durations to prevent too much discharge of sewage when in some years 60% of its water is utilized raw for fresh water pisciculture in the Bidyadhari, Bantala, Dhapa and Honakhali areas.

Table 2C. Monthwise Shape Factor at Different Centres During 1966*

Year	*Nowii-Sunti Junction*	*Haroahat*	*244 m. below O.F.*	*Malancha*	*Bermajore*
January	13.1	19.7	22.5	31.9	83.1
February	14.6	21.0	22.3	29.5	81.8
March	16.1	20.0	18.0	30.0	88.1
April	16.6	17.6	22.3	26.8	84.6
May	26.0	18.2	23.1	27.2	76.0
June	-	22.31	21.5	30.1	73.9
July	22.9	16.4	21.6	32.0	79.0
August	20.4	21.2	18.7	30.1	82.0
September	31.00	16.2	19.7	33.4	89.0
October	34.1	19.3	20.9	29.5	75.9
November	31.0	17.6	21.0	31.6	88.2
December	15.3	20.0	22.6	32.4	83.4

Higher values observed at different centres of the Kulti estuary during different months of the year indicating deterioration due to silting, these are-

Nowu-Sunti (Junction) : May-November

Haroahat : February, March, June, August and December.

Out fall : January, February, April, May, October-December.

Malancha : January-March, June-September, November and December.

Bermajore : January-April, August-September, November and December.

* Pramanik, H.R. and Roy, P.B. (1972).

Period of discharge at the outfall varies from month to month. Considering the discharge period of sewage on an average as 8 hours per day (4 hours during each tidal phase), the quantum of load disposed per day during each season is discussed as follows. Since the metropolitan wastes contain effluents of various chemicals and pharmaceuticals, electroplating, fertilizer, match factory, metallurgical (ferrous and non-ferrous), galvanizing, pulp and paper, soap detergents, textile, oil from petroleum pumps and a good number of tanneries, they are highly complex in character. The contribution of such effluents (industrial) in the total volume of metropolitan wastes was 0.23 m m^3. Seasonal (Summer, Monsoon and Winter) average load disposed per day (8 hours basis 4 hr. during each tidal phase), all based on the observations of David (1959), were 0.11664, 0.10206 and 0.059778 tonnes for suspended solids and 0.040969, 0.035866 and 0.040969 tonnes respectively for Bicarbonate alkalinity, expressed as $CaCO_3$ in the above order of seasons. Similarly, the total phosphous (PO_4) values were 0.000091, 0.000139 and 0.000071 tonnes and Nitrate-N values for summer and winter were 0.0000729, 0.00102 tonnes respectively. NEERI (1981) on the other hand observed values which were quite high. For example, the values calculated for total solids were 112.07, 56.24 and 45.49 tonnes : Suspended Solids 7.81, 6.427 and4.71 tonnes and for dissolved solids : 104.26, 49.8 and 40.785 tonnes respectively in the above order of seasons. Values for BOD_5 were 2.46, 4.327 and 3.15 tonnes; COD : 9.29, 9.45 and 8.499 tonnes respectively. The values of total Nitrogen were, however, of the order 11.61, 10.21 and 12.84 tonnes against total phosphorus (PO_4) values of 7.4358, 2.249 and 23.65 tonnes respectively in the above seasons. Ray *et al.* (1981) estimated total oganic load (BOD_5) disposed to be 234, 269 and 99.9 tonnes respectively. The waste also contained metals like zinc, copper, iron and manganese in concentrations of 3-56, 9-11, 110-164 and 13.9-60 ppm in the liquid fraction. The range of concentrations of metals and mean values in mg/kg in sludge of DWF canal at six points were 12.78-181.23 (89.03) for lead; 36.78-246.72 (113.92) for copper; 101.90-760.1 (366.1) for Zinc; nil for Cadmium; 23.19-50.84 (39.22) for Nickel; 10.95-16.32 (13.86) for Cobalt, 25.35-378.84 (197.05) for Chromium and 33.69-444.01 (333.24). The ranges of estimated values for the above parameters in the two km stretch of DWF canal from Lowgachi-Ghusinghata (Lock gate), which is agitated everytime by high tide water of the river and the suspension drawn back to the river, were 23.30.4 mg, 36.70-48.48 mg, 101.90-127.87 mg, nil, 33.28-37.49 mg, 14.7-15.14 mg, 47.27-68.57 and 420.38-444.01 mg/kg sludge.

The ebb flow starts carrying the sewage after two hours of commencement of the tide. With the progress of ebb tide more sewage is drawn into the estuary when ratio of water : sewage decreases. At the turn of tide the surcharged mass, now stretched upto 25 km down below travels back and covers 25-30 km above the outfall during a period of 3-4 hours from the onset of high tide. This back water gets more concentrated by accepting the fresh sewage ingress while crossing the outfall zone and enters into the brackishwater bheries through arterial canals. During each tidal play (High and low tide) the sewage ingress is for 4 hours only.

3 HYDROBIOLOGICAL FEATURES

No systematic information pertaining to hydrobiology are available on the Kulti estuary excepting stray investigations. The studies made so far can be classified as (1) Physicochemical and biological studies and (2) hydrological study in the context of its conservation and water quality survey for efficient sewage management.

Bose (1944) and Nair (1944) probably for the first time casually reported about the effect of Calcutta Metropolitan Wastes on the Kulti estuary. This was further augmented by David (1959) who studied biological as well as chemical conditions covering a stretch of 18 miles on either side of the outfall. Thakur (1975) studied the prospects of fish seed resources in this estuary. Subsequently, Ray *et al.* (1981) and the present author further intensified their studies on various aspects like hydrobiological, bacteriological, metallic contamination etc. Likewise, River Research Institute (1970) made some hydrological observations with a view to conserve the estuary. Lastly NEERI (1978) made some observations to find out the impact of Calcutta Metropolitan Wastes on the estuary purely from efficient sewage management point of view. Most of the investigations were carried out in the upper stretch of the estuary. The physicochemical and biological characteristics discussed below are based on the results of investigations made so far. Baseline data of the Kulti estuary for different seasons in different years are indicated in Table 3.

Table 3 : Baseline Data for the Kulti Estuary Expressed by Average Values for Different Seasons in Different Years

	1954–55						*1967**					
	Winter		*Summer*		*Monsoon*		*Winter*		*Summer*		*Monsoon*	
	U	*L*	*U*	*L*	*U*	*L*	*U*	*L*	*U*	*L*	*U*	*L*
Parameters												
Physical												
Temperature 0°C	–	–	–	–	–	–	–	–	–	–	–	–
Turbidity (units)	300 1000	–	500 900	–	–	–	–	–	–	–	–	–
Total Solids (ppm)	–	–	–	–	–	–	–	–	–	–	–	–
Suspended Solids ppm	–	–	–	–	–	–	88 319	–	95 490	–	105 358	–
Chemical												
pH	7.8-7.8	–	7.1-8.0	–	7.2-7.6	–	7.2-8.2	–	7.2-8.2	–	7.2-8.2	–
DO (ppm)	1.8-3.6	–	0.1-6.0	–	0.9-3.35	–	–	–	–	–	–	–
BOD (ppm)	–	–	–	–	–	–	–	–	–	–	–	–
COD (ppm)	–	–	–	–	–	–	–	–	–	–	–	–
OC (ppm)	15-28	–	14-22	–	11-27	–	–	–	–	–	–	–
Alkalinity (ppm) (as $CaCO_3$)	220 312	–	100 290	–	180 270	–	236 464	–	248 615	–	189 295	–
Salinity (‰)	0.8-7.0	–	10-17.1	–	1.1-12.95	–	2.6-4.1	–	8.2-21.2	–	1.66–6.1	–
Sulphate (ppm)	–	–	–	–	–	–	–	–	–	–	–	–
Sp. Conductivity (m.mhos/cm)	–	–	–	–	–	–	–	–	–	–	–	–

Contd.......

Table 3 – contd..

	1954–55						1967*					
	Winter		*Summer*		*Monsoon*		*Winter*		*Summer*		*Monsoon*	
	U	*L*	*U*	*L*	*U*	*L*	*U*	*L*	*U*	*L*	*U*	*L*
Ammonia–N (ppm)	–	–	–	–	–	–	–	–	–	–	–	–
Alb. Ammonia –N(ppm)	1.0-10.6	–	2.0-2.8	–	3.25-6.8	–	–	–	–	–	–	–
Nitrite–N (ppm)	–	–	–	–	–	–	–	–	–	–	–	–
Nitrate–N (ppm)	0.5-1.2	–	0.4-0.5	–	3.25-6.8	–	–	–	–	–	–	–
Inorganic P. (ppm)	0.22-0.8	–	0.16-1.0	–	0.268-1.48	–	–	–	–	–	–	–
Organic P. (ppm)	0.1-0.3	–	0.02-0.1	–	0.035-0.42	–	–	–	–	–	–	–
Parameters												
Physical												
Temperature 0°C	–	–	-	-	–	-	–	–	–	–	–	–
Turbidity (units)	–	–	–	–	–	–	–	–	–	–	–	–
Total Solids (ppm)	–	–	–	–	–	–	1430 2152	–	3728 12996	–	664 734	–
Suspended Solids ppm	53-378	–	47-924	–	61-344	–	124-704	–	192-648	–	52-84	–
Chemical												
pH	7.1-7.8	–	7.1-7.8	–	7.2	–	7.8-7.9	–	7.8-8.0	–	7.3 7.9	–
DO (ppm)	–	–	–	–	–	–	0.2-3.7	–	2.1-4.4	–	1.6-5.6	–
BOD (ppm)	–	–	–	–	–	–	14-21	–	4-10	–	9-20	–
COD (ppm)	–	–	–	–	–	–	56-92	–	60-103	–	19-56	–
OC (ppm)	10-25	–	6-16	–	11-31	–	–	–	–	–	–	–
Alkalinity (ppm) (as $CaCO_3$)	–	–	–	–	–	–	344-368	–	180-208	–	142 195	-
Salinity (‰)	1.78-5.36	–	6.135-27.1	–	1.07-3.08	–	0.86 1.14	-	1.39-11.25	–	0.15 0.27	-
Sulphate (ppm)	–	–	–	–	–	–	43-57	–	275-415	–	14-43	–
Sp. Conductivity (m.mhos/cm)	–	–	–	–	–	–	–	–	–	–	–	–
Ammonia–N (ppm)	–	–	–	–	–	–	8.9-11.6	–	1.12-1.4	–	0.6 1.61	–

Contd.......

Table 3 – contd..

	1954–55						*1967**					
	Winter		*Summer*		*Monsoon*		*Winter*		*Summer*		*Monsoon*	
	U	*L*	*U*	*L*	*U*	*L*	*U*	*L*	*U*	*L*	*U*	*L*
Alb. Ammonia-N (ppm)	–	–	–	–	–	–	–	–	–	–	–	–
Nitrite-N (ppm)	–	–	–	–	–	–	0.012 0.017	–	0.115 0.206	–	0.01 0.044	–
Nitrate-N (ppm)	–	–	–	–	–	–	0.135- 0.23	–	0.91- 3.5	–	0.05 0.265	–
Inorganic P. (ppm)	–	–	–	–	–	–	1.74 3.74	–	0.18- 0.33	–	0.2 1.09	–
Organic P. (ppm)	–	–	–	–	–	–	–	–	–	–	–	–
Parameters												
Physical Temperature 0ºC	–	–	30.3- 31.1	30.5- 31.0	28.1- 29.5	28.5- 29.2						
Turbidity (units)	–	–	–	–	–	–						
Total Solids (ppm)	–	–	–	–	–	–						
Suspended Solids ppm	–	–	–	–	–	–						
Chemical pH	–	–	7.3- 7.6	7.3- 7.6	7.3- 7.9	7.3- 7.9						
DO (ppm)	–	–	0.5- 1.48	0- 2.04	0.48- 1.87	0.5- 1.53						
BOD (ppm)	–	–	70- 125	100 220	19- 58	20- 50						
COD (ppm)	–	–	175- 250	150- 200	100- 180	60- 80						
OC (ppm)	–	–	60- 75	60- 100	40- 50	–						
Alkalinity (ppm) (as $CaCO_3$)	-	–	135- 150	104- 152	180- 172	150- 178						
Salinity (‰)	–	–	0.795- 1.05	10.8- 22.5	0.54- 0.756	2.6- 4.5						
Sulphate (ppm)	–	–	–	–	–	–		415				
Sp. Conductivity (m.mhos/cm)	–	–	973- 998	–	–	954- 4782						
Ammonia-N (ppm)	–	–	–	–	–	–						
Alb. Ammonia-N (ppm)	–	–	–	–	–	–						
Nitrite-N (ppm)	–	–	–	–	–	–						

Contd.......

Table 3 – contd..

	1954–55						*1967**					
	Winter		*Summer*		*Monsoon*		*Winter*		*Summer*		*Monsoon*	
	U	*L*	*U*	*L*	*U*	*L*	*U*	*L*	*U*	*L*	*U*	*L*
Nitrate-N (ppm)	–	0.45	0.45-2.5	0.4-1.2	0.05-0.25	–						
Inorganic P. (ppm)	–	–	– 0.45	0.08-0.56	tr.-0.84	0.08-						
Organic P. (ppm)	–	–	–	–	–	–						

U-Upper stretch (above Ghusighat) : L-Lower stretch (below Ghusigha-Nazat)

* Published in 1970

1967-1968 - River Research Institute. W.B. (1970) Publ. No. 36

1954-55 - David A. (1959)

1974-75 - NEERI (1978)

1982 - Ray *et al.* (1981)

3.1 Physicochemical Parameters

Water temperature : Except seasonal fluctuations no longitudinal temperature gradient has been observed. Seasonal fluctuation is not very appreciable as is observed in the case of the Hooghly estuary. A temperature range 25.5°C - 31.0°C during 1979-80 was recorded by Ray *et al.* (1981) in a stretch of 60 km between Beliaghata-Kharibari in the north and Bermazore in the south of the eatuary.

Turbidity : Sand, silt (Particles below 0.02 mm) and organic suspension are the factors responsible for causing turbidity in the water. These are washings from catchment area during the wet season and scouring effect. David (1959) recorded turbidity level between 300-600 and 900-1100, while Ray *et al.* (1981) observed similar ranges with the upper limit reaching as high as 1500.

Total Solids and suspended solids : The Kulti being a small estuary, its tidal impact is very much felt. This reslts in high scouring effect (maximum velocity 1.45 and 2.44 m per sec. recorded below outfall flow and ebb currents). Since the estuary is receiving the metropolitan wastes regularly, its total solids are estimated to be high, particularly during the monsoon months. Suspended solids ranged 88-490 ppm during 1967 and 47-924 ppm during the following year (River Research Institute report, 1970). NEERI (1978) observed a range 52-704 ppm during 1974-75 when the corresponding values for total solids recorded were 664-12996 ppm. From this it is evident that dissolved solids are significantly high.

pH : David (1959) recorded pH variation between 7.1 and 8.0 although the values mostly ranged between 7.1 and 7.3. During 1967, considerable improvement was noticed as the pH mostly varied between 7.2 and 8.2 but again followed by a tendency of fall in 1968 (pH mostly ranged between 7.2 and 8.0). NEERI (1978) observed considerable improvement again during the period 1974-775 as the pH values were mostly touching the level 7.9-8.0. Subsequently, Ray *et al.* (1981) again registered a slight depression (pH : 7.3-7.9).

Dissolved Oxygen : The fact that the ratio of dilution of non-conservative waste water of the metropolis to Kulti water in different seasons varies does not seem to hold good under the existing condition. David (1959) alhough observed a range of 0.8-6.0 ppm but concentration of DO was mostly registered below 3 ppm which on a number of occasions indicated values as low as 2.0 ppm or even less irrespective of any tidal influence. Twenty years later during 1974-75, NEERI (1978) observed some improvement as the concentration in general ranged 2-4 ppm ; but subsequently Ray *et al.* (1981) observed further deterioration as DO hardly could cross 2.0 ppm. Diurnal variation of DO (0.48-1.04 ppm) studied below the confluence clearly supports this statement.

To understand the nature of organics of the waste demanding oxygen, it has been observed that the time required for complete oxidation of the waste water. Samples of waste mixed with water from below the confluence were kept in a glass aquarium (25 lit. capacity) and exposed to the sun. Periodical sampling was done to estimate pH and DO. Result shows that during summer, DO after 84 hrs. attained 5.8 ppm from initial value of 1.28 ppm and pH from 7.6 to 8.0. In winter, however, DO attained is 5.16 ppm from 2.0 ppm in 88 hours. From the graphical interpolation the concentrations of DO in 12 hours comes to 1.7 ppm and 1.55 ppm respectively during summer and winter if the water remained undisturbed (cf. Fig. 6). In case of river water flows undisturbed and dilution unaffected, in 12 hours time, the gain in DO will be around 1.7 ppm and 1.55 ppm respectively during summer and winter months when high tide starts. By the turn of tide there may be some tendency of

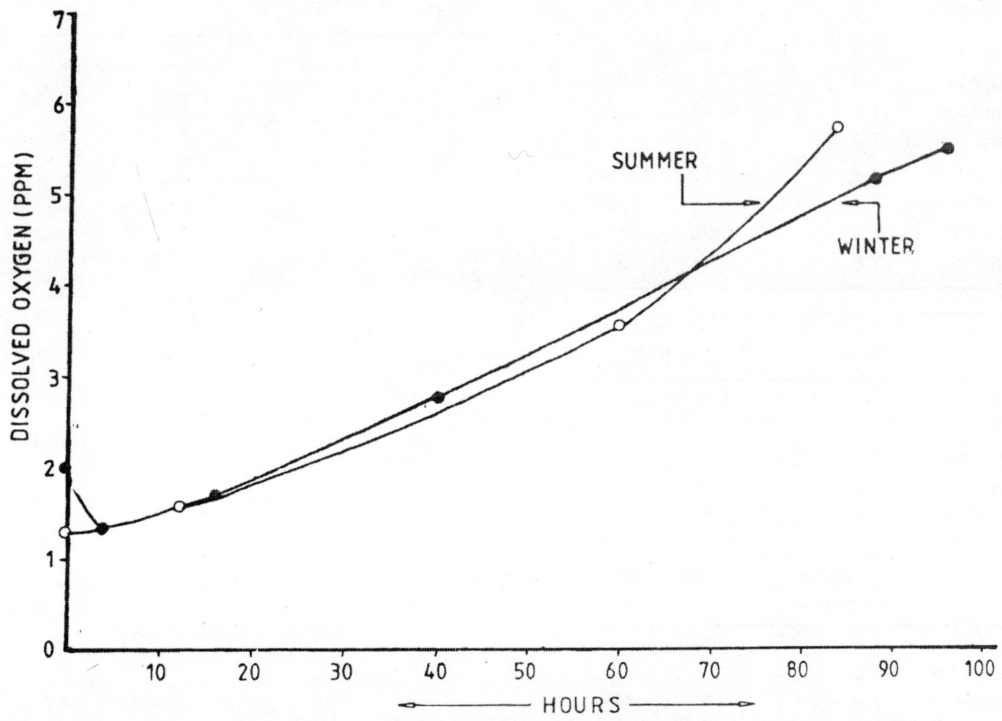

Fig. 6 : State of oxidation of reduced materials of the Kulti estuary as reflected on dissolved oxygen concentration

improvement in the DO content by dilution and aeration, but simultaneously the effect o addition of more of effluent in the water again prevents from increase in DO. Low value of DO in the estuary may thus be attributed to combination of several factors, *e.g.* ratio o dilution (sewage and river water), turbidity, Oxygen demand by non-conservative organi material and greater period of effluent retention. The above findings show that the estua hardly gets any chance of improving the DO content. The pH and DO profile (cf. Table 4 estimated in the Kulti estuary during monsoon when maximum dilution occurs justifies thes findings. Organic load (BOD_5) was estimated for the first time by NEERI (1981) during 1974-75 in a short stretch around the out fall. Values indicated higher in winter (14-24 ppm) an lowest during summer (4-11 ppm) : but subsequently Ray *et al.* (1981) registered much highe values (12-220 ppm) in the entire stretch of the estuary, summer value being higher (mean 70-220 ppm) estimated around the out fall only.

Table 4 : Hydrogen-ion concenration and dissolved Oxygen profile in the Kulti estuary during monsoon

	Beliaghata pH		*Haroa DO (ppm)*	
	HT	*LT*	*HT*	*LT*
Beliaghata	7.3	7.3	0.6	0.6
Palti	7.4	-	0.6	-
Sandalia	7.5	7.5	0.6	0.6
Batra	7.4	-	0.6	-
Ghosalabad	7.5	-	0.6	-
Tehata	7.6	7.6	0.56	0.64
Kharubata	7.6	7.5	0.56	0.56
Ganganaya	7.4	7.4	0.48	0.52
Dhanlat	7.6	-	0.52	-
Khaspur	7.4	7.4	0.72	0.52
Akandaberia	7.4	7.6	0.76	0.8
Sankarpur	7.4	-	0.76	-
Haroa	7.6	-	1.14	-

	Harda		*Above Outfall*	
Haroa	7.6	-	1.14	-
Laterbagan	7.6	-	1.2	-
Ranigachi	7.6	-	1.2	-
Kistopur	7.7	-	0.85	-
Above O.F.	7.4	7.3	1.53	1.32

	Below Outfall		*Malancha*	
Below O.F.	7.0	7.3	nil	1.12
Bamanpukur	7.5	-		
Minakhan	7.5	-		
Malancha	7.5	-		

	Kharibari pH		*Haroa DO (ppm)*	
	HT	*LT*	*HT*	*LT*
Kharibari	7.5	7.6	0.6	0.6
Gangalia	7.5	7.6	0.56	0.52
Belghata	7.6	7.6	0.6	0.52
Telighata	7.6	7.6	0.52	0.52
Haroa	7.6	-	1.14	-

Note-indicate no sampling was done.

Chemical Oxygen Demand (COD)

The oxidation of Carbonacious and other inorganic compounds present in the water and measured by chemical oxidation (dichromate method) seems to be quite high. Other than BOD_5, COD also is responsible for removing oxygen from the ecosystem. NEERI (1978) estimated during winter and summer respectively in 1974-75. Subsequently Ray *et al.* (1981) estimated higher values than what were observed by NEERI. In summer, the values were higher (150-250 ppm) than monsoon.

Oxygen Consumed (OC)

David (1959) estimated values of oxygen consumed between 11 and 32 ppm under all tidal conditions. The mean values at high and low tides were 16.16 and 20.75 ppm respectively. Comparatively higher values (mostly 16-32 ppm) were estimated by River Research Institute (1970) around the outfall. Such values were much lower (8-12 ppm) at Haroahat, about 8 km away from the outfall. Pramanik and Roy (1972) observed a range between 20 and 25 ppm. Subsequently Ray *et al.* (1981) estimated values ranging between 60 and 100 during summer covering the entire stretch of the estuary. Duing monsoon a slight fall was noticed. The above values clearly indicate the presence of high concentration of organic matter at all times.

Alkalinity

Moderatity high values of bicarbonate alkalinity were recorded in summer and monsoon while in winter the values further David (1959) observed low tide values to be slightly higher (mostly 250-290 ppm) than the high tide values (190-250 ppm). River Reseach Institute (1970) estimated values between 250 and 350 ppm during 1978, while NEERI (1978) observed values much higher during winter (340-368 ppm) and lower during monsoon (127-175 ppm) around the outfall. Ray *et al.* (1981) registered values ranging between 104 and 178 ppm in the entire stretch of the Kulti estuary. Seasonal effect was not very much appreciable.

Salinity

The Kulti estuary has a wide range of fluctuation in salinity. David (1959) observed ranges. Trace to 17,000 ppm (Tr-30.69%) and 880-11,700 ppm (1.59-21.1%) respectively during high and low tide around the outfall. River Research Institute (1970) estimated equally high values, 600-16,800 ppm (1.03-30.3%) around the outfall which ws lowered to 424-12,000 ppm (0.756-21.6%) at 8 km up the estuary near Haroa, while in the following year the values were 608-16,800 ppm (1.09-30.3%) and 780-16,000 ppm (1.41-28.88%) respectively in the above order. Pramanick and Roy (1972) recorded seasonal change ranging between 500 ppm (0.9%) (October) and 20,000 ppm (36.1%) (June). While NEERI (1978) reported high values (2215-6235 ppm) (3.99-11.25%) during summer, followed by winter, 312-670 ppm (0.56-1.21%) and monsoon. 82-150 ppm (0.148-0.271%). Such values undoubtedly were much lower than the previous records. Subsequently, Ray *et al.* (1981) recorded vlaues which were much higher than what was recorded earlier by NEERI (1978). The summer values in the entire estuary ranged between 795-22,500 ppm (1.43-40.6%) against 540-4500 ppm (0.97-8.12%) in monsoon. The lower values recorded were from the upper stretch.

Sulphate

Fairly high concentration of Sulphate (125-415 ppm) was estimated during summer, followed by winter (43-64 ppm) around the outfall by NEERI (1978).

Nutrient Status

Nitrogen and phosphorus are the important nutrients of a water body. The main sources are run off from the catchment, metropolitan wastes and decomposition of organic material. The different stages of nitrogen present in the Kulti estuary are ammonia, albuminoid ammonia, nitrite and nitrate. NEERI (1978) reported highest value for ammonia (8.5-11.6 ppm) around the outfall during winter, followed by summer (1.12-1.68 ppm) and monsoon (0.21-1.63). The same trend was observed in case of nitrate-N, 0.012-0.3 ppm in winter, followed by 0.044-0.206 ppm in summer and 0.01-0.87 ppm in monsoon. Nitrate-N, on the other hand registered highest value during summer, 0.45-3.5 ppm, followed by monsoon, 0.05-0.265 ppm and winter, 0.07-0.23 ppm. Earlier, David (1959) also recorded high values of albuminoid ammonia ranging 1.0-4.0 ppm during high tide and 2.0-10.6 during low tide around the outfall. Likewise nitrate-N values were 0.4-1.2 ppm and 0.26-1.1 ppm during high and low tide respectively. Subsequently, Ray *et al.* (1981) estimated concentrations ranging 0.40-2.5 ppm in summer and 0.05-0.25 ppm in monsoon in the entire streth of the estuary.

Phosphorus also indicated moderately high values. David (1959) observed concentrations ranging 0.16-0.4 ppm and 0.25-1.0 ppm during high and low tides respectively as organic phosphorus around the outfall. Values reported by NEERI (1978) for the year 1974-75 estimated were 1.74-3.74 ppm as P or 4.04-8.68 ppm as PO_4. This was followed by monsoon (0.2-1.09 ppm-P) of 0.464-2.53 ppm as PO_4 and summer (0.18-0.63 ppm-P) or 0.42-1.46 ppm-PO_4). Ray *et al.* (1981) on the other hand observed values ranging between Trace and 0.56 ppm, and 0.08-0.84 ppm as PO_4 during monsoon in the entire stretch of the Kulti estuary. The above values clearly indicate that the water of the Kulti estuary is highly rich in nitrogen and phosphorus fertilizers of cause eutrophic condition in the ecosystem. With all these it is still felt that there is a great dearth of data on the nutrient status of the estuary in time and space.

Heavy Metal

Concentration of heavy metals like Lead, Copper, Zinc, Cadmium, Nickel, Cobalt, Chromium, Manganese etc. were estimated in water and also in the precipitated soil of the estuary at certain points, *viz.*, Haroa (8 km above outfall). Below outfall, Malancha (5 km below outfal) and Nejat (25 km below outfall) during summer. The concentration of the above parameters were as follows (cf Table 5).

In the absence of any data for such metals in the water medium it is difficult to predict their actual impact on the aquatic biomass. However it is justified to presume that some percentage will remain in the water phase which is likely to increase or decrease depending upon the concentration of heavy metals in the sewage and scouring effect of tides and chilation. Adequate data on the above parameters both for water and soil phase would have provided valuable information of the ecosystem.

Table 5 : Concentration of metals in water and soil sediments of the Kulti estuary between Kharibari and Najat.

	Water Phase (ppm)	*Soil sediment (mg/kg)*
Lead	-	11.65-40.16
Copper	0.01-0.07	20.16-24.27
Zinc	0.036-0.73	62.24-130.09
Cadmium	-	ND
Nickel	-	24.00-27.63
Cobalt	-	11.46-13.44
Chromium	ND-0.2	41.93-53.48
Manganese	-	380.45-455.65

N.D. - Not detected.

\- - Data not available.

3.2 Biological Characteristics

Plankton form an important food of many small and large fishes in the estuary. The principal planktonic forms in the Kulti estuary are marine, brackish and freshwater forms (Table 6). David (1959) in his detailed qualitative study observed that plankton in general was poorly represented. He observed that among Protozoans, Rhizopoda was most frequently occuring. Two to three species of *Difflugia* were recorded almost round the year. Except one species, each of Foraminifera and Helizoa (marine origin) all other protozoans were conspicuously absent.

No copepods were recorded during July and December indicating a remarkable feature in the estuary, which is attributable to ill effects of pollution.

Cladocerans were entirely unrepresented. Zoea and Mysis were present only during June and July of 1954 and May and September (dead) of 1955.

Megalop larvae (*Varuna Literate*) were found dying in large numbers during low tides of July and August and perhaps were affected by pollution but remained unaffected during high tides.

Mycelial threads of *Beggiota* and *Zooglea* were found even in high tides. Numerous spares of sewage fungi and *Mycellium* were found in the polluted water during low tides. Besides, organisms of sewage origin *viz.*, Nematode worms, dislodged maggots etc. were collected. A number of aquatic insects from freshwater canals were washed into the estuary.

In the case of phytoplankton, various blue green algae were found time to time. *Anabaena, Spirulina, Microcystic, Oscillatoria* and *Nostoc* were abundant.

Not more than 11 genera of Diatoms were recorded and *Coscinodiscus* was the most frequently available genera. *Hemidismus*, a marine form was recorded on several occasions during high tides. Unstable water and its opacity associated with polluted water are inhibitory factors for an abundant production of Diatoms.

Green algae were represented by *Closterium, Pediastrum, Cosmarium* during monsoon. Desmids were flushed into the estuary during monsoon.

Zygotes of algae, zooglea of *Microcystis,* fungal spores etc. were recorded almost consistently in the Kulti. The seasonal appearance of plankton as studied by David is summarised in Table 6.

About 20 years after Ray *et al.* (1981) made a thorough study, both quantitatively and qualitatively, of plankton in the entire stretch of the Kulti estuary. It was observed that plankton concentration was poorly represented. Phytoplankon ranged between 52 and 149 unit/1 and Zooplankton 13-104 units per liter during monsoon but during winter this density further lowered to phytoplankton 11-31 units/1 and zooplankton 2-16 units/1. The dominant species being *Oscillatoria* and *Coscinodiscus, Spirulina, Nitzschia* etc. among phytoplankton and *Filinia* sp., *Disthyla* sp., *Brachionus* sp. and *Nauplii* etc. among zooplankton.

Table 6 : Plankton Organisms of the Kulti Encountered during Different Seasons (1954-1955)

Fauna	*Winter (Dec-Feb)*	*Summer (Mar-June)*	*Monsoon (July-Oct)*	*Flora*	*Winter*	*Summer*	*Monsoons*
Protozoa				Cyanophyceae-(Blue green-algae)			
Rhyzopoda							
Arcella spp.(2)		x	x^2	*Microcystis* sp.		x	x
Difflugia spp (3)	x^2	x^2	x	*Merismopedia* sp.			x
Nebela sp	x			*Coelosphaerium* sp.	x		x
Foraminifera		x^2	x	*Spirullna* sp.	x^2	x^2	
Heliozoa		x^4		*Oscillatoria* sp.	x		x^3
Mastigophora				*Nostoc* sp.	x	x	x
Euglena spp	x		x	*Anabaena*	x^2	x^4	x
Infusoria							
Vorticella							
Porifera				(P) Bacilaria-aceae (Dictoms)			
Sponge Spicules	x^2	x^2	x^3	*Coscinodiscus* sp.	x^3	x^4	x^4
Rotefera				*Melosira* sp.		x	
				Cyclotella sp.	x^2		x
Brachionus sp.		x^2	x^3	*Gyrosigma* sp.	x^2		
Notholca sp.	x			*Pluerosigma* sp.			
Conochilus sp.		x^2		*Navicula*-spp	x	x	
				Coconies sp.			
				Nitzschia sp.	x^2	x	
Annelida							
Polycheate eggs	x^2	x^3	x^2	*Fragillaria*	x		x
				Hemidiscus sp.			x
				Synedra sp.		x	
Crustacea							
Copepoda	x^2	x^4	x^3				
Nauplii	x^3	x^3	x	Chlorophyceae (Desmids)			
Mysids		x^2	x	*Closterium* sp.			x^2
Zoea		x^3	x^2	*Cosmarium* sp.			x
Mysis stages		x^3	x	*Spirogyra* sp.	x		x^2

Contd...

Table 6 – *contd...*

Fauna	*Winter (Dec-Feb)*	*Summer (Mar-June)*	*Monsoon (July-Oct)*	*Flora*	*Winter*	*Summer*	*Monsoons*
Megalopa stages		x	x^2	*Pediastrum* spp.			
Gammarus spp.	x	x^2	x^2	Miscellaneous :			
Sagitta spp.	x	x^2	x^2	Fungi-Sewage-fungas complex-(Strands and hyphae)	x^3	x^4	x^4
Hydrozoa							
Hydroid colonies		x		Zooglea of Micrococci-	x^2	x	x^2
Medusae		x		Clupeid Fish larvae		x	
Nematoda				Gobid Fish larvae		x	
Several spp.		x	x^2	Zygotes of Green algae	x		x
Insecta							
Notonecta				Peltate Hair of plants	x	x^3	x^2
Coryxa				Pollen	x		
Nepa	x		x				
Dyliscids			x				
Insect larvae			x				
Maggots (small)			x				
Mollusca							
Veliger larvae			x				
Minute Shells			x^2				

Power of x indicate number of months of occurrence.

Modified from David (1959).

Nair (1944) recorded plant life as scarce and no algal growth was observed up to 8th mile (13 km) beyond which, however, as few species of green algae (*Spirogyra* and *Spirulina*) were recorded as stray specimens which had probably come from the adjoining field.

Littoral and Burrowing Forms

Swarms of bright green *Euglena gigantia* were obtained from the greenish scum over the exposed mud flats at low tides close to the outfall (David, 1959). From his investigation David (1959) further reported that numeruos burrows of the Fiddler Crab (*Gelasimus* sp.) were found except during low sarine condition (August-October). Crabs like *Seasarma, Dotillopsis, Metaplax, Tympanomerus* and *Pachygropsus* are also recorded. They were abundant only below Minakhan. Due to sudden heavy discharge of waters during spring tides, these crabs crossed up stream with rising level. Similarly, *Varuna litterate* and *S. cyllaserrata,* the two important swimming crabs were found dead or in dying condition during low tides. Between May and July, once or twice every year an unusual mortality of crabs occured which adversely affected their breeding and recruitment considerably.

Another important feature is the absence of molluscan fauna. During numerous specific surveys carried out in 15 mile stretch only dead floating shells of freshwater gastropods e.g. *Pila, Belamya, Indoplanorbis* and *Melanoides* were obtained in the marginal debris. Thus extinction of molluscan fauna due to polluted water is not surprising. Oligachaetes were unrepresented for nearly a mile down stream of outfall.

The findings of David were further substantiated by Ray *et al.* (1981) who reported almost total absence of any biotic life in the estuary.

3.3 *Bacteriological Load*

In Water

Bacteriological load estimated as number/ml with regard to TPC. *Stapheureus, Pseudomona, E. coil* and *Vebriocholerac* in the estuary at several centres like Haoroa (HT), Ghusighata above outfall (HT) and below outfall, (L.T.), Malancha (LT) and Nazat (HT) during summer. TPC ranged 1×10^6–7×10^6. *Stepheureus,* 1×10^2–2×10^4 *Pseudomonas* 1×10^2–12×10^2. Below out fall it is absent *E. coli* 1×10^2–3×10^4 and *Vibriocholerac,* not found/doubtful.

In Soil

In the soil sediments around outfall TPC was in the range 1.8–2×10^6, *Stephcureus* sp. 4–7×10^4 and *Psuedonomas* 6×10^4.

In Fish

The estimation in fish was done both in surface washings and the mecerated kidney, liver and heart of *Tilapia mosambica.* TPC of surface washings was 1.8×10^6 while in the macerated kidney, liver and heart was only 7×10^4; Likewise, *E coil* was 2×10^2 and 1×10; *Pseudomonas* : nil and doubtful and *Salmonella* sp. absent in both the cases.

4. FISH AND FISHERIES

The notable species of fish like several species of sharks and rays and mullets, perches, thread fins, clupeids (Herringa), crockers and other prawn and crabs occurred till 1939, the first year of sewage admixture. But definite information on the presence of major fish *Hilsa ilisha,* either as fry or adult, was not available till that period. However, in the absence of any published account of fishes which occurred during prepollution days data on fish production was not available.

Nair (1944) probably, for the first time, observed fish moving around the outfall in distress in a stretch of 5 mils (8 km) on either side. Most of the fishes indicated below are freshwater species, brought from the local paddy fields which thrive in the existing low saline water of the estuary. The fishes found in distress are recorded below.

List of Fishes Recorded

Ophichhus bore	Dying in the zone of pollution
Mystus gulio	do
Pagasius pangasius	No dead specimen-found
Puntius sophore	Dying in the zone of pollution
Esomus danricus	do
Chela sp.	do
Xenantodan-cancila	do
Colisa fasciata	do
Anabas testudineus	Apparently living in comfort

Lates calcarifer	Dying in the zone of pollution
Mugil persia	do
Pseudapocryptes	do
lanceolatus Ophicephalus	Apparently living in sp. comfort.

Ten years later, David (1959) based on his observation of 1954-55 found alarming condition. The estimated total catch in 1954, on the basis of 18 days a month sampling, was 250 lbs (1.1334 t) for the entire 12 miles (19.2 km) stretch between Haroahat and Minakhan. By the end of 1955 the catch further declined and 75-80 per cent of it was *P. pangasius*, *M. gulio* and gobid genera, the rest being stray specimens of uneconomical fish and a few prawns. Table 7 indicates that out of the fishes listed at various times represent 13 per cent of fresh water species washed in from the surroundings of the stretch. Fig. 7 shows how sewage acts as a barrier and trap for fish ascent.

Fish Seed Resources

David (1959) recorded only larval and post larval stages of clupeids, sciaenids, gobids and some zoea and Mysis of crustaceans during high tides in almost dead conditions. During the wet months of July-September, when salinity decreases, *M. gulio* ascended in the fresh water through the Kulti in a fully mature state but young ones of this species (16-29 mm) were collected as dead in sufficient percentage in October and November. Besides post larvae and fry of *Thrissocles* spp., *Setipinna phasa, Mugil parsia, Amphipnous cuchia, Periopthalmus Schlosseri*, several prawns and crabs were also brought in by high tide flood. Estimated number of young bhetki (*L. calcarifer*) noticed in 1954 were found disappearing in 1955.

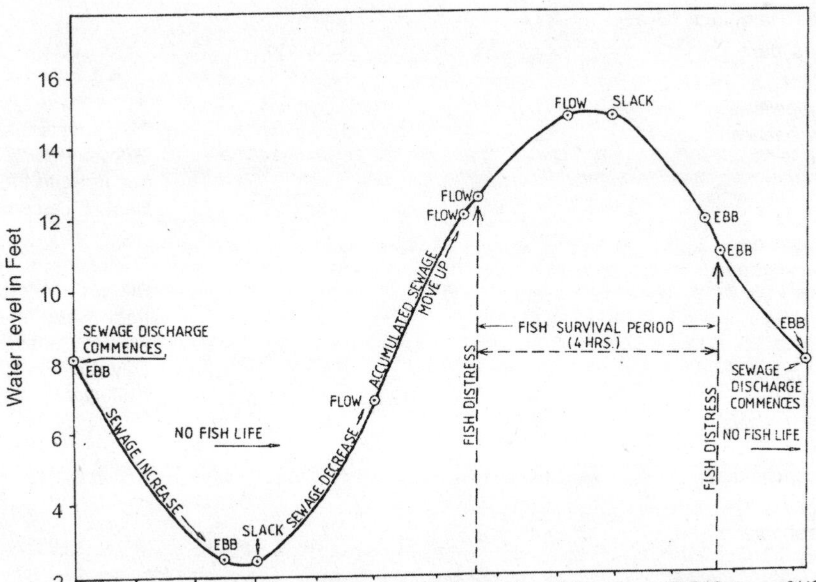

Fig. 7 : Drainage operation cycle under tidal influence acting as a barrier for fish migration.

Table 7 : List of Fishes of the Kulti Estuary.
(Stray-x, very rare-xxx, Rare-xx, Frequent-xxxx. Fry-F, Juveniles-J. Adults-Ad.

No.	*Scientific name of fish*	*How occuring*	*Stage Recorded*	*Remarks*
	Family *Megalopidae*			
1.	*Megalops cyprinoides* (Broussonet)	xxx	Ad	Few at a time noted in-'bheen' nets.
	Family *clupeidae*			
	Sub *family clupeini*			
2.	*Corica Soborna* (Hamilton)	xx	J.Ad.	Serial Nos 2-12 are the clupeids
3.	*Gadusia chapra* (Hamilton)	xxx	J.Ad.	Noted variously in the Kulti.
4.	*Hilsa toli* (Cuvier and valenciennes)	x	J.	(Hilsa ilisha was noted even as fry or Juveniles)
5.	*Pellona elongata* (Bennet)	x	J	
	Sub family Dorosomatini			
6.	*Anadontosoma chacunda* (Hamilton)	x	J	
	Family *Engraulidae*			
7.	*Anchoviella indica* (Van Hasselt)	xx	J.	
8.	*Coilia dussumeiri* (Day)	xx	J.	
9.	*Setipinna phasa* (Hamilton)	xxx	F.J., Ad.	
10.	*Setipinna taty* (Blacker)	xx	J.Ad.	
11.	Thrissocles Hamiltonii (Grey)	xxx	F.J.Ad.	
12.	*Thrissocles pura* (Hamilton)	xx	F., J., Ad	
	Family Scopelidae			
13.	*Harpodon nehereus* (Hamilton)	xxx	J., Ad.	This species is the well known 'Bombay Duck'. generally occurring in the Kulti between Sept. & December
	Family *Cyprinidae*			
14.	*Amblypharyngodon mola* (Ham.)	x	Ad.	These forms between Sl- Nos. 14-21 are the fresh water 'minnow carpa' ususally noticed between July and September when Kulti-water almost becomes fresh.
15.	*Chela laubuca* (Hamilton)	xx	Ad.	
16.	*Esomus barbatus* (Hamilton)	xx	Ad.	
17.	*Oxygaster bacaila* (Hamilton)	xx	Ad.	
18.	*Puntius conchonius* (Hamilton)	xx	Ad.	
19.	*Puntius sophore* (Hamilton)	xx	Ad.	
20.	*Puntius ticto* (Hamilton)	xx	Ad.	
21.	*Rasbora doniconius* (Hamilton)			

Contd.......

Table 7 – contd...

No.	Scientific name of fish	How occuring	Stage Recorded	Remarks
	Family *Tachysuridae*	x x	Ad.	
22.	*Osteogeniosus milltaris* (Linnaeus)	x x	J.	Serial number 22-28 are the Catfish
23.	*Tachysurus* spp.	x x	J.Ad.	species in the Kulti.
	Family *Plotosidae*			Mysius gulio and Pangasius
24.	*Plotosus canius* (Hamilton)	x x	J.Ad.	Pengastus are the only two species of economic value.
	Family Bagridace			
25.	*Mystus cavasius* (Hamilton)	x	J.	
26.	*Mystus gulio* (Hamilton)	xxxx	F.,J.,-Ad.	
27.	*Mystus vittatus* (Bloch)	x x	J.	
	Family *Schibeidae*			
28.	*Pangasius Pangasius* (Hamilton)	xxxx	J	
	Family *Heteropneustidea*			
29.	*Heteropneustes fossilis* (Bloch)	x x	J.Ad.	These two species 29 and 30 are the airbreathing forms often washed into the Kulti from the adjoining freshwater tracts.
	Family *Clariidae*			
30.	*Clarias batrachus* (Lunnaeus)	x x	J.Ad.	
31.	*Anguilla bengalensis* (Gray)	x x	J.	
32.	*Anguilla* spp.	x x	J.	Serial Nos. 31-34 are some of
	Family *Moringuidae*			the 'true' eels recorded.
33.	*Raitoborua raitoborua* (Hamilton)	x x	J. Ad.	Though Sl. No. 33 and 34 are airbreathing they did not survive
	Family *Ophichthydae*			in the polluted estuary.
34.	*Pisodonophys hijala* (Hamilton)	x x	J., Ad.	
	Family *Belonidae*			
35.	*Tylosurus strongylurus* (Van Hasselt)	x x	J.	These are 'Gar' fishes and Half beaks'. They were recorded
	Family *Hemirhamphidae*			mainly in outer channels
36.	*Hemirhamphus gaimardi* (Valenciennes)	x x	J.	connecting the Kulti.
37.	*Oryzias melastigma* (Mc Clelland)	x x	Ad.	
	Family *Mugilidae*			
38.	*Mugil cephalus Linnaeus = M Oeur* (Forskal)	x x	J.	These 'Mullets' have been noted mainly when there was no drainage of sewage.
39.	*Mugil parsia* (Hamilton)	xxx	F.,J.Ad.	
40.	*Mugil speigleri* (Bleeker)	x x	J., Ad.	
41.	*Mugil tade* (Forskal)	x	J.	
	Family *Polynemidae*			

Contd......

Table 7 – contd...

No.	Scientific name of fish	How occuring	Stage Recorded	Remarks
42.	*Eleutheronema tetradctylum* (Shaw)	x	J.	These 'thread-fins' have-occurred only as stray individuals; *P. paradiseus* in July-September only.
43.	*Polydactylus indicus* (Shaw)	x	J.	
44.	*Polynemus paradiseus* (Linnaeus)	x	J.,Ad.	
	Family *channidae*			
45.	*Channa gachus* (Hamilton)	xx	Ad.	These are 'Murrels', also-known as 'Snake heads'- and have occurred in monsoon months.
46.	*Channa punctatus* (Bloch)	x	J. Ad.	
47.	*Channa Striatus* (Bloch)	x	J.	
	Family *Amphipnoidae*			
48.	*Amphtpnous cuchia* (Hamilton)	xxx	F., J.Ad	This 'Mud eel' is recorded all the year round.
	Family *Centopomidae*			
49.	*Ambassis baculis* (Hamilton)	x	J.,Ad	
50.	*Lates calcarifer* (Bloch)	x	J.	This is the 'Bhatkl', a perch, cultivated in the connected 'bheris'.
	Family *Theraponidae*			
51.	*Therapon jarbua* (Forskal)	x	J.	
	Family *Sillaginidae*			
52.	*Sillago panijius* (Hamilton)	xx	J.Ad.	'Crocodile' fish recorded-during monsoon months.
53.	*Caranx Carangus* (Bloch)	xxx	J., Ad.	Serial No. 53-58 are Perch-like fishes occurring all the year round.
	Family *Lobotidae*			
54.	*Loboctes* sp.	x	J.	
	Family *Lieognathidae*			
55.	*Gerres* sp.	x	J.	
	Family *Leiognathidae*			
56.	*Leiognathus equulus* (Forskal)	xx	J.Ad.	
	Family Sciaenidae			
57.	*Pama pama* (Hamilton)	xxx	F.,J.,	These are 'croakers recorded from fry to juvenile stages in August-September.
58.	*Pseudo sciaena coltor* (Hamilton)	xx	F.,J.,	
	Family *Scatophagidae*			
59.	*Sactophagus argus* (Bloch)	x	F.	
	Family *Nandidae*			
60.	*Nandus nundus* (Hamilton)	x	J.	
	Family *Trichiuridae*			
61.	*Trichiurus savala* (Bleeker)	xxx	F.,J	Well-known 'Ribbon' fish-occasionally seen in the nets (marine form)

Contd.......

Table 7 – contd...

No.	Scientific name of fish	How occuring	Stage Recorded	Remarks
	Family *Stromoteidae*			
62.	*Stromoteus sinensis* (Day)	x	J.	Marine pomfrets occurring-'bheen nets'.
	Family *Anabontidae*			
63.	*Anabas testudineus* (Bloch)	xx	Ad.	'Climbing Perch' mainly a-fresh water form, recorded-in the Kulti occasionally.
64.	*Colisa fasciatus* (Bloch)	xx	J.Ad.,	
	Family *Eleotridae*			
65.	*Buttis butis* (Hamilton)	xx	Ad.	Sl.Nos. 65-73 constitute Gobid fishes and next to - Pangasius and Mystus-gulio together form sizeable fishery of some meagre-value.
66.	*Eleotris fusca* (Bloch)	xx	J.Ad.	
	Family *Goidea*			
67.	*Apocryptes bata* (Hamilton)	xxx	F.,J., Ad.	
68.	*Brachygobius nunus* (Hamilton)	xx	Ad.	
69.	*Glossogobius giuris* (Hamilton)	xx	F.J., Ad.	
70.	*Cobiopterus chuno* (Hamilton)	xx	Ad.	
71.	*Odontomblyopus rubicundus* (Hamilton)	xxx	F.,J., Ad.	
72.	*Pseudapocryptes lanceolatus* (Bloch)	xx	F.,J., Ad.	
73.	*Stigmatogobius sadanundio* (Bennet)	x	J.	
	Family *Periopthalmidae*			
74.	*Boleopthalmus baddaerti* (Pallas)	x	J.,Ad.	
75.	*Periopthalmus schlosseri* (Pallas)	x	J.	Common 'Mud Skipper'
	Family *Platycephalidae*			
76.	*Platycephalus insidiator* (Linnaeus)	xx	J.	
77.	*Cyanoglossus lingua* (Hamilton)	x	J.	'Flat Fishes'
	Family *Mastacembelidae*			
78.	*Macragnathus aculeata* (Bloch)	xx	J.,Ad.	
79.	*Mastacembelus armatus* (Lace'pe'de)	xx	F.,J., Ad.	
	Family *Tetrodontidas*			
80.	*Tetrodon* spp.	x	J.	

(No single species can be denoted as occurring 'Ver Frequently', 'Commonly' or 'Abundantly' during the studies); David (1959)

Thirteen years later in 1970 Thakur (1975) carried out investigations from September 1967 to May 1968, the main period of breeding season, to find out the seed potentiality of the Kulti estuary in a stretch of more than 20 miles (32 km) between Kharibaria in the north and Kalinagar in the south. The study revealed that the upper stretch (above outfall) did not contain any seed of commercially important fishes except some fry of *Glossogobius giuirs, Oryzias melasligma, Puntius ticto, Ambasis baculis, Panchax panchax, Mastacembelus armatus, Hemirhamphus gaimardii*, juveniles of *Tilapia* sp. and Mysids. In the lower stretch different brackishwater fishes recorded were *Clupeia tri, G. giuris, P. ticto, A. baculis, Esomus danricas, Amblyphoryngodon* sp. *O. melastigma, Peliona motius, Anguilla* sp. *Cynoglossus* sp., gobids, *Acetes indicus*, mysids, crabs and megalop larvae. Seeds of *P. indicus*. P. monodan, *L. styliferus*, and *Mugil parsia* were available in considrable quantity. The seed potentiality of Kulti along with some physicochemical parameters are indicated in Table 8. Thakur interpreted the availability of prawn coinciding with the increases in salinity during January to April in the lower stretch. Larvae and juveniles of *M. parsia* ranged from 9.0-25 mm. the author suggested that low DO alone or associated with comparatively low salinity in the super zone were probably the factors for non-availability of prawn fry.

Table 8 : Seed Potentialities of the Kulti Estuary along with some Physicochemical Parameters.*

Species of fish	*January*	*February*	*March*	*April*
P. indicus	H-2.88 L-3.45	H-14.80 L-9.06	H-17.37 L-16.82	H-15.69 L-23.16
P. monodon	H-6.61 L-5.24	H-6.02 L-5.36	H-4.61 L-3.63	H-2.10 L-2.94
L. styliferus	H-2.98 L-2.11	H-7.08 L-4.90	H-14.81 L-8.35	H-17.92 L-12.00
M. brevicornis	H-0.86 L--	H-0.80 L--	H-1.95 L-2.15	H-6.28 L-6.94
M. affinis	H-- L--	H-0.50 L--	H-1.46 L--	H-- L--
P. sculptilics	H-- L--	H-- L--	H-6.33 L-2.39	H-- L--
Palaemonid post Larve	H-12.67 L-10.16	H-10.85 L-7.40	H-1.09 L-1.92	H-0.52 L--
M. Parsia	H-16.67 L-15.60	H-8.95 L-6.60	H-1.86 L-2.37	H-6.04 L-4.48
Setipina Sp.	H-4.07 L-2.48	H-2.40 L-1.50	H-- L--	H-- L--
Polynemids	H-- L--	H-- L--	H-1.66 L-2.69	H-0.26 L--
P. Canins	H-0.80 L--	H-- L--	H-- L--	H-1.38 L--
M. gullo	H-- L--	H-- L--	H-0.79 L-1.30	H-- L--
I. elongata	H-- L--	H-- L--	H-- L--	H-1.92 L-0.92
H. nehereus	H-- L--	H-- L--	H-- L--	H-1.48 L-2.29

H = High tide; L = Low tide; * Data based on September and December 1967 and May 1968. Results are expressed in percentage by volume.

@ *Physico-Chemical Characteristics* (Recorded during 1968)

SS-(ppm)	72-340	104-382	102-896	48-1033
Depth (m)	6.0-7.6	6.1-7.8	7.0-8.0	7.0-7.7
pH	7.2-8.0	7.0-8.0	7.3-8.1	7.2-7.8
Alkalinity	295-402	295-335	295-375	342-362
Salinity (ppm)	2000-3280 (3.6-5.3%)	4000-5680 (7.2-10.2%)	5760-11440 (10.36-18.3%)	9520-15200 (17.14-27.36%)
O.C :- (ppm)	13-22	19-29	8-16	12-14

*Thakur, N.K. 1975 (*J. Inland Fish Sec.* India) Vol. VII. Dec.

@ River research Institute, 1970 Govt. of West Bengal Annual Report No. 36, Directorate of irrigation and water ways.

Ray *et al.* (1981) observed moderate number of juveniles of *P. pangasius* and *M. gulio* at the lower zone, 5-8 km below the outfall. Around the recovery zone, at Nazat, prawn seed collections were made during 1979-80 ranging between 1.906 million and 33.877 million of seed per four hours per month, the peak period being December (10.961 m) to June (17.310 m) attaining maximum in March. Centre-wise monthly total estimated number of prawn seed during 1979-80 are given in Table 9.

Table 9 : Centrewise Monthly Total Estimated Number of Prawn Seed in 1979-80 (in 1000) Centres

Month & Year		*Nazat Complex*	*Nazat local Kalinager and Ghoshpur*	*Percent*
November	1979	9,121	5937	65
December	1979	21,466	10,916	51
January	1980	16,669	13,609	82
February	1980	68,442	28,366	41
March	1980	96,154	33,877	35
April	1980	60,349	21,553	36
May	1980	45,685	22,826	50
June	1980	36,550	17,310	47
July	1980	10,921	5,537	51
August	1980	4,459	1,906	43
Total		3,69,816	1,61,837	44

The rapid survey made by Department of Fisheries, Government of West Bengal, in 1981 (Report of the Fact Finding Committee, 1984) revealed that the Bidyadhari river (including Chetokatatgachi, Ghatihar and Barrokalagachi) was the major source for prawn seed collection. During peak season about 48 lakhs of prawn seed were collected per by with the help of 3900 nets. The other sources were stretches from Hingleganj to Sahebkhali, Sahebkhali to Raimangal and Hasnabad to Barunhat of Kalindi river and from Chaktangramari to Kumarmari of Dasa river.

5. OVERALL IMPACT ON BIOTA

The industries contributing towards unfavourable complex condition of the waste water e.g. colour from electroplating, textile, paper, tanneries etc., suspended solids, from paper and tanneries; sulphate from tanneries textile, paper etc. and metals from electroplating, metallurgical-ferrous and non-ferrous galvanising, tannery, battery cell etc. are present in the metroplitan city. Such wastes when disposed into the Kulti create an unfavourable condition in the ecosystem and affect survival, growth, maturation, reproduction and migration of many biotic life.

Ray (1981) has correctly stated that the impact of pollution depends on littoral, vertical and longitudinal mixing as determined by turbulence. The turbulence is dependent on discharge, slope, depth, canal roughness, configuration and wind action. He further observed that the nature of toxins, dilution, mixing pattern, retention of effluents, duration of tide, tidal phase, natural regeneration and seasonal fluctuation are some of the important factors whose combined effect is very complex and is discussed in the context of pollution hazards of the Kulti estuary.

With the complete suspension of freshwater flow from upstream gradual deterioration occurred an account of silting in the Kulti estuary. Such siltation, particularly above the outfall region, was quicker. This resulted in shrinkage of the size of the estuary. Effluents accumulating in this region gradually increased in concentration. The saline water of flood could not penetrate upward, as it could do in early years, and resulted in pollution hazards. Such hazards gradually extended down stream beyond the outfall. The overall impact of such waters on pH is clearly indicated as the value of Kulti water ranging mostly between 7.3 and 7.6 and on a rare occasion could attain 8.0. Dissolved oxygen is very much affected especially in recent years, so much so that even in the arterial channels Nowii and Suntee (up stream) of the Kulti never rose beyond 1.0 ppm DO even during monsoon when highest dilution was experienced. Even in the main Kulti estuary beyond Malancha (down stream) on rare occasions the value remained just a little above 1.0 ppm. The pH and DO profile in the entire stretch of the Kulti (cf Table 4) recorded during both the tides coincides with the above findings. Perhaps the longer retention period of effluent has a major role in keeping the DO at a lower level. Such low oxygen level was also noticed in the years 1954-55 by David (1959) and pollution in the estuary was attributed to lack of oxygen and not from any bacterial, fungal or parasitic attack or from chemical coagulant which resulted in periodic fish kill (swollen nature of the gill and distended opercular cavities) and their fry and crustaceans at various stages brought up in swarms by currents contributing to a large amount of loss all round. Added to it was the high suspension due to scouring effect of ebb and flood water (Pramanick and Roy, 1972; Ray *et al.* 1981). Such factors were quite unfavourable for the survival, growth and reproduction of any microorganisms. Plankton were significantly reduced during 1954 (David, 1959) so much so that almost none of the brackishwater/marine forms indicated by Biswas (1927) and Sewell (1934) around Bidyadhari spill area or in arterial Channels of Bidyadhari was present (cf. Tables 18 & 20, in Culture Fishery).

Such condition persisted till recent years as is evidenced from plankton count and organic carbon production. Ray *et al.* (1981) opined that whatever plankton was estimated in the Kulti estuary were brought down from the neighbouring plankton-rich bheries when the bhery waters were released into the river prior to intake of nutrient–rich river water. The environmental condition of the river is not congenial either for plankton or biota to grow and reproduce. It is hence, suspected that all the plankters that come out with the discharged water from bheries could not survive and reproduce in the river although if contained enough of nutrients. Three hourly diurnal variation of pH and DO studied below the outfall indicated that pH was maintained at 7.6 level and DO ranged between 0.52 and 1.64 ppm. It has been experimentally observed under laboratory condition that the waste of the outfall area having initial DO 1.28 and 2.0 ppm during summer and winter respectively could attain DO values of 5.8 and 7.2 ppm in 84 120 hours. All the above findings clearly indicate that the existing condition is heading towards septic in the major stretch of the Kulti estuary and with the passage of time progressively approaching downstream.

The fact that the estuary is not congenial for fish to migrate and reproduce is evident from the observations of Nair (1944) as most of the fishes caught in the net were found to be dead or dying in a stretch of 8 km on either side of outfall. The freshwater species caught were brought from the adjoining paddy fields. The salinity of the estuary in the area being low and the fishes being hardy could tolerate such conditions and survived. David (1959) presented an interesting picture of fish movement along the flood tide (Fig. 3.3). At every 4/5 hours of continuous discharge of sewage in the estuary the supply got disrupted due to increase in the water level on account of flood tide. No fish could ascent till that time. After about 6 hrs. 45 mts. from initial time of discharge, with further rise in the water level and consequent dilution of sewage, the fish although ascended experienced some distress for a short while. With a further increases in the water level this problem disappeared. Fish could survive for a period of about 4 hours only. Thus, again after 10 hrs 45 mts. from the initial time when water preceded considerably due to ebb tide fish again indicated sign of distress until they could move elsewhere. Such artificial barrier has definitely restricted the up stream movement of fish to a great extent. Leaving *P. pangasius* and *M. gulio* this resulted almost in fishless conditions for a greater part of the estuary.

Nair (1944) conducted bioassay experiments with raw sewage using major carp fingerlings (Rohu, Catla and Mrigal) as test animals and showed that fishes proved to be able to live even in the raw sewage during July-August, provided the raw sewage was kept agitated and properly aerated. He further observed that the physiological responses of fishes depended upon the amount of DO present in the sewage and that the reactions took the form of respiratory embarrassment which, however, subsided as soon as the medium was mechanically aerated. He concluded that fishes provided with accessory respiratory organs tolerated the sewage throughout the canal. Hence, it is clearly that Calcutta sewage has no chemical toxic effect on fishes. Almost a similar result was achieved by Ray *et. al.* (1981) who recorded that the test fishes (Rohu, Catla and Mrigal) could survive for more than 10 days at 5 ppm DO level without any distress. But mere survival of fishes did not mean that they were not physiologically affected due to the presence of heavy metals.

In the water phase although the presence of zinc, copper and chromium was not alarmingly high, in the soil sediments, however, of the metals, zinc, lead, copper, nickel, cobalt, chromium and manganese were present in appreciable concentrations (Table 5).

Hence, bioaccumulation of such metals (even in sublethal level) in fish, especially in those provided with respiratory organs as observed in the case of bheri fish at Bantala (Part I), cannot be ruled out. Bloom and Ayling (1977) while investigating the heavy metal concentration in mussels, ediments and filtered water of the Derwent estuary (Tasmania) presented a unique example of this aspect. The physiological effect of such metals in the organism (mussel) is however, not indicated.

From the available data it is clear that Kulti estuary is loosing salinity; and there is a gradual increase in organic matter in the water which is responsible for the loss of salinity as evident from Fig. 8.

The decline in fish catch as recorded by Nair (1944) and David (1959) was mainly due to low DO in the water. Such statement was also supported by Ray *et. al.* (1981). Thakur, however, accounted low salinity for non-availability of prawn seed resources. The salinity in the upper reaches recorded during 1968 by River Research Institute (1970) was also quite high particularly during February, March and April when breeding was not an unusual feature and could have been possible. The author feels that in this case low DO may be another important factor which prevented prawn breeding. Prawn, however, have been found to breed profusely around Nazat, the recovery zone (cf. Table 9) where salinity was observed to be lower than the observed figure to Thakur (1975).

6. HEAVY METAL POLLUTION

The fact that heavy metal can cause harmful effect on aquatic biomass is well recognised. Except what has been furnished by Ray *et. al.* (1981, there is hardly any information available on heavy metals in the previous years either in the water or soil sediments of the Kulti estuary or nature of metals disposed along with the waste into the estuary. According to them the waste water disposed into the estuary contained zinc, copper, iron and manganese in the concentrations of 3-56, 9-11, 110-164 and 13.4-60 ppm and trace-5.61, 9.0-20, 100-160 and 0.5-6.5 ppm respectively in the solid fraction. The author in a subsequent survey reported that the ranges of average of lead; copper, zinc, nickel, cobalt, chromium and manganese of the Storm water canal carrying the effluent were 12.78-181.23, 36.78-240.72, 101.90-760.1, 23.19-50.84, 10.95-16.32, 25.35-378-84 and 33.69-444.01 mg/kg of the soil sediments at six centres. There was no information of such metals in the water medium. Although in the water phase of the entire Kulti estuary the metal concentrations were not alarmingly high, their presence in the soil sediments were quite high for some of the metals. Further, the same effluent drained into the bheries near Bantala indicated availability of copper, manganese and zinc in concentrations of trace - 1.5, trace-0.8 and 0.06-0.21 ppm (Culture Fishery A). This clearly indicates that the water possessed all these metals and that their concentrations in the upper stretch must have been higher than the lower zone, as soil sediments around Haroa contained higher values of such metals.

The fact that presence of different heavy metals in low concentration in the water is harmful for aquatic biomass either alone or in combination has been studied by some workers. Venugopalan (1979) from his experiments conducted to study the effects of copper and zinc on growth and photosynthesis of *S. costatum* and *N. longissima* (a marine species) remarked that copper concentration up to 10 μg cu/l (0.01 mg/l) increased growth and photosynthesis in both the species. Although cell multiplication was reduced at copper concentration of 25 μg/*l* (0.025 mg/*l*) photosynthesis was not affected even upto a concentration of 50 μg/*l* (0.05 mg/*l*). He further observed that unlike copper, relatively a higher concentration of zinc, 75 μg/*l* (0.75 mg/*l*) was tolerated by *N. congissima* without any detrimental effect on the growth and photosynthesis, on the other hand. *S. costatum* was

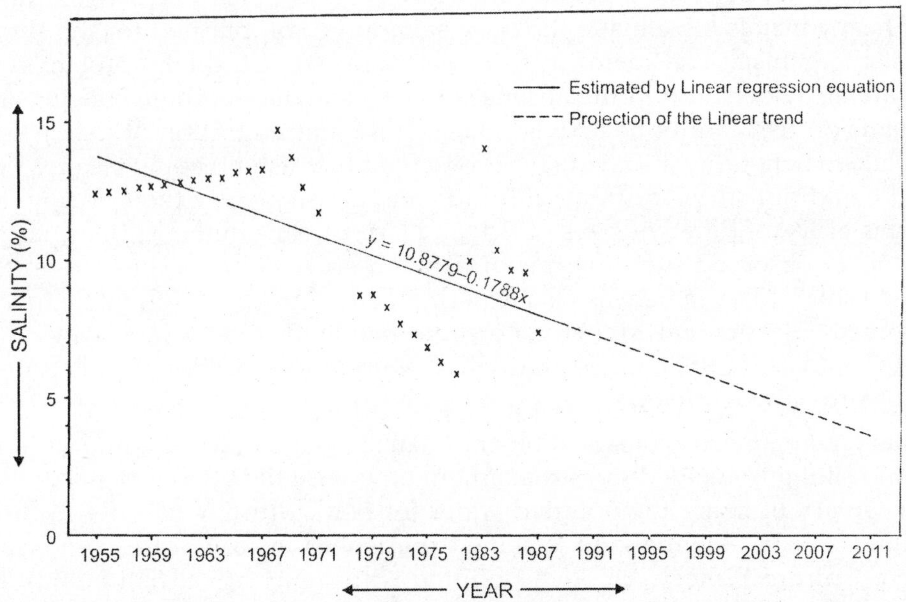

Fig. 8 : Linear trend of salinity during summer (March-June) in the Kulti Estuary (Upper Stretch).

able to tolerate only up to 10 μg/*l* zinc (0.01 mg/*l* zn). These results indicated the existence of ecologically significant intraspecific differences in the physiological and metabolic adaptations of the micro-algae, although they co-existed in the same niche and under same environmental conditions. He further observed that diatoms and other algal forms would revive after exposure to zinc pollution when transferred to a normal-medium, but chances of such revival were remote with copper pollution. Stones (1976) stated that copper, lead, mercury, nickel, silver and zinc if present in 1 ppm concentration can reduce BOD by 44, 23, 72, 25, 81 and 8 per cent or alternately if the water contains 0.2, 1.2, 0.07, 1.0, 0.03 and 7.0 ppm of above metals BOD can be reduced by 25 per cent. In case of fish food organism *Daphnia similis,* the toxicity values (LC^{50}) of Copper was found as 0.06 ppm in moderately hard water (Ray and Bandopadhyay 1984). Water flees *Daphnia magna* are killed within 5-10 days at 0.01 mg/l zinc. Sprague (1964a) observed that in soft water (Hardness, 18 mg/l as $CaCO_3$) the lowest concentration causing average fish to show significant avoidance were 0.053 mg/zn/l for zinc sulphate alone, and 0.006 mg/l for zinc in presence of 0.0004 mg cu/l. Brungs (1969), found that reproduction in fathead minnows was almost totally inhibited at zinc concentrations that had no effect on survival, growth and maturation. The number of eggs produced per female at 0.18 mg Zn/*l* was only 17 per cent of that produced at 0.03 mg Zn/*l*. In tests lasting about 4 months, during which rainbow trout were exposed to concentrations between 0.6 and 2.0 mg Zn/*l* in hard water, there was a small but significant mortality among the fish (Edwards and Brown, 1966). Individually, 8 ppm zinc and 0.2 ppm copper are harmful to fish, but when mixed together can cause synergitic action at a much lower concentration (1 mg Zn + 0.025 mg/*l* cu) (Water Quality Criteria, 1963).

In the estuary, a stretch of about 60 km is found to be highly polluted. From the above findings it would not be wrong to state that low DO, suspension due to scouring, and presence of heavy metals have made the ecosystem more complicated for the survival and growth of microorganisms considered as fish food. In case of fish similar problems including maturation are also likely. But in the absence of adequate data definite conclusions on the effect of metals on fish cannot be arrived at. The fact that migration of fish is very much restricted is adequately proved. David (1959) observed that fish breeding occurred profusely but the seed could not survive due to pollution and consequently there was an all round loss of swarms of fry, of different sizes of fish and Crustaceans in the Kulti estuary. But Ray *et al.* (1981) never observed such swarms of fish and crustaceans of any size in the larger section of the estuary. The same was also experienced by Thakur (1975) while investigating the seed resources. The gradual addition of organic load in the estuay, especially in the upper stretch acting as a vast septic tank, is gradually lowering its salinity and it is progressively extending towards down stream.

The estuary, hence, needs conservation and management so that the impact of pollution is reduced and silting is avoided to its maximum, otherwise the bheries flanked on both the sides of the estuary utilizing the polluted water for fish culture will suffer in future from lack of dilution. This is essential and is being experienced in case of freshwater bheries of Salt lakes (cf. Culture Fishery A). Such polluted water may affect fish health, maturity, recruitment, leading ultimately to loss of total production.

7. CONSERVATION AND MANAGEMENT OF THE ESTUARY

With the complete suspension of upland flow in the Jamuna, a tributory of the river Hooghly, due to silting the connecting small streams (Nowi, Suntee and Nanagang) have to depend entirely on local drainage during monsoon and the spill areas for preservation. Conditions being such and annual rainfall in the area being high (1043 mm in North 24 Parganas) enormous volume of silt is again brought into the estuary every year from the catchment area which instead of being washed down into the sea is pushed back upstream resulting in reduction of depth of water in the Kulti estuary. Added to it large volume of water which is being drawn in the bheries on either side of the estuary for pisciculture purposes, in the recent past, considerably reduces the impact of ebb tide and indirectly helps silting. The estimated higher values of shape factor at 5 centres stretched over the estuary (maximum being at Malancha followed by outfall at Ghusighata and at Nowii-Suntee junction clearly explain the phenomenon. Thus, it is apparently clear that sewage once discharged during ebb flow is again brought back by the turn of tide when some more effluent is added to it. This cycle continues during each tidal phase and as a result the concentration of pollutants gradually increases. Appreciable dilution is effected only during the spring tide phases of particular season of the year; but this is not sufficient to neutralise the toxins. This is how fishery has dwindled and only some species, provided with accessory respiratory organs, like *P. pangasius, M. gulio*, etc. could survive. Even fishery of these species is also greatly affected. Hence, conservation and management of the estuary have become essential to protect both the ecology and the fishery.

In this connection De (1929). Mazumdar (1953) and Maitra (1958) have made valuable suggestions for resuscination and conservation of the Kulti estuary. De's suggestions came when there was appreciable tidal flow in Bidyadhari-Kulti complex. Anticipating silting problem that might result in future, he suggested to construct two large reservoirs, about half a mile long each, on either sides of the estuary. The objective was during flood tide the reservoirs would be full and while ebbing such water would have scouring effect on the deposited silt in the estuary. But no attention was paid to these feeding channels and spill areas resulted. Subsequently Mazumdar (1953) offered another practical suggestion. According to him, the only connection the Kultigong had with the Hooghly river was through Nowii and Ichhapur Khal. Mere improvement of the Khal would not help as the heavily silt laden tides coming up both the Hooghly as also up the Kultigong would meet within the improved Khal and would result in deposition of silt. So, instead, with the help of Ganga barrage project if 20,000 cusecs was allowed to pass down and which was possible, it would have the effect of pushing down the flow tide. With the improvement of the Kultigong its tidal flow would be accelerated and it could reasonably be expected that the time of arrival of the flow tide upto the Kultigong would be advanced. As a result of these improvements leading to the retardation of flow tides in the Hooghly and its acceleration in the Kultigong the meeting ground of the tides could be shifted inside the Hooghly, the improved Kulti channel would maintain itself by natural action and the problem could be solved even without any diversion barrage across the Hooghly. The alternative would be to have a regulator at the off-take of the improved Kultigong. To avoid a tidal meeting ground, the regulator should be kept closed and nonfunctional during the flow tides but should be opened just when the Kultigong begins to ebb so that water from the Hooghly could be utilized for reinforcing the ebb flow with a view to scouring out the silt deposited during the flow tide. If this alternative has to be taken recourse to, consideration might also be given as to the feasibility of having a new connection link with the Kultigang taking off from a higher point on the Hooghly say near Bager Khal off-take (near Kanchrapara) which was really the off-take of the old Jamuna which is now dead. If we could get sufficient water for the purpose, this alternative would also enable us to resuscinate some of the other dead channels in this region which were originally spill channels from the Jamuna when it was active. The channels were Nowii, Suntee, Nonagon etc. Maitra (1958), however, suggested that by resuscitation of Sealdahgong and Nowii-Suntee. Kultigong might discharge its function of draining Calcutta and the surrounding areas. To provide the minimum upland water to the Kultigong he has recommended the introduction of 1000 cusecs drainage discharge taking into consideration the effect of tide lockage through each of the Channels-Nowii and Suntee in the beginning and increase the discharge gradually. This was in addition to the three discharge schemes viz., Nonagong, storm water canal and Bagjola already in operation. If the entire system can be made to work properly. It would improve the river condition to enable it to discharge its function. Under the present condition out of three suggestions the one made by Mazumdar (*op cit*) seems to be more practical, beneficial and economical.

After introducing necessary flow in the estuary enough fo scouring the deposited silt, attention has to be given for controlling the metallic contamination. It is difficult to treat the waste water at the source to make the effluent free from metals. Hence, it is suggested to construct two large reservoirs on either sides of the storm water canal and allow one to be filled with sewage water for settling purposes. Water hyacinth (*Eichhonia crassipes*) can be allowed to grow to absorb the metals, as discussed in the first part. The waste water after such treatment may be released into the river. The reservoirs can be alternately used. The estuary will then contain enough of nutrients and at the same time become free from toxic metals. No doubt some nutrients will be lost but metallic concentration once lowered to its minimum level, bacterial activity can be enhanced which will help release more of nutrients from organics. Fish then more safely can have a healthy growth, maturity and breeding more safely. The occasional discharge of water from the bheries, artificial tanks/reservoirs and spill areas, as is normally done, may bring down enough of plankton material in the estuary where they can survive and can be used by fish or other aquatic biomass. Besides, fish and prawn can migrate in shallow spilled swampy areas at different places adjoining the bheries (can be preserved as sanctuary) where biotic life can abundantly grow for use as food for fish and other crustaceans.

Since the quantum of metropolitan waste will further increase with the rise of population and industries, data on various aspects of the treated waste is necessary prior to it releases into the estuary. A simultaneous base line study on the physical chemical and biological characteristics of the estuary and the adjoining sanctuaries is needed for the purpose of regular monitoring of ecological condition. Periodical study of retention period of waste is also essential. In addition, metallic contamination both in the reservoir water before and during release in the estuary and in the estuary itself need thorough investigation. Bioaccumulation tests should be carried out in fish samples collected from different stretches of the estuary to determine metallic contamination. Avoidance test of treated sewage mixed water in various dilutions may be done using proper test fish to find out if the test fish of any size group has a dislike for any particular dilution. Water hyacinth and settled sludge of the reservoir can be used as fertilizer in the adjoining agriculture lands. Occasional survey of seed resource (fish and crustaceans) is required to ascertain if natural recruitment has improved with the changed ecological condition.

REFERENCES

Biswas, K.P. 1927. Flora of The Salt Lakes. University of Calcutta.

Bose, P.C. 1944. Calcutta sewage and fish culture. *Proc Nat. Inst. Sci., India.* 10 : 443-454.

Brungs, W.A. 1969. Chronic toxicity of Zinc to the Fathead minnow. (*Pimephales promelas* Rotinesque). *Trans Am. Fish. Soc.* 98 : 272-9.

Bloom H., and Ayling, G.M. 1977. Heavy metals in Derwent Estuary. *Environ Geol.* 2: 3-22.

De, B.N. 1929. *Kulti outfall scheme.* A report, Calcutta Corporation.

David, Augustine, 1959. Effects of Calcutta Sewage upon the fisheries of the Kulti estuary and the connected cultivable fisheries. Jour. Asiatic Soc. Bengal, 1 (4).

Director of Land Record and Surveys, West Bengal 1984. The problems of Paddy-cum-Brackishwater Fish/prawn cultivation and Forcible conversion of paddy lands into fisheries in parts of 24-Parganas District.

The Report of the Fact Finding Committee set up by the West bengal State Land Use Board. March.

Edwards. R.W. and Brown, V.M. 1966. Pollution and Fisheries. A Progress Report. *J. Inst. Water Pollut Contol. Land.*, 66 : 63-78.

Majumdar, S.C. 1942. *Rivers of Bengal Delta.* University of Calcutta.

Mazumdar, S.C. 1953. Ganga Barrage and the Bhagirathi-Hooghly River Problem. Indian *Jour. Pond River Valley Dev.* 3(3).

Mitra, A. 1954. *Census 1951, West Bengal District Hand Books-24 Parganas, West Bengal*, Govt. Press, Alipore, West Bengal.

Maitra, B. 1958. *Report on the Recent investigations and model studies on the Kultigong series of rivers.* River Research Institute, Govt. of West Bengal.

Nair, K.K. 1944. The effects of Calcutta sewage on fish life. *Proc. Nat. Inst. Sci.* India. 10 : 455-458.

National Environmental Engineering Research Institute, 1978. *Water quality survey of the Hooghly Estuary (1974-76) Part-III Survey of major Pumping Stations of Calcutta including their disposal system.* Sponsored by CMDA.

Pramanick, H.R. and Roy, P.B. 1972. Conservation of River Kultigong. *Seminar on estuarine pollution.* March, 9-10 sponsored by CMDA and CPHERI, Nagpur.

Ray, P., and Bandyopadhyay, S.K. 1984. *Utilization of Municipal wastes of Calcutta Metropolis in the Freshwater Bheries for Aquaculture Programmes and Problems Associated with its Recycling.* Final Report 1982-84 CIFRI (ICAR), Barrackpore, W.B.

River Research Institute 1970. *Annual Report 1968 Govt. of West Bengal. Directorate of Irrigation and water ways.* Publication No. 36.

Ray, P., S.B. Saha and R.K. Banerjee 1981. A case study of use of Calcutta Municipal wastes for fish culture in the Bidyadhari-Kulti complex, West Bengal.

International Symposium on water Resources, Conservation, Pollution and Abatement. Dec. 11-13, 1981. Univ. of Roorkee.

Ray, P. 1993. Aquaculture in Sundarban Delta its perspective. International Books and Periodicals Supply Service.

Ray, P. 1998. Ecological imbalance of The Ganga river system – its impact on aquaculture, Daya Publishing House. Delhi.

Sewell, R.B.S. 1934. A study of Fauna of The Salt lakes. Calcutta. *Rec. Ind. Mus.* 36 : 45-121.

Stones, T. 1976. A discourse on the Biochemical Oxidation of Sewage. *Effluent water treatment Journal,* 16 : 352-355.

Sprague. J.B. 1964a. Avoidance of copper-zinc solutions by young salmon in the laboratory. *J. Water Pollut Control Fed.* 36 : 990-1004.

Thakur, N.K. 1975. Observations on the availability of brackishwater fish seed in the Kulti estuary, W.B., *J. Inland Fish. Soc. India,* Published in 1976 March.

Venugopalan, V.K. 1979. Effect of Pesticides and heavy metals on growth and primary production of a few species of marine algae. Expanded Abstract of Marine Environment and Related Eco-systems. *National Seminar on Protection* of Marine Environment and related ecosystems India. National Institute of Oceanography. Goa, November, 13-15. Vol. 1.

Water Quality-Criteria 1963. State Water Pollution Control Board. Publ. 3-A, U.S.A.

4
The Bhagirathi-Hooghly Estuary

1. METAMORPHOSIS OF THE BHAGIRATHI-HOOGHLY ESTUARY

There was a time when the river Ganga had contributed considerable amount of fresh water to Ichhamati and Bidyadhari through the major tributary Jamuna, and the Bhagirathi-Hooghly for which the entire estuarine system Bidyadhari, Matla and other important estuaries lying criss-cross the delta were alive, maintained the salinity to a particular level. Biodiversity of Sundarbans with rich flora and fauna was thus maintained. With the diversion of the main flow of Ganga to thus maintained. With the diversion of the main flow of Ganga to Padma, Damodar flowing further down and subsequently construction of Farakka Barrage the ecosystem has completely changed (Fig. 1a, b).

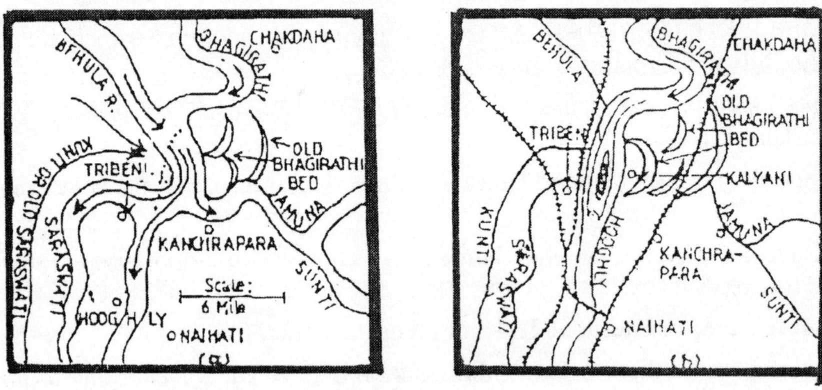

Fig. 1 : (a) Active Ganga, Jamuna and Saraswati during seventeenth century. (b) Conditions of these rivers after the Ganga (Hooghly) and the Damodar (Behula) shifted their Courses.

Source : Ray, P. (1993).

The Bhagirathi which flows through Behrampur, recorded about 140,762 cusecs and still further down the stream near Jangipur it was 140,160 cusecs with velocity of water 1.8-2.31 m/sec in now considerable reduced during dry months with the result that Bhagirathi converted into a small spill channel (Sarkar, 1988).

This fact was also evidenced when people could even wade across the river during 1972. The inland water-ways Authority of India recorded least available water depth between Tribeni and Farakka to be two metres (Mukherjee, 1993).

Near Calcutta, the Hooghly is wider, (about 609.6 m) as a result velocity of water decreases. During monsoon huge quantity of alluvial soil is brought alongwith the water. Bera (1984) is of the opinion that if the flow of water is above 1.52 m/sec the silt moves along with the water, but if it is below, silt settles, further; if the velocity exceeds about 2.1 m/sec, effect of erosion takes place.

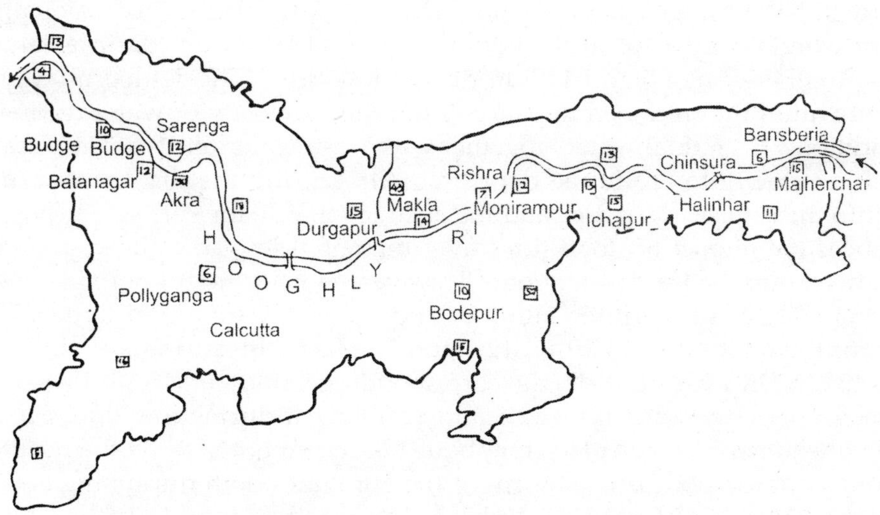

Fig. 2 : Location and number of brick field cluster flanked on both sides of the Hooghly estuary under Calcutta Metropolitan District.

Source : Bhattacharya, S.K. and S.M. Chatterjee (1972).

The discharge @ 40,000 f^3/sec comes to 345 crore f^3/day. According to Bera (1984) during monsoon the quantity of alluvial soil brought daily along with the water is about 17.25 lakh f^3. In four months of monsoon the quantity comes to 20 crore f^3. Ealier study (Chatterjee, 1972) revealed that 90-100 million f^3 of silt were carried alongwith the water every year into the Bhagirathi-Hooghly, three fourth of which settles in the river bed. This comes to 60-66 m f^3/per year. In the present observation 3/4 of 20 crore f^3 that settles in four months comes to 13.3 crore f^3 which is more than double the earlier value in course of 12 years. Also in 1947-48, the dredging quantity was limited to about 7 million tonnes and in 1967-68 it was in the order of 21 million tonnes. Again according to ministry of irrigation and Power an estimate, about 2.8 million m^3 of silt (about seven million tonnes) enters the Bhagirathi-Hooghly channel annually. About 2/3 of silt is deposited below the Bansberia point. In peak rainy season it is carried further down and deposited 20 km south of Calcutta. In addition the river receives about 8 million tonnes of sand from the sea which chokes up the stream flow (Mukherjee, 1993). To be precise the quantum of silt deposited annually, is undoubtedly alarming. The effect of such huges silt load is shown from the slope (cf. Table 2). Also 279 brick fields flanked on both sides of the river between Kalyani and Achipur (100 km) are the outcome of such silt load (Bhattacharjee and Chatterjee, 1972). (Fig. 2 and Table 1). The reduction of flow velocity near Calcutta causes such huge mass of silt to settle.

In the last 10 years as stated by Bera (1984), the depth of water is reduced from 3-5 ft (0.9-1.51m) in the main channel. Decrease of depth of river bed results increase of tidal range. In 1947, the river experienced 70 bore days which in 15 years time rose to 161 days. (Banerjee, 1972). Fig. 3 indicates the gradual increase in annual frequency of bore tides in the river Hooghly.

Chakraborty and Banerjee (1966) who studied the depth conditions of the Hooghly estuary between Bally bridge (down stream) and tribeni (up stream) occasionally observed uneven muddy and sandy areas at the bottom of the estuary. This confirms large variation of depth ranging between 4.5 and 14.5 m during low tides. The settled muddy bed shifts up and down within the river bed depending upon the velocity of water during flood and ebb tide conditions. Even tidal waters together with fresh water release from Farakka cannot create suitable velocity to erode the deposited silt. During the dry season, discharge of freshwater into the estuary decreases and the abstraction rate for various purposes increases which enhances the impact of flood tides over the ebb tides. This allows the saline water to penetrate deep into the freshwater zone. The average velocities of boretide and neap tide during different seasons *i.e.* winter, summer and monsoon are found to be 27.7, 28.3 and 27.9 km/hr for the high tide and 10.16, 12.1 and 10.96 km/hr during low tide of bore tidal phase respectively. During neap tide, the corresponding values are 18.18, 18.5 and 20.3 km/hr at the time of high tide and 10.15, 12.6 and 12.5 km/hr during low tide. On an average, the bore tide phenomenon takes place for about 12 days/month, while the neap tidal phase lasts for about 14 days/month. The height of the bore tides both during high and low tides is always higher than that of neap tidal phase. The ebb tidal flow lasts for a longer period than the flood tide. Duration of ebb tide flow is 6.5-8.5 hr during dry months and 7.0-11.5 hrs in the wet months. The flood tide flows last for 4.0-6.0 hr and 2.5-4 hr in dry and wet months respectively. Both bore and spring tides have appreciable scouring effect on the

Table 1. Brick field clusters of Calcutta Metropolitan District flanked on both sides of the Hooghly Estuary

Area	*No. of Brick fields*	*Area*	*No. of Brick fields*
1. Achipur	4	11. Kalyani, Majherchar	15
2. Budge Budge, Pujali	10	12. Uluberia upto Bauria	13
3. Batanagar, Nangi	12	13. Sarenga, Manickpur	12
4. Akra	30	14. Podra, Radhadasi, Benkra	18
5. Bhangakhal, Badartala	18	15. Durgapur (Howrah)	15
6. Tollygunge	6	16. Makla, Raghunathpur	45
7. Barrackpore, Manirampur	12	17. Kotung	14
8. Garulia	3	18. Rishra, Serampur	7
9. Ichapur	15	19. Champdani, Chandemagore Chinsura	13
10. Halisahar, Kanchrapara	11	20. Bansberia	6

Source : Bhattacharya, S.K. and S.M. Chatterjee (1972).

littoral zones of the estuary. The mouth of the Hooghly-estuary is wide and deep (Kulpi, 4,59 km, Diamond Harbour, 4.65 km. and Falta, 1.2 m). The corresponding maximum depth at high tide during monsoon are 16.9, 27.5 and 20 m. The maximum depth at low tide during winter being 10, 20 and 13.9 m respectively. The banks taper into a broad funnel admitting the tides to extend high up into the river. They often form 'bores' in the narrow part of the estuary and maintain average height of 4.0-4.6 m with strong scouring action. (Ray, 1981). Additional discharge from Farakka simply extends the freshwater zone further down.

Table 2. Slope characteristics in different stretches of the Bhagirathi-Hooghly with corresponding mean rate of flow.

	Stretch	*Section*	*Length in Km.*	*Average slope*	*Mean annual rate of flow $M^3/sec.$*
1.	Farakka to Nabadwip	Deltaic non-tidal plain.	230	1 in 23000	1000-1300
2.	Nabadwip to outfall	Deltaic tidal plain	240	1 in 24000	Variable due to tides.

Source : Das Gupta, S.P. (1984).

2. LIMNOLOGICAL STUDIES

In the estuarine zone conditions are totally different and the Hooghly waters remain comparatively clear during winter. Thereafter turbidity increases steadily until it attains its maximum value during July and August. Climatic and seasonal variation in the tidal traslatory movements effect turbidity significantly. It is also influenced by the lunar phases. The coincidence of the perigee position of the moon and the spring tide condition, as observed by Bose (1956) has the maximum mechanical *effect* in pushing the neritic waters and mixing of column water with the soil particles that had sedimented during the earlier neap phase of the tide. On the other hand, the coincidence of apogee position of the moon and the neap condition of tide is found to have maximum effect on turbidity. Bose further reported increase in turbidity due to daily tidal movements of water in areas where there are curves in the estuary. Further, the turbulence occurred on account of occasional movement of ships also promotes turbidity.

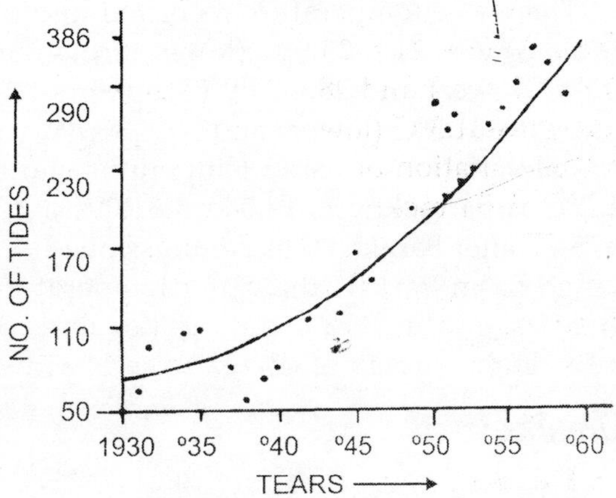

Fig. 3 : Annual Frequency of Bore Tides in Hooghly River (1930-1960).

Source : Banerjee, B.N. (1972)

Substantial work is also done in this stretch by CIFRI both before and after the commission of Farakka barrage. Aspects covered are qualitative and quantitative studies of physicochemical factors, phytoplankton and other biological features and their seasonal pattern in a small stretch of the river near Palta initiated by Roy (1949 and 1955) This was followed by a survey (Dutta *et. al.*, 1954) conducted in a 108 km. stretch from Palta to

Diamond Harbour. Later, Shetty *et al.* (1961) and Saha *et al.* (1975) reported variations in the plankton of the Hooghly-Matla estuarine system throughout the 290 km tidal stretch of Hooghly. Susequently, extensive studies were made related to the effects of industrial and domestic effluents (Seth and Bhaskaran, 1950; Basu 1966; Ghosh and Basu 1968; Basu *et al.*, 1970; Basu and Ghosh 1970; Gopalakrishnan *et al.*, 1970 and 1978; Ghosh *et al.*, 1973, 1976, 1977, 1979a, 1979b, 1980a, b; Ray *et al.*, 1977, 1979 and Ray 1980) in and around the outfall of some industries and their overall impact in the river. The above findings are based on the studied made before Farakka barrage was commissioned in 1975. Subsequently after Farakka was commissioned ecological changes in the Hooghly was studied by Nandy *et al.* (1983), for a period of two years from 1975 to 1977. The overall impact of industrial and municipal wastes on the trend of fisheries was highlighted by Ray and Laha (1981). This was followed by an attempt to study the heavy metals in the estuarine water, bottom sediments, fish and shrimp together with their bioaccumulation in different organs by Ghosh *et al.* (1983), and pesticide pollution by Joshi (1985). Carbon assimilation in the estuary was studied by Babulal (1990), and finally the spawning activity of hilsa (De and Sehgal, 1989).

2.1 *Physico-chemical Characteristics*

Water Temperature

The water temperature recorded in the estuary in the different seasons during 1967-71 (Table 3) was 21.6-23.9°C (lower stretch) and 21.9-24.3°C (upper stretch) in winter; 27.1-32.8°C (lower) and 28.4-31.9°C (upper stretch) in summer and during monsoon the values were 28.8-31.3°C (lower) and 29.5-31.7°C (upper) respectively. Nandy *et al.* (1983) recorded annual variation of water temperature ranging between 17.5 and 32.0°C at Kalna, 19.5 and 31.2°C at Barrackpore, 19.5 and 31.0°C at Uluberia and 18.5 and 33.5°C at Geokhali during 1975-77 after Farakka was commissioned & (Table 4) Das Gupta (1984) documented in "The Ganga Basin Part II" the water temperature of 1981-82 ranging between 26 and 32°C with an average value 30.2°C during May and 13-22°C with an average value 18.9°C) during July in the entire stretch of Ganga from Haridwar to Diamond Harbour.

Turbidity

The maximum turbidity of 1000-2500 JTU and the minimum value of 100 JTU were reported by Bose (1956) between Uluberia and Kakdwip. Discharge from the rivers Damodar and Rupnarayan, which join the Hooghly in tis area and the river Haldi which joins it further down, also contributes in increasing the turbidity. High turbidity values were observed in the lower stretches during the non-monsoon seasons from 1967 to 1971 (Table 3).

Hydrogen-ion-Concentration

Roy (1949) recorded pH 8.3 and 8.4 during 1945-48. Bose (1956) recorded between 7.9 and 8.4. The pH value did not show any yearly variation during 1967-71 (Table 3). The average value in the upper estuary was 8.1 in winter, 8.1-8.3 in summer and 8.1 in monsoon. In the lower zone however, the pH ranged 8.1-8.2 (Av. 8.1), 8.2-8.3 (Av. 8.2) and 8.1-8.2 (Av. 8.1) during the above seasons (Ray, 1981). The above values pertain to pre Farakka release. After Farakka was commissioned a tendency of slight rise in values towards upper level (8.5)

Table 3. Baseline date for Hooghly Estuary expressed by average values for different seasons during the period 1967-71

Period	Season	Temp (°C)		turalidity (cm)		pH		DO (ppm)		BOD_5 (ppm)		ALK ($CaCO_3$)		Salinity (%)		Phyto, (u/1)		Zoo (u/1)	
		U	L	U	L	U	L	U	L	U	L	U	L	U	L	U	L	U	L
	W	24.3	23.9	21.5	7.2	8.0	8.1	6.8	5.0	2.1	1.7	259	194	0.07	0.25	78	66	62	27
1967	S	31.9	32.8	-	-	8.2	8.3	4.3	5.1	-		-		0.17	9.75	-		-	
	R	30.6	31.3	4.9	7.4	8.1	8.2	4.8	4.9	-		122	154	0.06	7.07	-		-	
	W	23.8	-	16.8	11.2	8.1	8.2	5.9	6.3	1.9	2.5	218	218	0.06	0.46	352		153	
1968	S	29.2	27.9	5.9	3.3	8.1	8.2	5.2	5.4	3.0	2.9	237	212	0.74	7.66	144	88	49	25
	R	31.7	30.9	3.5	2.6	8.1	8.2	5.5	5.3	1.8	1.9	71	66	0.06	0.06	55	49	21	18
	W	22.6	21.9	19.5	9.4	8.1	8.1	6.3	5.6	2.4	2.3	280	251	0.06	0.45	908	211	209	77
1969	S	28.5	27.1	9.5	3.9	8.3	8.2	6.3	6.4	3.0	3.8	232	216	0.07	3.0	91	300	28	28
	R	30.3	29.6	3.3	3.9	8.1	8.1	6.3	5.5	3.0	2.2	122	144	0.07	0.37	137	103	21	23
1970	W	22.6	21.8	11.4	9.0	8.1	8.1	6.3	5.4	2.5	2.1	278	252	0.11	0.64	184	70	95	17
	S	29.7	28.8	7.0	5.9	8.2	8.2	5.5	5.0	2.0	2.0	229	231	0.18	5.3	107	41	49	7
	R	29.8	29.6	2.5	3.8	8.1	8.1	5.8	5.3	2.1	2.1	105	119	0.06	0.19	103	61	5	3
1971	W	21.9	21.6	-	6.5	8.1	8.1	8.1	6.3	1.8	1.4	289	267	0..06	0.22	-		-	
	S	28.4	28.8	5.7	4.8	8.1	8.2	5.8	5.6	2.8	1.9	190	184	0.165	1.39	-		-	
	R	29.5	28.8	3.0	3.1	8.0	8.1	5.1	5.1	1.5	1.5	32	37	0.33	0.11	-		-	

W - Winter Season. U - Upper Stretch

S - Summer Season L - Lower Stretch

R - Rainy season or monsoon.

Source : Ray, P. (1981).

was noticed (Nandy *et al.*, 1980). The annual average of 1975-77 recorded at the 5 centres stretched over the estuary, Kalna-Kakdwip, was 7.6-8.5 with Kalna 7.6-8.5 Barrackpore, 7.8-8.5, Uluberia, 8.0-8.5, Geokhali, 7.8-8.5 and Kakdwip 8.1-8.5 (Table 4).

Dissolved Oxygen

In the estuarine stretch DO around Calcutta was more than 4.0 ppm excepting during summer months (Bose, 1956). Between Barrackpore and Kulpi minimum DO values ranged from 8.4 and 5.1 during 1953-55. Subsequent study revealed that the average DO was 5 ppm in all the seasons. At the outfall zones it ranged from 3.8-4.8 ppm (Ray and Ghosh, 1976). Near industrial outfall region the DO was either absent or present in traces. Such low values of DO may be due to combination of several factors *e.g.* high turbidity, great demand of oxygen by decomposing organic material in summer and grater period of effluents retention during the neap tidal phase. The annual variation of DO during the post Farakka release 1975-77, as stated by Nandy *et al.* (1983) ranged 5.1-10.8 ppm at Kalna, 5.0-10.0 ppm at Barrackpore, 5.0-10.2 pm at Uluberia 5.2-8.6 ppm at Geokhali and 4.7-10.6 ppm at Kakdwip (Table 4) and is certainly an indication of improvement over the years of 1967-71 attributable to release of fresh water.

Table 4. Annual Variation of some Physico-Chemical parameters in the Hooghly estuary after Farakka discharge

Parameters	*1975-1977*				
	Zone-I		*Zone-II*		*Zone-III*
	Kalna	*Barrackpore*	*Uluberia*	*Geokhati*	*Kakdwip*
Water Temp °C	17.5-31.0	19.5-31.2	19.5-31.0	18.5-33.5	20.1-34.5
Suspended Solids (ppm)	220-2300	120-3400	80-2240	140-3800	50-3125
pH	7.6-8.5	7.8-8.5	8.0-8.5	7.5-8.5	8.1-8.5
Dissolved Oxygen (ppm)	5.1-10.8	5.0-10.0	5.0-10.2	5.2-8.6	4.7-10.6
T. alkalinity (ppm)	80-175	68-195	80-180	72-180	40-158
Sp. Conductivity at 25 °C x 10–6 (mhos)	194-425	143-479	199-799	211-7199	11999-18275
Nitrate as NO_3-N (ppm)	Trace-0.224	Trace-0.18	Trace-0.2	Trace-0.27	- -
Phosphate as PO_4 (ppm)	0.037-0.61	Trace-0.41	0.03-0.31	Trace-0.265	0.05-0.15
Silicate (ppm)	3.0-25.0	3.0-27.0	1.0-25.0	2.0-25.0	1.0-5.0

Source : Nandy, A.C., M.M. Bagchi and S.K. Majumdar (1983).

Biochemical Oxygen Demand (BOD_5)

The extensive work done by CIFRI (Barrackpore) in the Hooghly estuary in different years prior to Farakka discharge indicate that the organic load of the Hooghly estimated or the first time by Ray during 1967-71 at the lowest dilution level. The average BOD_5 never exceeded 5 ppm during the period in the main body of the estuary (Table 3) except for slight rise in the summer months between Naihati and Calcutta (Ray and Ghosh, 1976). However, around the outfall the values were significantly higher.

Table 5. Chemical Characteristics of some Industrial, Domestic and Municipal Wastes Expressed as average Values During the Period 1968-71

Nature of waste water	pH	Alkalinity	Sp. Cond. m mhos	DO	BOD_5	$KMnO_4OC$ mg/l			Nitrogen			Lignin
						3 mt	4 hr	30 mts 100° C	Free & saline	Alb Amn	Nitrite-N	
Pulp and Paper Industry Sulphate Process	6.5-9.4 (7.4)	50-940[b] (286)	479-8779 (2191)	Nil-6.8 (2.3)	12-700 (144)	Nil-8.0 (4.0)	90-225 (145)	4-228 (120)	Tr.-0.9 (0.4)	Tr.-1.6 (0.33)	Nil Nil	265[d]
Sulphite process	5.3-6.9 (6.2)	30-450 (169)	654-79993 (9095)	Nil	400-1920 (986)	6-985 (370)	145-2144 (1201)	240-3300 (1899)	0.5-1.4 (0.8)	0.11-1.36 (0.89)	Nil	1000[d]
Soda process[a]	7.1-9.5 (8.5)	126-2070 (6.08)	287-3789 (1214)	Nil-7.7 (3.45)	13-450 (60)	3-180 (34)	14-360 (84)	5-520 (140)	-	-	-	200[d]
Textile Industry	5.6-10.5	160-1896[c] (564)	514-7199 (2946)	Nil-8 (2.1)	17-1335 (380)	1.2-72 (30)	4-250 (98)	6-1060 (162)	1.4-2.35 (1.87)	0.45-0.5 (0.47)	Nil	
Natural Synthetic	2.0-4.9 (2.8)	Nil-14 (1.0)	912-10284 (3697)	Nil-2.3 (0.3)	26-1175 (331)	18-32 (24)	4-23 (16)	40-438 (141)	-	-	Nil	
Distillery	5.0-6.1 (5.5)	502-040 (1135)	7199-10284 (5160)	Nil	5760-21600 (15687)	1600-2000 (1806)	1040-8500 (4918)	1025-31300 (8502)	Nil-0.1 Tr.	Tr-0.25 Tr.	Nil	
Yeast	5.4-7.7 (6.6)	432-650 (564)	1565-2399 (1866)	Nil	1460-5760 (3387)	25-85 (45)	79-1200 (790)	130-1793 (1216)	0.6-19.2 (9.63)	4.5-2.9 (16.75)	Nil	
Tannery	5.2-8.1 (7.2)	42-500 (292)	2057-21838 (7173)	Nil-3.2 (0.6)	60-1280 (404)	25-65 (44)	65-750 (120)	88-1020 (339)	5.6-25.5 (10.03)	0.9-8.0 (2.9)	Tr-0.12 (0.04)	
Rubber	6.6-8.0 (7.3)	240-320 (271)	604-835 (697)	1.5-6.7 (4.8)	10-106 (40)	2-72 (11)	10-40 (19)	9-89 (51)	1.48-3.3 (2.2)	0.80-0.8 (0.34)	Nil	
Oil refinery[e]	5.5-9.0	110-142	2181-3999	2.6-6.0	7-30	-	-	-	-	-	-	
Miscellaneous	6.1-10.5	104-920	757-4499	Nil-9.6	8-457	Nil	6-25	8-376	0.3-6.0	03-1.5	Nil-56	
Paints and vamish	(8.1)	(370)	(1513)	(3.00)	(137)		(11)	(88)	(2.0)	(0.43)	(.12)	
Safety match	6.8-8.1 (7.5)	187-396 (308)	3428-4388 (3345)	Nil-6.1 (3.0)	16-76 (39)	- -	- -	20-118 (79)	2.4-6.22 (4.30)	.85-1.5 (1.17)	Nil	

contd....

Table 5 – contd...

Nature of waste water	pH	Alkalinity	Sp. Cond. m mhos	DO	BOD_5	$KMnO_4OC$ mg/l			Nitrogen			
						3 mt	4 hr	30 mts 100° C	Free & saline	Alb Amn	Nitrite-N	Lignin
Shellac	5.0-7.3 (6.3)	832-1500 (1137)	1285-19452 (9250)	Nil	310-1600 (963)	50-250 (75)	225-800 (375)	320-1420 (900)	-	-	Nil	
Metal and Steel	6.1-80 (7.2)	11-250 (194)	299-4234 (1161)	Nil-6.2 (2.4)	10-74 (46)	- -	- -	8-76 (40)	Nil-0.1 (Trace)	Nil-0.4 (Trace)	Nil	
Combined Wastes (Bagh Canal)	6.9-8.2 (7.8)	360-480 (407)	4056-9506 (6969)	Nil-2.4 (1.6)	126-156 (136)	10-35 (25)	32-65 (55)	54-118 (80)	0.12-4.24 (2.78)	Tr.-8.4 (2.9)	Tr.-0.2 (0.12)	
Domestic and Municipal Wastes	6.7-8.6 (7.7)	204-660 (412)	1531-8999 (4048)	Nil-7.4 (0.4)	24-560 (187)	15-95 (45)	20-156 (120)	31-540 (198)	0.12-12.5 (6.6)	0.14-4.0 (0.83)	0.02-0.2 (0.15)	

Results are expressed in mg/1 except otherwise stated
a-Data pertain to 1970-1973
b-Contains carbonate alkalinity
c-Contains hydroxyl alkalinity
d-Basu (1966)

e-Data provided by B.B. Ghosh, Central Inland Fisheries Research Institute, Barrackpore

Further data : TSS (mg/1) : 18-55
TD (mg\1) : 1200-1420
Sulphide (mg/1) : 0.2-2.0
Phenolic Compound (mg/1) : Trace-1.0
Oil (mg/1) : 2-12

Ghosh, Ray and Gopalakrishnan (1973)

Source : Ray, P. (1981).

Alkalinity

In the estuarine region, alkalinity of the Hooghly water remains the same except for some seasonal variation between freshwater zone–Kalna and saline zone Diamond Harbour. September onwards alkalinity increases until it attains maximum value during March and June. It abruptly decreases with the advent of monsoon (Dutta *et al.*, 1954). Bose (1956) observed that from a minimum 2 mg/1 (102 ppm) alkalinity increased to 4.5 and 7.0 mg/l (230 to 357 ppm) in January/February. After March it declines but salinity increases. Observations made during 1967-71 by the author shows that the maximum salinity value of alkalinity gradually declined. The highest values were observed during November-February which declined in March-June and were the lowest during the rainy season (Table 3). Subsequent study made by Nandy *et al.* (1983) during 1975-77 after the Farakka discharge shows that the excess freshwater release has brought in considerable dilution to the constituents resulted in the variation 40-195 ppm in the tidal zone. Their split up at 5 centres were–Kalna, 80-158 ppm, 68-195 ppm at Barrackpore, 80-180 ppm at Uluberia, 72-180 ppm at Geokhali and 40-158 ppm at Kakdwip.

Salinity

Salinity fluctuations in the estuarine zone is quite significant. In general, observations were made from April to early June due to the intrusion of the sea water. Salinity varies between 8-12 ppm during January-Feb., 10-120 ppm in March, 10-208 ppm in april, 26-180 ppm in May, 18-48 ppm in June and 4-8 ppm in July-December (David, 1954).

He also observed wide fluctuations, i.e. 72-432 ppm during the course of same day (25th April 1948). During the pre-monsoon period salinity remains high and shows erratic fluctuations. The highest salinity of 4000 ppm was recorded at Palta in 1972. Perhaps there are no vertical gradients of salinity, as it was not recorded.

During summer salinity increases 3 per cent in the brackishwater zone at Kakdwip. It decreases during the monsoon to 1.6 per cent and may become almost like freshwater (Bose, 1956). During April saline water penetrated deep into freshwater zone on account of reduced flow from upstream. Upstream of Barrackpore the water remains fresh throughout the year. Around Barrackpore maximum salinity was 2 per cent during May/June (Bose, 1956). In Brackishwater zone near Diamond Harbour, difference of 2 per cent between surface (20 per cent) and bottom waters (22 per cent) at a depth of 18 m during maximum high tide was recorded. The difference was marginal during the lowest water level. Variations in salinity between surface and bottom waters around Calcutta may be due to local vertical eddy currents at the time of high tide (Bose, 1956).

The Hooghly estuary is free from salinity zones such as marine tidal and gradient zones (Bose, 1956). Between Kakdwip and Palta lies the brackishwater zone and upstream of Calcutta is freshwater zone. The area between these two zones possesses transitional characteristics. During 1956-58 Shetty *et al.* (1961) recorded salinity in traces (less than 0.1 per cent) between Medgachi and Khusigali, the upper zone; trace-15 per cent, in the middle or gradient zone between Konnagar and Diamond Harbour, and 6.18-32.85 per cent in the lower zone upto Fresergan j. Observations made during 1967-71 indicated almost similar trend (Table 3). Seasonal fluctuations in salinity before and after Farakka discharges both in fresh and brackishwater zones around Barrackpore and Kakdwip are shown in Fig. 4. A sizable drop in the salinity occurred all over the estuary. At Kakdwip, the Salinity dropped from the range 1.5-32.8 per cent in 1960-61 to 3.2-29.2 per cent in the first year of release (May 1975-April 1976) and further to 2.5-18 per cent in the second year of release (May 1976-April 1977) Nandy *et al.* (1983). At Barrackpore the values came down to traces from 0.2 per cent.

Specific Conductivity

Nandy *et al.* (1983) estimated conductivity to range 143-18275 mhos (at 25°C×10^{-6} mhos) between Kalna and Diamond Harbour. At Kalna, the upper end of estuary, the range was 194-425, further down at Barrackpore, 143-479, Uluberia, 199-799, Geokhali, 211-71289 and Kakdwip 11999-18275 mhos thereby indicating that the water was more of brackish in character (Table 4).

Bacterial Density

Around Calcutta bank-bacterial density studied by NEERI (1978) during 1975-76 in a stretch, Pathak to Chintamony (about 20 km). during low and high tide of summer and monsoon, indicated right bank to be more polluted (Left bank, LT : 26.7×10^5, HT : 30.29×10^6 during dry season against the right bank, LT : 37.6×10^5, HT : 11.253×10^6 during monsoon). Some of the values mentioned by Das Gupta, 1984 appears to be quite high (higher vlues 53,533) (post-monsoon) and 9,16,667 (winter). It is clear from all the studies made so far that all the values shown are abnormally high especially below Farakka and are much above the drinking water standard (5000 MPNV/100 ml) and for this no explanation is offered. Incidentally it may be mentioned that data presented by Das Gupta (1984) are based in the data of WPCB, West Bengal.

Orgnaic Pollution indicated by Bacterial density are shown in Tables 6 and 7.

Table 6. Seasonal Variations of the mean Concentrations of indicators of organic Pollution of the lower plain during 1980-82.

BOD mg/l			*D.O. mg/l*			*Coliform Bacteria (MPN) no/100 ml.*		
Post monsoon	*Winter*	*Summer*	*Post-monsoon*	*Winter*	*Summer*	*Post-monsoon*	*Winter*	*Summer*
1.0	2.82	3.07	6.26	8.27	6.46	28859	271579	260509
0.53-1.23	0.5-15.23	0.45-15.58	4.53-7.80	7.47-9.10	5.40-6.90	2110-53533	8833-916667	22600-902333

Source : Das Gupta, S.P. (1984) Table modified.

Coliform bacteria an indicator of Organic Pollution were also studied by various workers/Organisations which indicate very high bacterial count. Their values no/lit are indicated in Table 7 for different stretches.

Nutrient Status (Nitrate, Phosphate)

The main sources of nitrogen and phosphorus are from the run off of the catchment area, industrial and sewage effluents and decomposition of organic waste which settles on the substratum. Around Palta 0.001-0.0035 ppm phosphate (PO_4) has been reported by Roy (1955). It is the highest in spring and summer and the lowest in winter. The phosphate (PO_4) content in the estuary is low during monsoon but relatively high in summer possible due

Table 7. Coliform bacteria (no/lit.) studied in different stretches or Bhagirathi-Hooghly.

Centre	*Total count*	*Season*	*Source*				
1.	Pathak-Chintamony (20 km.)	Low tide-26.7 × 10^5 High tide-30.29 × 10^6	Summer Summer	Left Bank	37.5 × 10^5 (L.T.) 11.253 × 10^6 (H.T.)	R. Bank	1975-76 NEERI.(1978)
2.	Nabadwip	Bank-21000-110000 (855500) Mid River-4600-46000 (25300) 62% less	April & May				Das Gupta (1984)
	Katwa	Bank-110000-240000 (175000) Mid River-21000-46000 (80% less) (33500)					
3.	Nabadwip 58500 Dakshineswar 240000 Diamond Harbour 8250		... ...	... ...	... ...	... ... Das Gupta (1984)	WPCB (West Bengal 1981, 82.

	No.\100 millilitre (in lakh)	*Source*
Palta	14	Thakur Debdoot
Dakshineswar	85	5.7.2004
Shibpur	9.5	Ananda Bazar Patrika.
Garden Rich	14	In : WPCB (WB).
Uluberia	9	
Diamond Harbour	16	

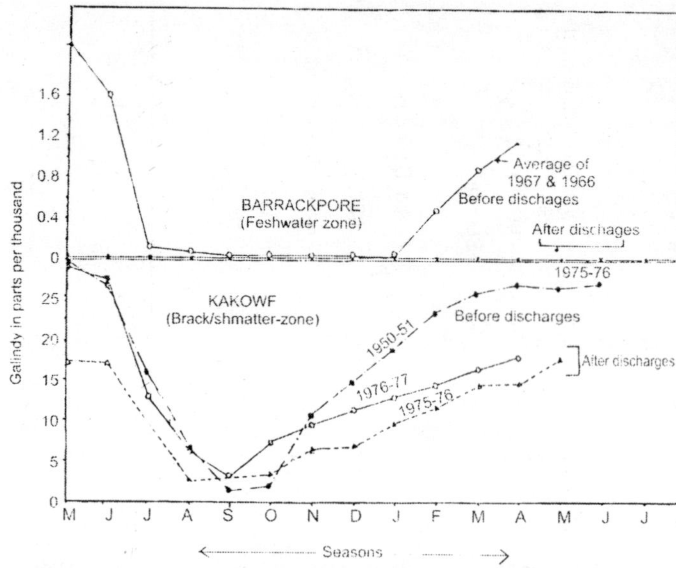

Fig. 4 : Seasonal fluctuations in salinity before and after Farakka discharges.

Source : Ray, P. (1981)

to its regeneration from the bottom sediments (Bose, 1956). He has also found that the phosphate content between Barrackpore and Uluberia were relatively high, varying from 3.0-4.0 mg atom p/m^3 (0.9201-1.2268 ppm PO_4 compared with 1.5-3.0 mg atom p/m^3 (0.46005-0.9201 ppm PO_4) at other centres. The lower values varied between 0.1 and 0.4 mg atom p/m^3 (0.3067 and 0.12268 ppm PO_4). Silicates, however, varied between 5.25 and 22.2 ppm (SiO_2) (Roy, 1955). The maximum concentration is reached during the winter season and is maintained till late spring or extended upto early summer. Bose (1956) recorded 400-540 mg atom Si/m^3 (0.856-1.556 ppm SiO_2) towards freshwater zone (Uluberia and above); and 180-300 mg atom Si/m^3 (0.3852-0.428 ppm SiO_2) in the lower zones. However, the values of Silica gradually increased from 40-160 mg atom Si/m^3 (0.0856-0.3424 ppm SiO_2) from brackish to freshwater zones. In general, there is a dearth of data on the nutrient status of the estuary on a time specified basis. The nutrient status during the post Farakka release as reported by Nandy *et al.* (1983) ranged between trace and 0.94 ppm for nitrate-N, trace and 0.75 ppm for phosphate P (PO_4) and for silicate it ranged between trace and 31 ppm, between Kalna and Kakdwip. Centrewise the values of nitrate were trace-0.224 ppm at Kalna; Trace to 0.18 ppm at Barrackpore; Trace-0.2 ppm at Uluberia and further down trace-0.27 ppm at Geokhali. In the case of Phosphates the vlaues were 0.037-0.61 ppm, trace-0.41 ppm, 0.03-0.31 ppm trace-0.265 ppm and 0.05-0.15 ppm for the centre in the above order, and finally silicate ranges 3-25 ppm, 3-27 ppm 1-25 ppm, 2-25 ppm and 1-5 ppm at the above five centres (Table 4).

**Sulphates* : In the lower Ganga plain values were 17.7-27.6 ppm (av. 21.4) 21.5-28.7 ppm at. 23.3) and 16.9-28.5 ppm (av. 24.3) for the seasons, post-monsoon, winter and summer.

**Calcium* : In the estuarine zone, the values were 83.6 ppm (70-100), 95.6 ppm (62.5-113.3) and 80.1 ppm (63.6-145.7) for the seasons, postmonsoon, winter and summer.

**Source* : Modified from Das Gupta S.P. (1984).

Metals and Pesticides

Ghosh *et al.* (1983) while studying the pollution status of the Hooghly with reference to heavy metals estimated upper level of Zinc and Chromium above the outfall ranges as Zinc (0.4 ppm) and Chromium (0.4 ppm). Near the outfall the upper level Zinc ranged between 00.5 and 7.2 ppm, and Chromium 0.2 and 20 ppm, Copper and lead could not be detected. Interestingly concentration of metals in sediment registered much higher values than the wastes Zinc was estimated to be 381.6 ppm (outfall of Rayon industry) followed by outfall of Paints and varnishes (160 ppm) and oil refinery (80 ppm). Chromium level was recorded at 9.6 ppm. Zinc contamination in sediments was found to extend beyond 500 m.

Table 8A. Accumulation of metals in some organisms of the Hooghly estuary at different outfall areas

Industry	Organism		Size		Concs of metals (ppm)			
			Length	Weight	Zinc	Copper	Chromium	Lead
Rayon	*Fish*	*Ailia Coila, Mastacembelus armatus* Apocryptes bato, Rita rita, Setipina phasa, Clupisoma garua Glossogobius giuris.*	60-185	1.9-16.5	2.5-105.8	–	–	ND
	Shrimp	*Macrobrachium Villosimanus Macrobrachium rude*	51-80	2.5-4.0	29.2-112.2	–	–	–
Dunlop Rubber (Sahaganj)	*Fish*	*Puntius ticto, Odontamblyopus rubicundus, R. rita, Mystusmenoda, G. giuris, Cynoglossouslingua. S. Phasa, Eleotris-macrodon.*	54-115	2.0-5.8	12.4-120.4	2-20	ND-4.5	ND
	Shrimp	*Macrobrachium-mirabile, Leander fluninicola, M. rude*	40-85	0.3-5.5	14.8-40.4	2-10	ND	ND
Paints & Varnishes (Garifa)	Fish	*O. rubicundus*	115	5.0	7.2-62.4	–	ND-0.26	ND
Paints & Vanishes (Shalimar)	*Fish*	*A. bato, Silonia silondia, C. garua, Pama Pama, Pangasius pangasius, P. ticto, Colisa Fasciatus, O. rubicundus.*	50-149	1.5-30.4	8.4-46.4	ND	ND-4.5	ND
	Shrimp	*M. rude, M. mirabile*	42-85	1.8-5.0	20.4-40.0	14-38	ND	ND
	Crab	*Scylla seratta*			24.4	14.0	ND	ND
Cotton Textile (Garden Research)	*Fish*	*S. Phasa, Chela bacaila, Hilsa ilisha*	98-116	5.0-5.2	26.4-64.6	ND	ND	ND
Tannery (Batanagar)	Shrimp :	*M. Villosimanus*	74-80	2.5-4.0	24.0-54.0	–	–	–
Oil Refinery (Haldia)	*Fish :*	*Boleophthalmus boddacrti, Boleophthalmus viridis, P Pangasius*	82-100	5.0-5.0	10.0-24.0	ND	ND	ND
	Shrimp :	*Parapenaopsis sculptilis*	66.0	1.2	27.0	8.4	ND	ND
	Mollusc :	*Pila globosa*	–	–	83.0	20.0	ND	ND

*Adult fish; ND - Non-detectable

Source : Ghosh, B.B., M.M. Bagchi and D.K. De (1983).

Table 8B. Accumulation of heavy metals in tissues of some fishes of the Hooghly estuary around Barrackpore

Location	Species of Fish	Size		Nature of tissue	Concentration of heavy metals (ppm)			
		Length (mm)	Weight (g)		Zinc	Copper	Chromium	Lead
Barrackpore	*Silago panijus*	272-274	115-120	Gill	40-60 (48.5)	ND	ND	–
(Dhobighat)				Muscle	12.5-ND (15.7)	ND	ND	-
				Liver	-	-	-	-
				Kidney	30-40 (5.5)	ND	ND	-
				Gut (intestine)	25-65 (47.5)	ND-30 (11.5)	ND	-
				Gonad	-	-	-	-
	Rita rita	420-430	900-920	Gill	11-15 (13.3)	ND	15-19.2 (17.4)	ND
				Muscle	10-13 (11)	2-4(3)	ND	ND
				Liver	30-60 (28.7)	ND	ND	ND
				Kidney	17.5-25 (21.5)	5-15 (10)	ND	ND
				Gut (intestine)	17.5-25 (18.7)	ND	ND	ND
				Gonad	137.5-146.8 (114.5)	ND	ND	ND
Nawabganj	*Hilsa ilisha*	379-493	650-1475	Gill	20-24* (21.4)	ND	ND	ND
					22.4-26 (24.1)	ND	ND	ND
				Muscle	7.5-9.5 (8.0)	ND	ND	ND
					6.4-8.0* (7.2)	8-11 (8.4)	ND	ND
				Liver	13.8-16.8 (15.1)	2-8 (5)	ND	ND
				Kidney	17-24* (21.4)	11-17 (14.4)	ND	ND
				Gut (intestine)	14-19 (18)	2-8 (5)	ND	ND
					4.5-8.5* (7.4)	ND	ND	ND
				Gonad	5-35 (10)	ND	ND	ND

*Smaller Size group (379 mm/650g); ND-Non-detectable. figures in the parenthesis denote average values.

Source: Ghosh, B.B, M.M. Bagchi and D.K. De (1983).

Metals were found to accumulate in fish, shrimp crab and molluscs (cf. Table 8A) The upper level of zinc was found to accumulate in fish (24-120 ppm); in shrimp. (24-112.2 ppm); in crab, (40 ppm) and in molluscs (83.0 ppm). Similarly, the upper level of copper was 20 ppm; Chromium 0.26-4.5 ppm and lead could not be detected.

Fish tissues are also found to accumulate metals appreciably (Table 8B). In gill zinc was found to accumulate in the range 11-60 ppm, in muscle 7.5-22.5 ppm, in kidney 17.5-40 ppm, in gut (intestine) 16-65 ppm and in gonad, 5-146.8 ppm. Copper accumulation was comparatively less and its accumulation in muscle is 2-11 ppm, in liver, 2-8 ppm, in kidney, 5-15 ppm, and in gut, 2-30 ppm respectively. Chromium was found to be present in gill of only one fish (15-19.2 ppm) and lead was absent.

Thus it is clear that zinc is present in highest level and copper although present is in a much lower concentration. Incidence and magnitude of DDT and BHC-V residues in fishes of tidal stretch of the Ganga were of higher order in the industrial zone compared to riverine (13 per cent) and estuarine agricultural zone (9 per cent). Greater acumulation of DDT was

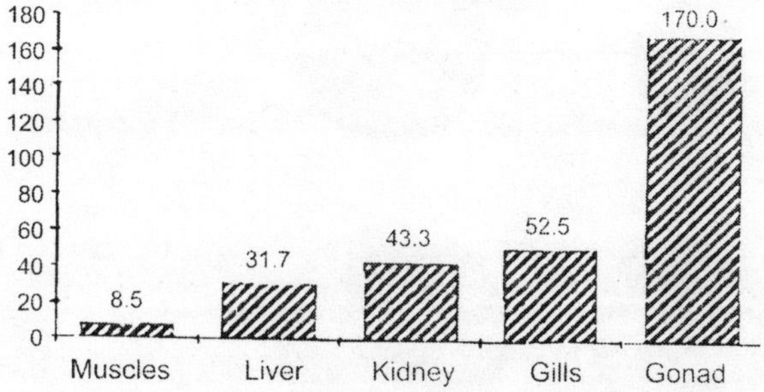

Fig. 5 : Bioaccumulation of zinc (in mg/gl) in tissues of Rita rita in Hooghly estuary.
Source : Research Highlights, CICFRI, Barrackpore, 1989-90.

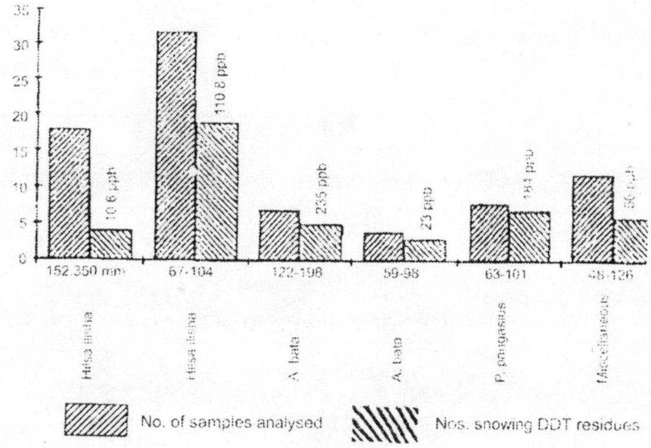

Fig. 6 : DDT Residues in Fishes in the Hooghly Estuary.
Source : Joshi, H.C. (1986).

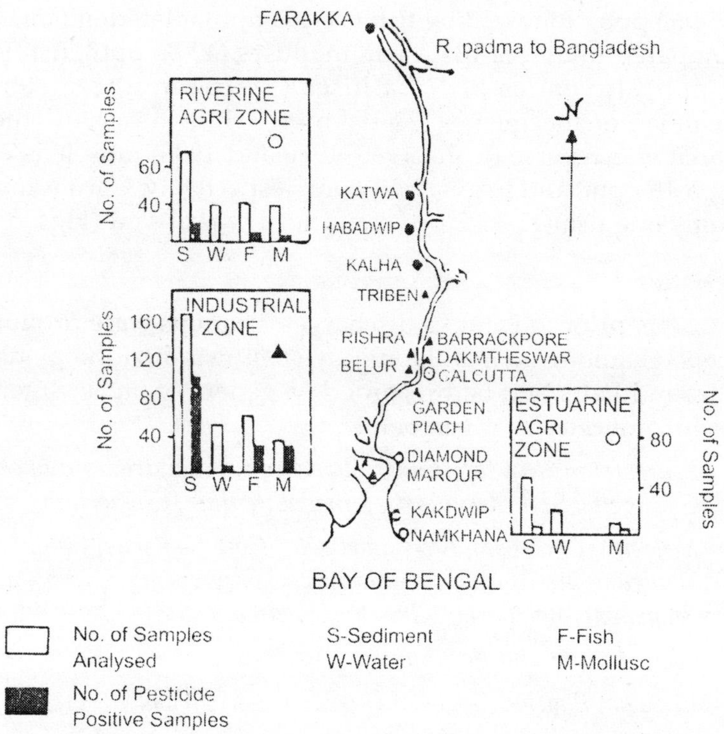

Fig. 7 : Pesticide Monitoring profile in the Hooghly Estuary.
Source : Joshi, H.C. (1985).

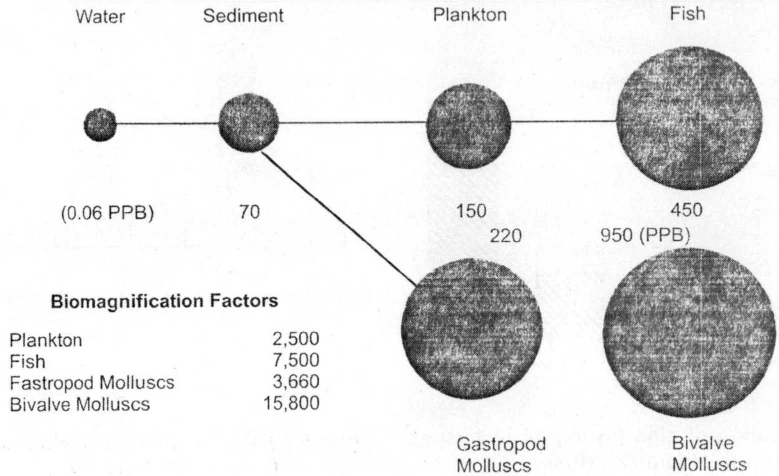

Fig. 8 : Biomagnification of DDT in Aquatic Food Chain in Hooghly Estuary.
Source : Joshi, H.C. (1988).

recorded in molluses (65-963 ppb), followed by fish (31-460 ppb), plankton (15-150 ppb) and sediment (17-80 ppb) compared to that of BHC-Y in molluses (40-86 ppb), fish (46-210 ppb) and sediments (21-70 ppb). Juveniles of *Hilsa ilisha* (67-104 mm) registered greater accumulation of DDT than larger specimens (122-198 mm), while in case of other estuarine fishes like *Apocryptes bato* it was reverse (Joshi, 1985, 1986 and 1988) muscle is 2-11 ppm; in liver, 2-8 ppm; in kidney, 5-15 ppm; and in gut, 2-30 ppm, respectively. Chromium was found to be present in gill of only one fish (15-19.2 ppm) and lead was absent (Figs. 5-8).

2.2 *Biological Characteristics*

In the brackishwater zone, prior to Farakka discharges, Diatoms were prominent among phytoplankton and copepods among zooplankton, Roy (1949) studied the plankton species of the Hooghly estury around Palta and had recorded 38 genera of algae, 4 genera each of protozoa, rotifera and schizomycetes as given below.

Chlorophyceae	:	*Pediastrum, Micractinum, Tetraedron, Ankistrodesmus, Actinastrum, Crucigenia Scenedesmus, Geminella, Penium, Closterium Cosmarium*
Bascillariophyceae	:	*Melosira, Cyclotella, Stephanodiscus, Coscinodiscus, Fragillaria, Synedra, Navicula, Stauroneis, Pleurosigma, Gamphonema, Cymbella, Encyonema, Rhopalodia Bacillaria, Nitzschia, Surirella.*
Dinophyceae	:	*Glenodinium and Peridinium*
Englenophyceae	:	*Euglena, Phacus, Trachelomonas and Ascoglena*
Myxophyceae	:	*Oscillatoria, Phormidium, Lyngbia, Anabaena, Aphanizomenon.*
Protozoa	:	*Sarcodina-Difflugia* *Mastigophora-Monas* *Infusoria-Codonella*

Rotifera : *Bdelloida-Rotifera*

Schizomycetes or higher bacteria

Chlamydobacteriales or Iron Bacteria– *Leptothrix, Didymohelix* and *Crenothnx*

Thiobacteriales or Sulphurbacteria– *Beggiatoa*

Subsequently Dutta *et al.* (1954) also found that diatoms are dominant plankton followed by green algae especially between Palta and Barrackpore. These freshwater forms were conspicuously absent between Palta and Diamond Harbour. The blue green algae although present throughout this section were not so significant as diatoms and green algae. They were found in the following sequence : diatoms 72, green algae 18, blue green algae 9; euglenoids 3 and dinoflagellets 3. zooplankton mostly consisted of freshwater and brackishwater species. A few marine forms were also recorded from Palta and Diamond Harbour (lower stretch) during March-June, when salinity was high. Freshwater species were predominant in the stretch between Palta and Diamond Harbour which had high turbidity. At Palta and Calcutta, zooplankton were maximum during November-March when turbidity was the lowest. The same was true at Palta and Diamond Harbour during October-January. (Table 9).

During 1956-58 in zones* I and II diatoms were found to dominate throughout the year and they were followed by green and blue green algae. Dominance of diatoms in 1959-62 has been reported by Saha *et al.* (1975). Shetty *et al.* (1961) and Saha *et al.* (1975) observed that protozoa, rotifera, copepoda, cladocera were dominant zooplankton in different zones during 1956-1967. Shetty *et al.* (1961) observed that there was a gradual fall in plankton from 1956 to 1968 in freshwater, intermediate and brackishwater zones. The yearly plankton fluctuations in the three zones are given in Fig. 4.9.

Effect of freshwater release from Farakka Barrage on the ecological condition of the Hooghly shows tht plankton population density has increased by 3-5 times in different zones of the stretch during 1975-77 (Nandy *et al.*, 1983) (cf. Table 11) as compared to those reported by Shetty *et al.* (1961). It now consisted mainly of freshwater forms. Marine forms were encountered from the lower zone. The upper zone from Kalana to Barrackpore, the chlorophyceae members were encountered in large numbers. The zooplankton consisted mainly of freshwater and brackishwater species in the upper and the middle stretch of the Hoohly and Roopnarayan estuary.

Diatoms and copepods wre found to dominate the phyto and zooplankton respectively. Phytoplankton mainly belong to two major classes with Bascillariophyceae having the maximum representation in terms of species (18) followed by chlorophyceae (6). The fauna belong to two major phyla of which copepoda was the most abundant and having the maximum representation. Copepod of calanoid and cyclopid forms with various copepodid and nauplii stages. Plankton population during pre and post Farakka discharges at various distances in the Hooghly from seaface are indicated in Table 11. The zonewise distribution of main plankton groups during pre and post Farakka discharges are also shown in Table 12. Besides a list of phytoplankton identified earlier during 1949-51 and also their Zonewise distribution are indicated in Tables 9 and 10.

*Zone I : Medgachi-Konnagar; Zone-II : Konnagar Diamond Harbour.

Studies on Benthic fauna in the estuary, revealed that prior to Farakka discharge the fauna was contributing poorly towards productivity. Benthic organisms are mostly confined to the freshwater zone of the estuary which is comparatively shallow and mostly free from tidal influence.

The most common forms were chironomid and oligochaete larvae, gastropods : *Melanoides tarebia* (Müller), *M. lineatus* (Grey), *M. Luteola f. ovalis* (Grey), *M. granifera* (Lam), *M. tuberculata* (Müller), *Thiara scabra* (Müller), *vivipara bengalensis* (lam), *V. dissimalis* (Müller), *Assiminea francissii* (wood); megalapa larvae; Stenothyra deltae (Bensono); *Macrobrachium Spp. Palaemon rude*, etc. (Ray, 1980). their density was low except at Hazinagar which represents the recovery zone of pulp and paper mill effluents. Benthic biota remains absent around the outfall areas.

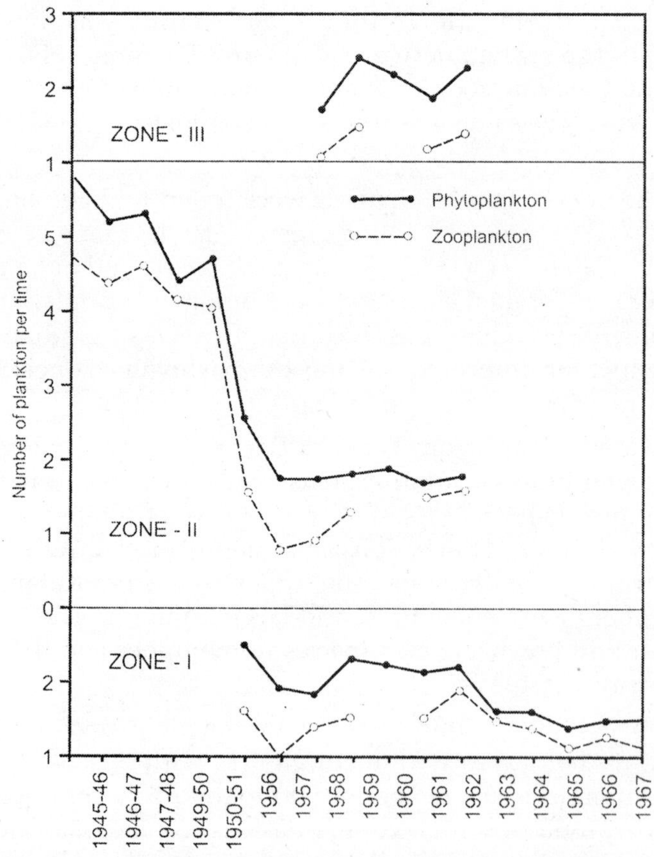

Fig. 9 : Average yearly fluctuation of phyto and zooplankton population on logarithmic scale in three broad zones in the Hooghly estuary.

Source : Ray, P. (1981).

Table 9. A list of Phytoplankton identified at the four Stations on the Hooghly during 1949-51

I. Diatoms

1. *Melosira granulata* (Ehr.) Ralfs
2. *M. granulata var. angustissima* (Ehr.) Ralfs. Mull.
3. *Skeletomema Costatum* (Grev.) Cleve
4. *Thalassiosira marinata* Cleve
5. *Cyclotella Meneghiniana* Kutz
6. *C. Kutzingiana* Thwaires
7. *C. striata* (Kutz.) Grunow
8. *Coscinodiscus Gramu* Gough
9. *C. radiatus*
10. *C. commutatis*
11. *C. gigas* Ehr.
12. *Coscinodiscus 5* spp.
13. *Asteromphalus cleaveanius* Grunow
14. *Planktoniella Sold* (Wallich) Schutt
15. *Bacillana paradoxa* Gmelin
16. *Actinocyclus Ehrenbergu* Ralfs
17. *Corethron hystrix* Hensen
18. *Laudena* sp.
19. *Leptocylindrus danicus* Cleve
20. *L. minimus* Gran
21. *Guinardia* sp.
22. *Rhizosolenia robusta* Normau
23. *R. Stolterfothii* H. Peragallow
24. *R. imbricata* Brightwell
25. *R. styliformis* Brightwell
26. *R. alata Brightwell forma indica* (Peragallo) Ostenfeld
27. *R. alata Brightwell forma inermis* (Castracane) Hustedt
28. *Bacter astrum hyalinum* Lauder
29. *B. vanans* Lauder
30. *B. elegans* Lauder
31. *Chaetozeros orientalis* Schiller
32. *Ch. perwvianus* Brightwell
33. *Ch. affinis* Lauder
34. *Ch. curvisetus* cleve
35. *Eucampia zoodiacus* Enr.
36. *E. comuta* (Cleve) Grunow
37. *Climacodium* sp.
38. *Ditylum Brightwellii* (West) Grunow
39. *Triceratium* sp.
40. *Biddulphia mobiliensis* Bailey
41. *B. sinensis* Greville
42. *B. rhombus* (Enr.) W. Smith
43. *Cerataulina* sp.
44. *Hemiaulus sinensis* Greville
45. *Hemidiscus Harmannianus* (Greville) Mann.

contd....

Table 9 – contd...

46.	*Navicula pygmaeea* Kutz
47.	*Navicula* sp.
48.	*Synedra ulna* (Nitzsch) Ehr.
49.	*Lithodesmium* sp.
50.	*Rhopalodia gibba* (Ehr.) O. Mull.
51.	*Thallassionema Nitzschiodes* Grunow.
52.	*Thallassiothrix longissima* Cleve and Grunow
53.	*Asterionell japonica* Cleve
54.	*Gyrosigma* sp.
55.	*Pleurosigma angulatum* (Quekett) W. Smith
56.	*Pleurosigma* sp.
57.	*Streptotheca* sp.
58.	*Diploneis* sp.
59.	*Pinnularia* sp.
60.	*Amphora* sp.
61.	*Nitzschia vitrea* Norman
62.	*N. seriata* Cleve
63.	*N. closterium* (Ehr.) W. Smith
64.	*N. longissima* (Brebisson) Ralfs
65.	*N. sigma* (Kutz) W. Smith
66.	*Nitzschia* sp.
67.	*Fragillaria intermedia* (Grun.)
68.	*Surriella elegans* Ehr.
69.	*S. tenera* Gregory
70.	*Cymbella cistula* (Hemp.)
71.	*Gomphonema* sp.
72.	*Achanathes* sp.
II. Green Algae	
1.	*Volvox* sp.
2.	*Eudorina elegans* Ehr.
3.	*Tetraedron* sp.
4.	*Ankistrodesmus* sp.
5.	*Pandorina* sp.
6.	*Pediastrum clathratum* (Schroeter) Lemmerman
7.	*P. duplex* Meyen
8.	*Crucigenia* sp.
9.	*Scenedesmus quadricauda* (Turpin) Brebisson
10.	*Spirogyra* (3 spp.)
11.	*Tribonema* sp.
12.	*Closterium* sp.
13.	*Cosmarium* sp.
14.	*Coelastrum* sp.
15.	*Mougeota* sp.
16.	*Oocystis* sp.
17.	*Zygnema* sp.

contd....

Table 9 – contd...

18.	*Selenastrum* sp.
III. Blue-Green Algae	
1.	*Microcystis aerugionosa* Kutz
2.	*M. flos-aquae* Kutz
3.	*Merismopedia* sp.
4.	*Lyngbya* sp.
5.	*Oscillatoria* (3 spp.)
6.	*Phormidium* sp.
7.	*Spirulina* sp.
8.	*Anabaena* sp.
9.	*Aphanocapsa* sp.
IV. Euglinineae	
1.	*Phacus* sp.
2.	*Euglena acus* Ehr
3.	*E. oxyuris* Ehr.
V. Dino Flagellates	
1.	*Peridinium* sp.
2.	*Ceratium hirundinella* O F Muller
3.	*C. tripos* O. F. Müller

Source : Dutta. N., J.C. Maihotra B.B. Bose (1954)

4 Stations are : I-Palta, II-Calcutta, III-Falta & IV-Diamond Harbour

Table 10. List of identified plankters and their zonewise distribution ('x' = Present, .. = Absent)

		Zone		
		I	*II*	*III*
(a) Diatoms				
1.	*Synedra ulna*	x	x	x
2.	*Coscinodiscus granii*	x	x	x
3.	*C. radiatus*	..	x	..
4.	*C. commutatis*	..	..	x
5.	*C.* spp.	x	x	x
6.	*Surirella elegans*	x	x	x
7.	*Melosira granulata*	x	x	x
8.	*M.* spp.	x	x	x
9.	*Nitzschia vitrea*	x	x	x
10.	*Navicula* spp.	x	x	x
11.	*Pleurosigma* spp.	x	x	x
12.	*P. elongata*	..	..	..
13.	*Amphora* sp.	x	x	..
14.	*Climacodium* sp.	x	x	x
15.	*Asterionella japonica*	x	x	x
16.	*Stephanodiscus* sp.	x	x	x

contd....

Table 10 – contd...

			Zone	
		I	*II*	*III*
17.	*Cyclotella* spp. ..	x	x	x
18.	*Cymbella cistula* ..	x	x	x
19.	*Chactoceros* spp. ..	x	x	x
20.	*Diploneis* sp. ..	x	x	x
21.	*Bidhulphia mobiliensis* ..	x	x	x
22.	*Bacillaria paradoxa* ..	x	x	x
23.	*Mastogloea* sp. ..	x	..	..
24.	*Meridion* sp. ..	x	x	..
25.	*Planktoniella* sol ..	x	..	x
26.	*Lithodesmium* spp. ..	x	x	x
27.	*Thalassiothrix longisima* ..	x	x	x
28.	*Ditylum brightwellii* ..	x	x	x
29.	*Rhopalodia* sp. ..	x	x	..
30.	*Gomphonema* sp. ..	x	..	..
31.	*Diatoma* sp. ..	x	..	x
32.	*Eucampia* sp. ..		x	x
33.	*Actinocyclus* sp. ..	x	x	..
34.	*Hemidiscus hardmannianus* ..	..	x	x
35.	*Hemiaulus sinensis* ..	..	x	x
36.	*Sceletonema costatum* ..	..	x	..
37.	*Triceratium* sp. ..	..	x	x
38.	*Epithemia* sp. ..	..	x	..
39.	*Rhaphoneis* sp. ..	..	x	..
40.	*Gyrosigma* spp. ..	..	x	x
41.	*Rhizosolenia* spp. ..	..	..	x
42.	*Bacteristrum* spp. ..	..	..	x
43.	*Lauderia* sp. ..	..	..	x
44.	*Cocconeis* sp. ..	..	..	x
45.	*Cerataulina* sp. ..	..	..	x
46.	*Thalassionema* sp. ..	..	..	..
47.	*Thallasiosita marginata* ..	..	..	..
48.	*Aulacodiscus* sp. ..	x	x	x
49.	*Fragilaria* sp. ..	x	x	..
50.	*Pinnularia gibba* ..	x	x	..
(b) Green Algae				
1.	*Pediastrum simplex* ..	x	x	..
2.	*P.* spp. ..	x	x	x
3.	*Mougeotia* sp. ..	x	x	..

contd....

Table 10 – contd...

			Zone I	II	III
4.	*Closterium* sp.	..	x	x	..
5.	*Ankistrodesmus* sp.	..	x	..	x
6.	*Cosmarium* sp.	..	x	..	..
7.	*Spirogyra* spp.	..	x	x	x
8.	*Eulorina elegans*	..	x	x	..
9.	*Pleodorina* sp.	..	x	..	x
10.	*Actinastrum* sp.	..	x	..	..
11.	*Tribonema* sp.	..	x	..	x
12.	*Euastrum* sp.	..	x	..	..
13.	*Chlorella* sp.	..	x	x	..
14.	*Scenedesmus quadricauda*	..	x	x	..
15.	*Zygnema* sp.	..	x	..	..
16.	*Desmidium* sp.	..	x	x	x
17.	*Pandorina morum*	..	x	x	..
18.	*Platydorina caudatum*	..	x	..	..
19.	*Borgea planktonica*	..	x	x	x
20.	*Treubaria* sp.	..	x	..	..
21.	*Chlorococcus* sp.	..	x	x	..
22.	*Closteriopsis* sp.	..	..	x	..
23.	*Sturastrum* sp.	..	x	..	..
24.	*Oedogonium* sp.	..	..	x	..
25.	*Ulothrix* sp.	..	x	..	..
26.	*Micrasterias* sp.	..	..	..	..
27.	*Microspora* sp.	..	..	x	..
28.	*Volvox* sp.	..	x	x	..
29.	*Coelastrum* sp.	..	x	..	..
30.	*Kirchneriella* sp.	..	x	..	..
(c) Blue-green Algae					
1.	*Microcystis* spp.	..	x	x	..
2.	*Phormidium* spp.	..	x	x	x
3.	*Nostoc* spp	..	x	x	x
4.	*Oscillatoria* spp.	..	x	x	x
5.	*Anabaena* spp.	..	x	x	x
6.	*Nostochopsis* sp.	..	x	..	..
7.	*Gloeotrichia* sp.	..	x	x	..
8.	*Apharizomenon* sp.	..	x	x	..
9.	*Cylindrospermum* sp.	..	x	..	..
10.	*Merispmopedia* sp.	..	x	..	x

contd....

Table 10 – contd...

			Zone	
		I	*II*	*III*
11.	*Spirulina major* ..	x	x	..
12.	*Lyngbya* sp. ..	x	x	x
13.	*Aphanocapsa* sp. ..	x	..	..
14.	*Coelosphaerium* sp. ..	x	..	..
15.	*Plectonema* sp. ..	..	x	..
16.	*Trichodesmium* sp. ..	..	x	x
17.	*Rhizoclonium* sp. ..	x	..	..
18.	*Wollea* sp. ..	x	..	..
(d)	**Flagellates**			
1.	*Euglena* sp. ..	x	x	x
2.	*Dinobryon* sp. ..	x	x	..
3.	*Ceratium hirudinella* ..	x	x	x
4.	*C. tripos* ..	..	x	x
5.	*Phacus* sp. ..	x	x	x
6.	*Peridinium* spp. ..	x	x	x
7.	*Noctiluca miliaris* ..	x	..	..
8.	*Trachelomoas* sp. ..	x	..	..
(e)	**Protozoa**			
1.	*Difflugia* spp. ..	x	x	x
2.	*Arcella* spp. ..	x	x	x
3.	*Bursaria* sp. ..	x	x	..
4.	*Vorticella* sp. ..	x	x	..
5.	*Nuclearia* sp. ..	x	..	..
6.	*Centropyxis* sp. ..	..	x	..
7.	*Proteomyxa* sp. ..	..	x	..
8.	*Tintinnidium* sp. ..	..	..	x
9.	*Globigerina* sp. ..	..	..	..
(f)	**Rotifera**			
1.	*Brachionus* spp. ..	x	x	..
2.	*Keratella* spp. ..	x	x	..
3.	*Trochosphaera* sp. ..	x	..	..
4.	*Tetramastix* sp. ..	x	..	..
5.	*Trichocerca* sp. ..	x	..	..
6.	*Asplanchna* sp. ..	x	..	..
7.	*Filinia* sp. ..	x	..	..
8.	*Hexarthra* sp. ..	x	x	..
9.	*Nothoica* ..	..	..	x
10.	*Ploesoma* sp. ..	..	..	..

contd....

Table 10 – contd...

		Zone I	Zone II	Zone III
(g) Copepoda				
1.	*Pseudodiaptomus* spp. ..	x	x	x
2.	*Diaptomus* spp. ..	x	x	x
3.	*Microsetella* sp. ..	..	x	x
4.	*Acartiella torianiformis* ..	x	x	..
5.	*A. major* ..	..	x	..
6.	*Paracalanus* sp. ..	..	..	x
7.	*Cyclops* sp. ..	x	..	..
(h) Cladocera				
1.	*Bosmina* sp. ..	x	x	..
2.	*Bosminopsis* sp. ..	x	x	x
3.	*Ceriodaphnia* sp. ..	x	x	..
4.	*Daphnia* sp. ..	x	x	..
5.	*Holopedium* sp. ..	x	..	..
6.	*Monia* sp. ..	x	x	..
7.	*Diaphanosoma* sp. ..	..	x	..
(i) Miscellaneous				
1.	*Mesopodopsis orientalis* ..	x	..	..
2.	*Sagitta* sp. ..	..	x	x

Zone I : Medgachi, Khasbati and Khusigoli (upper zone).

Zone II : Konnagar, Fuleswar and Diamond Harbour (Middle zone).

Zone III : Kakdwip, Namkhana and Fragerganj (Lower zone)

N.B. *Source :* Shetty, H.P.C; S.B. Saha and B.B. Ghosh (1961).

Table 11. Distribution in the plankton population of the Hooghly during pre and post Farakka discharges

Kalna 225 km. from sea face		*Barrackpore 171 km. from sea face*		*Uluberia 126 km. from sea face*		*Nurpur 86 km. from sea face*		*Kakdwip 40 km. from sea face*	
Pre- 1960-61	*Post- 1975-77*	*Pre- 1960-61*	*Post- 1975-77*	*Pre- 1960-61*	*Post- 1975-77*	*Pre- 1960-61*	*Post- 1975-77*	*Pre- 1960-61*	*Post- 1975-77*
Phyto plankton	219.38 154.8 (FW)	32.18	162.16 73.88 (FW)	48.70	91.24 7.02 (FW)	18.16	64.16 20.63 (FW)	39.81	48.70 Nil (FW)
No./1. 68.04	2.20 (BW) 62.29 (C)		24.45 (BW) 63.83 (C)		56.31 (BW) 27.91 (C)		24.0 (BW) 19.52 (C)		44.34 (BW) 4.36 (C)
Zooplankton	84.99	7.97	60.62	11.51	53.55	7.88	89017	6.90	7.97
No/1. 26.88	55.32 (FW) 8.31 (BW) 21.36 (C)		30.42 (FW) 19.13 (BW) 11.07 (C)		1.90 (FW) 46.10 (BW) 5.55 (C)		24.80 (FW) 55.32 (BW) 9.05 (C)		Nil (FW) 5.33 (BW) 2.64 (C)

FW-Freshwater ; BW-Brackishwater ; C-Common

Source : Nandy, A.C., M.M. Bagchi and S.K. Mazumdar (1983).

Table 12. Zonewise distribution of the main Plankton Group (No/l) during pre and post Farakka discharge.

A. Phytoplankton Organisms	*Zone–I* Pre-1960-61	*Zone–I* Post-1975-77	*Zone–II* Pre-1960-61	*Zone–II* Post-1975-77	*Zone–III* Pre-1960-61	*Zone–III* Post-1975-77	*Zone–IV* Pre-1960-61	*Zone–IV* Post-1975-77
Diatom	55.48	121.08	30.54	133.33	42.78	89.75	17.16	40.83
		FW 71.08		FW 53.23		FW 6.61		FW 10.13
		BW 2.00		BW 24.10		BW 56.29		BW 23.41
		C 48.00		C 56.00		C 26.85		C 7.29
Green algae	10.05	93.25	0.92	23.83	4.51	1.41	0.26	16.50
		FW 80.76		FW 19.27		FW 0.41		FW 7.90
		C 12.49		C 4.06		C 1.00		C 9.00
Bluegreen algae	2.42	5.01	0.69	5.50	0.17	0.08	0.75	6.83
		FW 3.01		FW 1.38		FW Nil		FW 3.00
		BW 0.20		BW 0.35		BW 0.02		BW 0.60
		C 1.80		C 3.77		C 0.06		C 3.23
Englenoids	0.09	0.04	0.03	-	1.24	-	0.01	-
TOTAL...........	63.04	219.38	92.18	162.16	48.70	91.24	18.16	64.16
		FW 154.89		FW 73.88		FW 7.02		FW 20.63
		BW 2.20		BW 24.45		BW 56.31		BW 24.01
		C 62.29		C 63.83		C 27.91		C 19.25

FW-Freshwater ; BW-Brackishwater ; C-Common.

B. Zooplankton Organism

B. Zooplankton Organism	*Zone–I* Pre-1960-61	*Zone–I* Post-1975-77	*Zone–II* Pre-1960-61	*Zone–II* Post-1975-77	*Zone–III* Pre-1960-61	*Zone–III* Post-1975-77	*Zone–IV* Pre-1960-61	*Zone–IV* Post-1975-77
Copepod	3.49	46.17	2.52	48.11	5.67	46.50	4.32	64.25
		FW 28.4		FW 22.04		FW 1.40		FW 10.15
		BW 5.03		BW 18.00		BW 43.10		BW 48.10
		C 13.10		C 8.07		C 2.0		C 6.00
Cladocerans	11.48	24.33	0.42	7.00	0.04	1.50	0.47	13.00
		FW 23.28		FW 6.00		FW 0.50		FW 10.00
		BW 1.05		BW 1.00		BW 1.00		BW 3.00
		C Nil		C Nil		C Nil		C Nil
Rotifers	0.43	4.66	0.24	5.51	0.17	0.07	0.26	7.25
		FW 3.0		FW 2.38		FW Nil		FW 3.05
		BW 0.10		BW 0.13		BW Nil		BW 2.15
		C 1.56		C 3.00		C 0.07		C 2.05
Others	11.48	9.83	4.79	-	5.58	5.68	2.93	4.67
		FW 1.00		FW Nil		FW Nil		FW 1.16
		BW 2.13		BW Nil		BW Nil		BW 2.51
		C 6.70		C Nil		C 3.58		C 1.00
Total :	26.88	84.99	7.97	60.62	11.51	53.55	7.88	89.17
		FW 55.32		FW 30.42		FW 1.90		FW 24.80
		BW 8.31		BW 19.13		BW 46.10		BW 55.32
		C 21.36		C 11.07		C 5.55		C 9.05

Source : Nandy, A.C.M., M. Bagchi and S.K. Majumdar (1983).

2.3 *Pollution Status of the Estuary*

The deltaic tidal stretch encounters heavy discharges from industrial effluents (430 MLD) from more than 96 factories such as, pulp and paper, distillery and yeast, viscose rayon, cotton textile, tannery, paints and varnishes, rubber and cycle rim, metal and steel safety match, Chemicals, Thermal power plants, pesticides (BHC) unit hydrogenated vegetable oil and soap, phosphate fertilizer, oil storage. Chloro-alkyl plant, antibiotic etc. flanked on both the banks between Nayasarai and Birlapur (about 92 km) and oil refinery and pesticide (dimethoate) manufacturing unit at Haldia (150 km down stream of Calcutta) beside pesticide residues through urban and agricultural run off. Organic components of industrial wastes (240 MLD) in the tidal stretch make a total BOD_5 demand of 55.1t/d of which lignocellulosic matter makes a BOD demand of 28.6t/d (Ghosh *et al.*, 1973). Further total pollution load separately, their broad classification and chemical characteristics of some industrial domestic and municipal wastes are also indicated in Tables 13, 14 and 15. The stretch also receives oil (0.31t), heavy metal like zinc, copper and chromium are drained from viscose rayon (8 mg/l), CottonTextile (19 mg/l) and Tanneries (28 mg/l) respectively. Wide spectrum of heavy metal contaminants from various industries released into the basin are given in Table 16. Industries Causing Pollution in the Bhagirathi-Hoghly are indicated in Table 13.

Table 13. Industries causing Pollution in the Bhagirathi-Hooghly.

Lower Ganga Plain	
Pulp and Paper	Miscellaneous
Sulphite Process	Paints and Varnishes
Kraft Process	Polythene
Soda Process	Hydrogenated vegetable oil and soap
	Safety match
Textile	Chlorine
	Rubber chemicals
	Insecticides
Natural Cotton	Phosphate fertilizer
Synthetic rayon	Shellac
	Pesticides
	Antibiotics
	Chloro-alkyl plant
Tannery	
Rubber	
Distillery	
Power (thermal)	
Oil	
Petroleum	
Refinery	

Ghosh *et al.* (1983) while studying the pollution status in the Hooghly estuary with reference to heavy metals disposed through industrial wastes have revealed valuable information on the behaviour of metals in the ecosystem. For example, metal content were studied in the industrial effluents, their concentration in sediments and water, bioaccumulation in some fishes and in fish tissues, shrimps, molluscs and crabs, biomagnification of metals (based on lowest ambient concentration of some metals). DDT residues in fish an their biomagnification, impact on plankton and benthos etc. The results are given in Table 8A, B and Figs. 4.5-4.8.

Table 14. Pollution Load Contributed by Various Industries Estates and Municipalities with their Corresponding Percentages.

Nature of waste water	*Vol. of waste Water (m^3/day)*	*BOD_5 tonnes/day*	*Total solids tonnes/day*	*Suspended Solids tonnes/day*	*Dissolved Solids tonnes/day*
Industrial Wastes					
Pulp & Paper Industry	178612.34 (15.5%)	28.59 (26.90%)	547.40 (23.61%)	299.20 (28.29%)	248.20 (19.68%)
Textile Industry	20366.08 (1.77%)	7.26 (6.83%)	71.80 (3.10%)	41.82 (3.95%)	29.98 (2.38%)
Distillery & yeast	675.08 (0.06%)	10.29 (9.69%)	5.242 (0.22%)	0.402 (0.04%)	4.84 (0.38%)
Tannery	11365.00 (0.99%)	4.54 (4.27%)	86.70 (3.74%)	13.50 (1.28%)	73.20 (5.81%)
Rubber	10910.40 (0.95%)	0.43 (0.4%)	8.70 (0.38%)	1.70 (0.16%)	7.0 (0.56%)
Oil Refinery	6600.00 (0.49%)	0.158 (0.15%)	9.976 (0.43%)	0.215 (0.02%)	9.761 (0.07%)
Miscellaneous	129728.29 (11.24%)	14.92 (13.95%)	164.235 (7.08%)	55.305 (5.23%)	108.93 (8.64%)
Domestic and Municipal Wastes	79555.00 (68.95%)	40.08 (37.72%)	1424.36 (61.44%)	645.40 (61.03%)	778.96 (61.78%)
Total	1153807.19	106.268	2318.413	1057.542	1260.871

Source : Ray P. (1981)

Table 15. Broad Classification of the waste waters and their Pollution Load/contributed into the Hooghly Estuary with Percentages.

	Vol. of waste Water (m^3/day)	*BOD_5 tonnes/day*	*Total solids tonnes/day*	*Suspended Solids tonnes/day*	*Dissolved Solids tonnes/day*
A. Wastes Containing Chiefly Organic Materials :					
Biological wastes[1] (wastes from processing of biological materials and/or from biological process)	196471.30 (73.49%)	45.91 (85.10%)	680.642 (87.77%)	346.522 (93.75%)	334.11 (82.33%)
Miscellaneous organic Chemical wastes[2]	32770.41 (12.26%)	2.598 (4.82%)	41.281 (5.32%)	9.520 (2.58%)	31.761 (7.83%)
Hydrocarbon wastes[3]	10910.40 (4.08%)	0.43 (0.80%)	8.70 (1.12%)	1.70 (0.46%)	7.0 (1.72%)
B. Wastes Chiefly Mineral in Nature or Partly Mineral and partly Organic[4]	27185.08 (10.17%)	5.01 (9.29%)	44.83 (5.78%)	11.90 (3.22%)	32.93 (8.11%)

1. Pulp and paper, textile (cotton), tanneries, distillery and yeast.
2. Paints and varnishes, shellac, oil refinery and hydrogenated vegetable oil and soap.
3. Rubber.
4. Metal and Steel, textile (rayon) and match.

Source : Ray, P. (1981).

Table 16. Wide spectrum of heavy metal contaminents released ino the environment

Industry	*Zn*	*Cu*	*Fe*	*Mn*	*Pb*	*Cr*	*Ni*	*Cd*	*Sn*	*CO*	*Sn*	*CN*	*Hg*	*As*	*Se*	*Ag*	*B*	*Phenol*
Electroplating	•	•				•	•	•					•					
Textile Waste	•	•				•	•	•										
Iron and Steel			•	•														
Iron and Ferro manganese	•	•	•	•	•													
Synthetic rubber PVC	•				•			•	•									
Soap detergents																	•	
Pulp and Paper	•	•			•	•	•						•					
Organic Chemical and petrochem	•		•		•	•		•			•		•			•		
Fertilizer	•	•	•	•	•	•	•	•					•	•				
Petroleum Refinery	•	•	•		•	•	•	•										
Leather tanning and finishing						•												
Alkalies, Chlorine Inorganic Chemicals	•		•		•	•		•			•		•					
Paints and Varnishes	•				•	•		•										
Perfumery, drug pharmaceuticals & Chemicals			•		•	•		•					•					
Galvanising	•							•										
Cosmetics																		
Battery/Battery Plant.	•				•		•	•				•			•			
Fly ash	•				•	•	•	•					•	•	•		•	
Cement factory	•														•			
Domestic Waste	•	•			•	•	•	•						•		•		
Spoil Heaps								•		•				•				•
Photography development & X-ray plates.								•								•		
Fish Processing	•	•				•	•	•										
Meat Processing	•	•				•	•	•										
Bakery	•	•				•	•	•										
Brewery, soft drinks and flavouring	•	•				•	•	•										
Ice Cream	•	•				•	•	•										
Laundry						•	•											
Soft drink																		

1. Highly toxic to bacteria, soil flora and plants. 2. Powerful herbicide.

Table 17. Concentration of some heavy metals estimated (mg/l) in various industrial wastes. Ghosh *et. al.* (1983).

	Zinc	*Copper*	*Chromium*	*Lead*	*Remarks*
Rayon (Kuntighat)	0.06-8.50 (3.22)	ND	ND	ND	ND-Non detectable
Rubber & Cycle Rim (Sahaganj)	ND-0.35 (0.12)	ND	ND-1.92 (0.68)	ND	Fig. in the parenthesis denote average values
Paints & Varnish (Garifa)	ND-0.42 (0.22)	ND	0.03-0.30 (0.14)	ND	
Paints & Varnish (Shalimar)	ND-5.59 (1.01)	ND-0.20 (0.03)	ND-0.60 (0.14)	ND	
Cotton Textile (Garden Reach)	ND-0.22 (0.12)	ND	ND-1.23 (0.28)	ND	
Tannery (Batanagar)	ND-0.50 (0.25)	ND-0.50 (0.26)	0.2-18.50 (6.85)	ND	
Oil Refinery (Haldia)	ND-0.10 (0.01)	ND	ND-0.12 (0.02)	ND	

Table 17 indicates presence of Zinc, Copper and Chromium in the effluents of tannery, and paints and varnishes; (Shalimar) Zinc and chromium in rubber and cycle rims, Cotton textile and oil refinery effluents, Zinc in Rayon effluents, while led was non-detectable. As high as 8.5 ppm of zinc in Rayon effluents (during summer), 18.5 and 0.5 ppm of chromium and copper respectively in tannery effluents were estimated.

Outfall of Rayon industry registered the highest zinc accumulation (381.6 ppm) in the sediments followed by paints and varnishes (160 ppm) and oil refinery (80 ppm), chromium level to the extent of 9.6 ppm was estimated at the outfall of paints and varnishes while outfalls of the upstream registered a non-detectable amount of chromium and copper and ND-10.7 ppm zinc. Zinc contaminated stretches extended upto 500 m below the outfalls of Rubber, paints and varnishes and oil refinery and chromium below the outfall of paints and varnishes. Copper was non-detectable both in sediment and the water. Higher values of zinc were recorded in the surface water of Rayon outfall and chromium at the outfall of Tannery.

Ghosh *et al.* (1983) have intensified the work further by estimating the accumulation of metals in different species of fish small and larger sizes, shrimps, crabs and molluces etc. around the outfall of some industries disposing wastes on the Hooghly estuary (Table 18).

High accumulation of copper was observed in shrimps; mollusc registered more zinc and copper accumulation than crab, while all the fishes showed zinc accumulation. Accumulation of copper was registered in fishes like *Hilsa ilisha, R. Corsula, P. pama, S. Panijas, R rita, G. giuris E. macrodor* and *M. menoda*. Chromium accumulation was indicated in a few fishes like *R. rita* and *O. rubicundus*. In shrimps, copper was estimated in *M. rude, M. Mirabile* and *P. sculptilis* while highest zinc was estimated in *M. Villosimanus*. Lead was not detected in all the test arrivals. Marked bioaccumulation of zinc (maximum 120.4 ppm) and copper (maximum 20 ppm) in fish like *E. macrodon* and chromium (maximum 4.5 ppm) in *O. rubicundus* was estimated. Although copper was non-detectable in rubber effluents fish and shrimps caught around the outfall of rubber effluents indicated its presence. This was possibly brought down from the effluent of Tribeni Tissue (Saha, 1980) located further up or partly due to migration of fish *E. macrodon* and shrimp, *M. rude.* The entire body of some juvenile fishes registered higher levels of zinc copper and chromium as compared to those in the adult fishes like *M. armotus* and *C. bacaila.*

Table 18. Bioaccumulation of metals in fish, shrimp, crab and mollusc around the outfall of the Hooghly estuary.

Industry	Zinc	Chromium	Copper	Lead
Rayon (Kuntighat)	(F.S)+	ND	ND	ND
Rubber and Cycle rim (Sahaganj)	(F.S)+	F+	(F.S)ND	ND
Paints and Varnish (Garifa)	F+	F+	ND	ND
Paints and Varnish (Shalimar)	(F.S.C.)+	F+	(S.C)$^{+}$	ND
Tannery (Batanagar)	S+	+	+	ND
Oil Refinery (Haldia)	(F.S.M.)+	+	(S.M.)ND	ND
Cotton Textile (Garden Reach)	F+	+	ND	ND

+ Indicates presence and ND not detected of metals in the Corresponding wastes of industries disposed in the river, Metals are found to be accumulated in Fish (F), Shrimp (S), Crab (C) and Molluscs (M) as the case may be

Source : Based on data of Ghosh, B.B., M.M. Bagchi and D.K. De (1983)

Table 18 shows bioaccumulation of copper in fish and shrimp at the outfall of Rubber and cycle rim where copper is absent. Bioaccumulation of copper in shrmp and crab at the outfall of Shalimar paints where copper was not present. Zinc was present in fish at outfall of cotton textile were zinc found absent. Finally in outfall of oil refinery, copper was found to accumulate in shrimp and molluscs where copper was absent. Such anomalies were possibly due to different types of tidal influence during different seasons that might be the cause of movement of organisms like fish, shrimp, molluscs and crab.

Ghosh *et.al.* (1983) further intensified the study by estimating the metals in fish tissues like, gill, muscle, liver, kidney, gut (intestine) and gonad of some fishes like *Silago panijus, Rita rita* and *Hilsa ilisha* collected from in and around Barrackpore (Table 17 for the four metals studied). Accumulation of zinc was common in all the three species *Silago Punijus, Rita rita* and *Hilsa ilisha,* highest value was estimated in the gonad of *Rita rita.* Copper was found to accumulate in muscle, liver and gut (intestine), while in *Rita rita* it was present in muscle and kidney and in case of *Silago Panijus* only gut content showed its presence. Chromium was found to be accumulated only in the gill of *Rita rita.* Lead was non-detectable either in Rita or hilsa fish. From the available data it is apparent that the problem of metallic pollution was mainly from zinc.

Estimation of Biomagnification of metals, based on the lowest ambient concentration of zinc (0.02 ppm) and chromium (0.03 ppm) show that the former was of the order of 7340 in gonad, 6020 (whole fish), 5610 in shrimp, 4300 in molluscs and 1400 in crab. For chromium, the values were 640 in gill, 150 (whole fish) and 8 in case of shrimp. In earlier observation of Ghosh and Bagchi (1981) the highest zinc concentration factor for the whole fish and shrimp was reported as 3527 and 3740 respectively.

Apart from industrial and muncipal wastes there is another source which contributes substantial metallic pollutants from the washings of images at the time of immersion in the Ganga every year Ghosh (1993) has revealed some valuable information on this aspect in the article "The alarming tale of the Ganga pollution." His data is based on three years during October to January around Calcutta (Between Barrackpore and Diamond Harbour) when all the immersion of major seasonal images takes place.* According to him normally the water contains 0.1 mg chromium, 0.05 mg zinc and 0.1 mg Copper per litre. During October to January the values of Chromium and lead increases 10 to 20 times and Zinc and Copper to 1.5 mg per litre. He has concluded that such progressive increase of the metallic pollutants is capable of upsetting the ecological balance requires to be controlled immediately. Ghosh (1993) however did not indicate the stretch he studied.

Recently Somendra Ghosh and Sudipta Bhattacharya of the Environmental Scientific organisations (S.M.G. High and Saviors and Friend of Environment) have revealed valuable informations on the immersion of Durga idol of Kolkata. According to them at least in 10,000 idols, chemicals are used for their shining. Analysis of samples collected from 100 ft away from the ghats (Outram ghat, Baghbazar, Kashipur, Ariadaha and Barrackpur reveals-suspended matter-1600-2000 mg/l; D.O.-2.5-3 mg/l; oily matter-17-28 mg/l and heavy metals-zinc, lead, chromium (hexavalent) concentrations were much above the standard level (Bartaman, 19.10.2005).

Besides various cours (both in the form of liquid and perfumed colour powders) that are largely used during the Holi festival, contain appreciable concentration of various metals are finally drained into the river through Municipal wastes. Such water may cause concern to the aquatic life of the river and hence should be properly treated.

Under such ecological condition planktonic concentration were poorly represented. Similar observations were recorded by Ray (1981) during March to June in the Hooghly estuary. In the present case upstream of various outfall regions indicated a range 2-43 U/l of plytoplankton and nil-16 U/l Zooplankton while at the outfall their ranges were lowered viz. nil-14 and nil-8 U/l respectively. (Ghosh *et al.*, 1983). The Chief organisms being *Coscinodiscus* sp. *Melosira* sp. and *Spirogyra* sp. among phyto-plankton and *Keratella* sp., *Brachionus* sp. and *Cyclops* sp. among Zooplankters. Comparing their data with Ray (1981) it is distinctly clear that concentration of plankters in the later period is much less than the former years. This indicates that even after Farakka release the concentration declined sharply when stability was attained.

In the Hooghly estuary average Zinc accumulation in fish (50-185 mm) ranged between 11.1 and 70.6 ppm; Shrimp (40-85 mm), 23-61.6 ppm; crab, 24.4 ppm and molluscus, 83 ppm; Copper 2-20 ppm in fish (54-115 mm); Shrimp (40-85 mm) ranged 8.1-24 ppm; molluscs, 20 ppm and chromium in fish (50-149 mm) ranged nd-3.1 ppm in the Hooghly estuary. Besides, in longer specimen, *Silago panijus* (272-274 mm), Zinc accumulation was recorded in gill, muscle, kidneys, gut (intestines) and gonad exept in liver, with highest concentration in the gill (40-60 ppm) and copper was present only in gut (nd-30 ppm). In Rita rita (420-430 mm) zinc accumulated in all the tissues stated above, the highest concentration recorded being in the gonad (137.5-146.8 ppm); copper was recorded in muscle (2-4 ppm) and in Kidney (5-15 ppm) and Chromium in gill only (15-19.2 ppm). In *Hilsa ilisha* (374-493 mm) except kidney in all other bissnes and organs stated above accumulated zinc, the highest concentration recorded was in gill (22.4-26.0 ppm); copper was recorded in muscle, liver, and gut, the highest concentration recorded was in liver (11-17 ppm).

It is interesting to note that zinc content in the Rayon waste was highest which was around 8.5 ppm and in water around its outfall, the concentration came down to 7.2 ppm. Its accumulation recorded in small sized fishes, shrimp, crab and molluscs was several times more than what was estimated in water. Even copper which could not be detected either in water or sediment was found to be very high both in small sized fish, shrimp and molluscs and in different tissues of large size fish. Lead was not detectable in all the test organisms. It is interesting note that Chromium accumulation indicated lowest value when it was abundant in the water around tannery waste.

Bottom dwelling animals like crab, shrimp and molluscs however accumulated a very high concentration of zinc, possibly some metal enriched sediments are ingested along with food material. This fact is revealed in two groups of fishes, plankton and bottom feeding crustaceans of average size 102 mm (6.9 gr.) in the case of former, the accumulation of zinc and copper were respectively 40.8 ppm and 14.0 ppm (average value); while in case of later, the bottom feeders of average size 67 mm (2.9 grs) only, the concentrations were almost similar i.e. 40.2 ppm and 14.4 ppm respectively. This shows the bottom feeding crustaceans even though were quite smaller in size, indicated accumulation of zinc and copper similar to larger sizes (plankton feeder).* This finding is based on the results obtained by Ghosh, *et al.* (1983). Possibly they might have ingested orally some precipitated from the river bed as opined by Whitley (1968).*

Of late Joshi (1985) in his pesticide residue monitoring in the Bhagirathi-Hooghly has made some interesting observations in water, sediment, fish, plankton, mollusc and suspended particulate matter. According to the author DDT, BHC, Dimethoate, Endosulfan residues were detected in 190, 57, 11 and 3 nos. of samples. DDT and BHC residues were not detectable in water samples. Molluscs showed maximum pesticide detection rate PDR (50 per cent) followed by suspended particulate matter (50 per cent), fish (43 per cent), sediment (37.7 per cent), plankton (28 per cent) and minimum in water (3 per cent). Pesticide positive samples showed presence of DDT in mollasses (65-963), fish (31-460), plankton (15-130) and sediment (17-89 ppb) while lower values were observed in case of BHC-y in molluscs (40-86), Fish (46-210) and sediment (21-70 ppb). Juvenile *H. ilisha* (67-104 mm) registered greater accumulation of DDT than those of the larger specimens (122-198 mm) while in case of estuarine fish like *Apocryptes bato* it was reverse (Joshi, 1986). Industrial zone registered highest pesticide residues (PDR 51 percent) PDR's separate values for riverine and estuarine agricultural zones were 13 per cent and 9 per cent respectively. Pre-monsoon (February-April) values for PDR were always higher than the monsoon (July-September) values. Biomagnification of DDT in plankton, fish, gastropods and bivalves has been 2,500, 7,500, 3,600 and 15,800 times of its ambient level in water (0.06 ppb) Joshi (1985, 1986, and 1988).

Joshi (1985) remarks that as a food commodity these levels of DDT and BHC in fish are far below the tolerance limits prescribed by WHO (1976). However their adverse effects on the health of fish have remained to be seen.

---

* In *Metal Pollution in aquatic environment* by Förstner, U. and G.T.W. Whittmann, 1979.

GLC Estimation of Pesticide residues in sediment and biotic samples

Type of Samples	No. of samples analysed	Pesticide residues in ppb	
		DDT*	BHC-V
Sediment	17	17-89	21-73
Fish	57	31-460	46-210
Molluscs	5	65-963	40-85
Plankton	3	15-130	Not detected

Source: Joshi, H.C. (1985) *DDT - DDT+DDE+DDD

Further down to Haldia near Diamond Harbour conditions are equally polluted due to waste water from various complex industries like oil refinery, fertilizer plant and other petrochemical industries. Besides mercury, the waste waters of Haldia port also contains other carcinogenic heavy metals, like lead and chromium above permissible levels. This was revealed from a pilot survey conducted recently by the Indian Institute of Health and Hygiene, Department of Marine Biology, Calcutta University and also by pollution control Board Laboratories, Calcutta (Simi Kamboj, 2003). Lately, from a study conducted by the Centre for Science and Environment (CSE) opioned fish in Coastal waters of the country have been diagnosed to have a high quantity of mercury in them. (*Statesman News Service*, 2003).

Marine specialists in the city said - as a result of discharges, the entire range of food chain in the water was in jeopardy. The rising mercury levels in the waters causes a process called bio-magnification which ultimately affects the whole food chain. The Government of West Bengal also accepted the high toxic levels in the waters.

On the basis of the findings by the surveys, the state Environment Department indentified 65 different factories which were discharging toxic effluents into the Haldia waters. The water contamination thus caused by the Haldia Industrial Project has assumed alarming proportions, and lives of thousands of fish-eating inhabitants in the area may be at grave risk. The world Bank who was interested in the Meen Dwip Project believes that Meen Dwip Island off Haldia by 3 km. may soon turn into another *Minamata* (Japan) where about 1000 people in these Japan Islands lost their lives and thousand more suffered brain damage, paralysis and loss of hearing and sight due to high mercury level in the waters. Mercury poisoning there was caused by the effluents discharge into the sea be a chemical Fetilizer company called "Nippon Chisso Hiryo" (Seema Paul, Delhi and Tapas Chakraborty, Kolkata, 1994).

In a recent study carried out in the vicinity of Sagar Island and Canning to assess the potential of orgainsms for accumulation of mercury and its environmental impact in the Hooghly estuary by the Calcutta University's Department of Marine Science of Jadavpur University's Department of Metallurgical Engineering, it was revealed that, the mercury content in the liver and kidney of Bhola Bhetki (*Lates Calcarifer*) and "Bombay Duck" (*Harpodan nehereus*) were found to range between 1.18 to 1.20 mg/kg, when the WHO's limit is 1 mg/kg.

The mercury concentration in the muscle of both these varieties of fish hovered around 0.70 to 0.80 mg/kg. and in rare cases, even stretched to 0.90 mg/kg. Dr. S.K. Sarkar, reader,

Department of Marine Science, Calcutta University claims that "Mercury compounds enter the body of the fish primarily through the food chain and the metal accumulation is steadily rising over the years as pollution increases. The survey further reveals a high level of metal concentration in the muscle of Carnivorous fishes, the metal accumulation staying consistent with the top predators.

Sea grasses like *P. Coarclota* also showed a higher mercury concentration in the levels explaining the mercury deposit in fishes like "Bhola Betki", lates calcarifer a herbivore.

The Hooghly has several sources of mercury contamination ranging from industrial discharge (Paper factories using phenylmercuric acetate, chlorine factories using mercury electrodes), Agriculture run off composed of fungicides containing mercury and sewage sludge.

A considerable amount of toxic and hazardous contaminants are also being released from Haldia Petrochemical Refinery Complex (Simikamboj, 2003).

Ray (1998) has discussed overall stressed environment and biotic life in his valuable document "Ecological Imbalance of the Ganga River system - its impact on Aquaculture" which has highligted pollution problems in details.

Similar findings are alos reported by Bhattacharya (2005) in his recent article "Sundarbans endangered - trouble in tide country" appeared in the *Statesman*, South Bengal Plus. The findings are enclosed in Addendum-I.

Addendum-I

Bengal known for the pristine natural beauty of the Sundarbans home, home to Royal Bengal Tiger and its fish sadly is under threat, courtesy the accumulation of metallic and toxic pollutants not only in its land and water but also among fish are reported in the following facts. (Bhattacharya, Kolkata, 2005).

1. Studies by Scientists from the School of Oceanographic studies of Jadavpur University in collaboration with Central Pollution Control Board under the union ministry of forest and environment found that high level of toxic metallic wastes have invaded not only the land and waters of the Sundarbans but also the food chain and bodies of different varieties of fish swarming in its waters. It is reported that concentration of heavy metals including lead, chromium, copper, cadmium have been found along with the right bank of the Hooghly down stream. Such deposits have also been found at the bottom of the river, especially on its right bank.
2. Toxic elements and heavy metals such as lead, cadmium, mercury, chromium have been found in the muscle tissue of fish inhabiting the waters of Sundarbans.
3. Fertilizers and insecticides, including DDT and endosulphate have also left deposits.
4. The contaminats mainly come from fish feed used in aquaculture farms. The contamination enters the food chain vectored by feed used by fish farms. Aquaculture farms especially of prawns have mushroomed in the Sundarbans of late.
5. Paint and oil pollution is also a major natural hazard found in the region. The researchers found that the ship-wrecking industry along the dock yards of Kolkata,

Falta and Kulpi in down stream of Hooghly add to the toxic pollution in the Sundarbans area.

6. According to the Central Pollution Control Board, the original project started in 1998 which dealt with climatic change, global warming, coastal hazard and other factors in the Sundarbans delta when the researchers stumbled upon the facts of pollution - as reported by Dr. Sugata Hazra, Director of the School of Oceanographic studies.
7. Apart from metallic and toxic wastes, arsenic and bacterial pollution found by the researchers has put to risk not only the ecosystem but locals as well. Arsenic pollution has been discovered in the areas under Sandeskhali Block-II at 60 metre aquifar depth, and bacterial pollution in Diamond Harbour mud points revealing high levels of coliform bacteria.
8. Only 25 per cent of the wastes generated in the city goes to the East Kolkata dumping ground. The rest is dumped into the Hooghly. The waste then travels downstream and accumulates in Sundarbans.
9. Certain pollutants are naturally controlled by the mangrove plantations, but fertilizers and pesticide residues accumulated over the years in the ecosystem of Sundarbans may pose a problem in the future. Lead found in the fish is hazardous for health and cadmium affects kidneys.

3. IMPACT ON BIOTA AROUND THE OUTFALL AND THE MAIN RIVER – THEIR POLLUTION STATUS

Physiochemical and biological studies around the effluent discharge points of industries are important in determining implication of pollution. The effluents discharged have different types of chemical composition. The composition include mineral acids, bases and many other toxic substances of non-biodegradable in nature (Ray, 1981).

In the estuarine stretch of Bhagirathi-Hooghly which is caused from the effluents of synthetic rayon pollution studies were made industry by Ghosh *et. al.* (1976); Pulp and paper mills (soda process) by Basu *et. al.* (1973) Ray *et al.*, (1979) and Ghosh *et al.* (1980 b); Rubber industrial complex by Ghosh *et al.* (1979 a); Pulp and paper industry (sulphite process) by Basu (1966); Ghosh *et al.* (1977) and Ray and Mitra (1980); Distillery by Ray *et al.*, (1977); Tannery by Ghosh *et al.* (1980 a) etc. (Most of these studies were made for 1-1 1/2 hour prior to high tide while in few cases, observations were made during the high tide also). Besides these additional studies were made for heavy metal pollution by Ghosh and Bagchi (1981) and Ghosh *et al.* (1983), Joshi (1985-86) further investigated accumulation of pesticide residue specially DDT in fish flesh of the estuary.

Any alteration in turbidity, PH, DO, CO_2, SS, immediate oxygen demand (IOD), BOD and heavy metal concentration in the water affecting the survival, growth and reproduction of aquatic organisers or that which impaires the legitimate use of water is generally referred to as a aquatic pollution. Sand, clay, silt, chalk, gypsum and highly resistant organics like cellulose, liguin, fly ash are conservative pollutants as they do not decompose easily. Domestic sewage and effluents from tannery, distillery, breweries, yeast, shellac, pulp and paper, rubber, oil, fat etc. are biodegradable. These substances contribute towards the pollution. The impact of pollution depends on lateral, vertical and longitudinal mixing of river water which is determined by turbulence. The turbulence is dependent on discharge, slope depth canal roughness, configuration and wind action.

In the estuary however conditions are different because of different tidal influences. The effluents discharged into the estuary have extremely variable chemical compostion. The include mineral acids, bases and many other toxic substances of non-biodegradable and biodegradable nature. Some studies on the impact of industrial effluents on the Hooghly have been done and are discussed below. Generation and discharge of waste water (domestic and Industrial) around the Hooghly estuary are indicated in Fig. 10 a,b

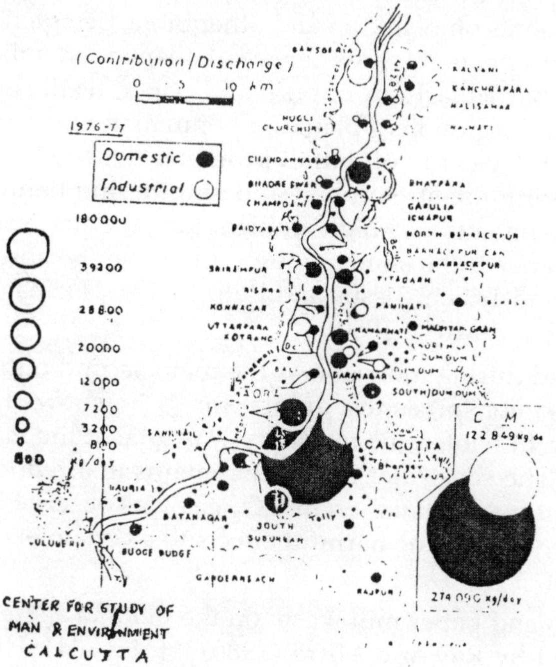

Fig. 10(a) : Generation and Discharge of Wastewater.

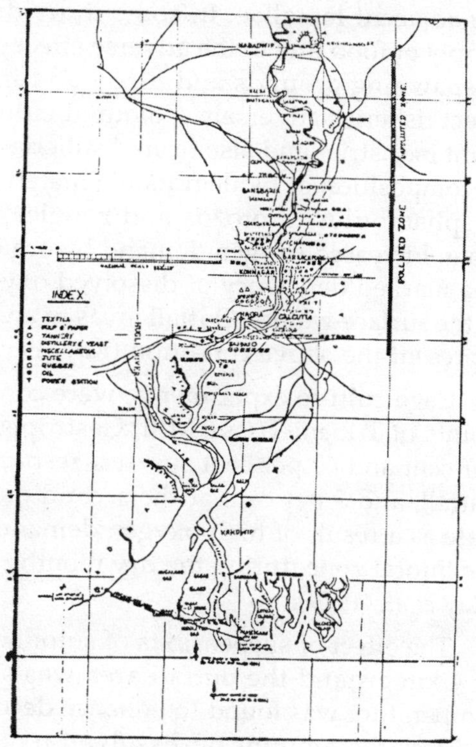

Fig. 10(b) : Industries Causing Pollution in the Hooghly Estuary.

Source : Ray, P. and G.C., Laha, (1981)

About 240,152 m^3 of organic waste is discharged every day which amounts to 89.82 per cent of the total liquid effluents discharged into the estuary. The corresponding loads are 90.72 per cent for BOD_5; 94.21 per cent for TS;96.79 per cent for SS and 91.88 per cent for DS. Pollution around the outfall of different factories releasing organic wastes such as effluents from pulp and paper mills (sulphite, sulphate and soda process) textile, tannery, distillery, yeast, shellac, rubber, etc. contains highly decomposable organic matter along with non-biodegradable materials like sand, silt, clay, chalk, gypsum, lignin, etc. Lignin content in pulp and paper waste is rather high, 200 ppm and 1000 ppm for soda and sulphite wastes respectively which corresponds to 2.2 t and 16.6 t per day (Basu, 1966). Continuous discharge of these substances raise the bed of the estuary, reduces plankton, eliminate food producing riffles, spawning grounds, fish eggs and larval populations. High turbidity in and around the outfall areas is another important contributory factor.

Finely suspended particles, possibly lignin clog the gills of fish and have synergestic effect due to asphyxia and poisoning of gastro-intestinal tract (Ray, 1961). Filtered effluents did not show any ill effect on the test fish. Experiments done using china clay (particle size 0.13-5 microns) and diatomaceous earth in 270 ppm concentration resulted in more than 50 per cent mortality of the test fish within 2-12 weeks (DSIR, 1961). Biopsy of the test fish revealed pathological changes in the gills, thickening of epithelial cells and after fusion of the adjacent lamellae. In some individuals even fishrot was observed. Reduction in the supply of food due to the adverse effect of effluents on plankton and other biota, destruction of spawning grounds and direct and indirect effects (by avoidance) of suspended solids affect fisheries adversely. Coloured effluents, particularly from pulp and paper, textile and paint industries and also from distilleries affect the growth of plankton population adversely. Decomposition of biodegradable materials in the vicinity of outfall promotes dense growth of sphaerotilus, protozoa and zooglea, covering the stream bed and smothering benthic biota. Increased oxygen demand by organic sediments and high ambient temperature result in a marked deficiency of dissolved oxygen beneath the slime, though it may not be absent in the surface waters. Outfall areas in and around Tribeni Tissue and Pulp Mill at, Hazinagar, represent the above condition (Ray, 1980; Ray and Mitra, 1980).

Cage culture experiments, were conducted during monsoon at the main second outfall points of Tribeni Tissue Mill, Gastropods (*Lymnea* sp.) as test animal for 71 hr. showed 50 per cent and 60 per cent survival respectively. In winter, the mortality was total at the main outfall, and 5 per cent at the second outfall. The soil water interface remains in a reduced state as a result of high oxygen demand. Continuous emission of CO_2 gas was observed in the littoral zone during the dry months, silting reduce the harmful effects to a certain extent (Ray *et al.*, 1979).

The effect of suspensiods of sulphite pulp and paper mill waste on the biota in a stretch of 4 km around the outfall area was studied by Ray and Mitra (1980) (cf. Fig. 11). High temperature was found to enhance decomposition of organic matter in the sediments which reduces the amount of dissolved oxygen at the soil water interface. Carbon dioxide is evolved more vigorously during March-May but ceases completely with the monsoon floods due to the settling of the suspended materials. The estimated carbon dioxide value in water ranged between 3.5 and 15.5 mg/l and 13.5 and 25.5 mg/l above and below the outfall point respectively. The sediment few mm below the surface were completely black generating H_2S and CO_2 within the top few mm of the sub-stratum. H_2S gets oxidised leaving behind an anaerobic zone of unsaturated inorganic and organic (carbonaceous) material at the soil water interface. This process continues even during the low tide when large areas below the outfall are exposed. These changes alter the equilibrium of the system. The suvival of invertebrates is seriously affected. Benthic organisms are wiped out completley and the equilibrium of the system changes drastically. However, the onset of monsoon changes the situation by diluting the pollutants. Common organisms of the estuary were conspicuously absent from the outfall areas. The impact of pollution was more severe during summer. The numbers of species recorded were 2.4 and 6 at 1.5 km above the outfall in summer, monsoon and winter respectively 3.4 and 5 at 0.7 km above the outfall, 3.2 and 4 at 1.5 km below the outfall and 4, 8 and 7 at 2.5 km below the outfall physical and chemical characteristcs of the substratum in relation to living organisms are given in Fig. 11. The following species formed the biotic community *Macrobrachium* spp, *Melanoides Lineatus* (Grey), *Melanoides granifera* (Lam), *Assiminea francessii* (Wood), *Stenothyra deltae* (Bensen), eel, *Acetes* sp, *M. Tuberculata* (Muller), *Thiora scabra* (Muller), *Vivipara dissimiles* (Muller), *Vivipara bengalensis* (Lam), *Lymnea Luteola f. Ovalis* (Gray), megalopa larvae, chironomid and oligochaetes.

Emergence and propagation of plankton from the soil samples collected from 1.5 and 0.7 km above the outfall, and 1.5 and 2.5 km below the outfall using pollution free water of the Hooghly estuary were studied. The dominant species recorded at each location are as follows :

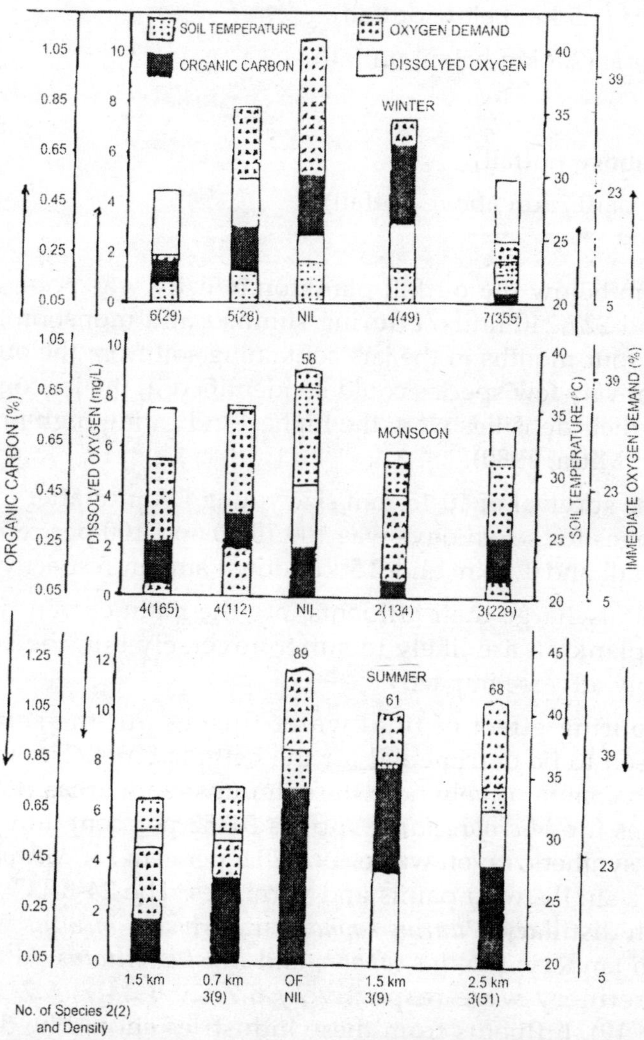

Fig. 11 : Physical and chemical parameters of soil and water in relation to biotic life.

Source : Ray, P. and E. Mitra (1980)

In the summer season

Ankistrodesmus, Oscillatoria, Euglena and *Closterim* (1.5 km above outfall). *Navicula, Oscillatoria, Cymbilla* (0.7 km above outfall)

Euglena (Outfall)

Oscillatoria (1.5 km above outfall)

Oscillatoria (2.5 km above outfall)

In the monsoon season

Euglena, Pinnularia, Closterium (1.5 km above outfall)

Cymbella, Closterium (0.7 km above outfall)

Closterium, Navicula (outfall)

Closterium, Euglena (1.5 km below outfall)

Navicula, Oscillatoria (2.5 km below outfall)

In the winter season

Euglena (1.5 km above outfall)

Oscillatoria, Euglena (0.7 km above outfall)

Euglena (Outfall)

At 1.5 and 2.5 km below the outfall plankton density was poor. Maximum plankton count was 20,360 and 227,240 units/l during summer and monsoon respectively after an incubation period of four months in the jars containing soil from the outfall point. Although in the laboratory relatively few species could be identified in the jar containing soil collected from the outfall but their densities were the highest indicating high nutrient content in the substratum (Ray and Mitra, 1980).

During winter the survival of 10-15 mm size young Lymnea after exposing them at soil-water-interface in cages for seven days was 100,10,60 and 100 per cent at 1.5 km at the up stream, near the outfall and 1.5 km and 2.5 km down stream respectively.

Other inudstries discharge their effluents into the main current where water is fairly deep. Accordingly plankton are likely to suffer adversely but, the substratum generally remains free from any adverse impact.

Further, the scouring effect of tidal water flushes out the settled material which otherwise is supposed to be detrimental for the bethnic biota. The tidal wave carries the young gastropods to comparatively near the calm upstream areas (Ray and Ghosh, 1976). The 96-h LC_{50} values for *Microbrachium* spp. as 31-52 per cent and 1.75-7.5 per cent for *Daphnia similes* with synthetic rayon waste, 19.0-20.0 per cent for *Microbrachium* spp. and 6.0 per cent for Dephnia similis with paints and varnishes. The 24-h LC_{50} values recorded for *Puntius sophore* with distillary; *Puntius sophore* and *Mystus vittatus* with pulp and paper; *Puntius sophore* with tannery; *Puntius sophore* and *Mystus vittatus* with textile and *Daphnia lumholzi* with oil refinery were respectively 6.7, 20.6-34.7, 6.7-10.03, 0.7-11.75 and 10.0 per cent (Table 19). Effluents from these industries should be diluted or adequately treated before releasing into the receiving water. On account of high oxidizable organic matter the immediate oxygen demand becomes high. This results in nil or low DO, high BOD and OC (3 m and 30 m at 100°C by permaganate) in the receiving water. Such a condition becomes unfavourable for the growth and propagation of plankton below the outfall (Table 20). Major damage to plankton results during the transifional period of low and high tide and vice versa at each tidal phase, particularly during the neap tide in summer when dilution ratio of effluents to receiving water was minimum and effluents spread over a much wider area. The condition deteriorated further by the nature of toxin, dilution, mixing pattern, duration of tides, flow of water, tidal phase, seasonal influence and retention period of the effluents (Ray and Ghosh, 1976). Self-purification capacity was greatly affected. On several occasions, around the outfall points plankton were either absent or had very low density. Rapid fluctuation in the chemical characteristics of waste (rise or fall of PH) also affect plankton adversely.

Table 19. Toxicity Tests of Some Selected Industrial Effluents

Industry	*Test animal*	*Period of experiment*	*Temprature range °C*	*LC_{50} (%) by Vol.*	*References*
1. Synthetic rayon	*Macrobrachium* sp	96 hr	27.5-34	31-49	
	-Do-	-Do-	-Do-	52	
	-Do-	-Do-	-Do-	33	Ray Ghosh
	Daphnia similis	-Do-	-Do-	1.75	Bagchi
	-Do-	-Do-	-Do-	5.6-7.5	and Mazumder (MS)
2. Paints and	*Macrobrachium* sp.	-Do-	29.5-32.0	19.0-20.2	
Varnishes	*Daphnia simils*	-Do-	-Do-	6.0	
3. Rubber	*Macrobrachium*	-Do-	29.0-31.0	32	
4. Distillery	*P. Sophore*	24 hr	19.0-25.0	6.7	
5. Pulp and					
paper	-Do-	-Do-	21.0-25.0	25.4	Ray (1961)
(a) Pulp	*M. vittatus*	-Do-	-Do-	34.7	
(b) Paper	*P. sophore*	-Do-	-Do-	24.5	
	M. vittatus	-Do-	-Do-	26.5	
(c) Combined	*P. sophore*	-Do-	-Do-	20.8	
	M. vittatus				
	-Do-	-Do-	-Do-	20.6	
6. Tannery					
(a) Sulphide	*P. sophore*	-Do-	21.5-22.0	7.7	
	-Do-	-Do-	30.6-32.8	10.2	
(b) Vegetable	-Do-	-Do-	21.5-22.0	10.03	David and Ray (1960)
	-Do-	-Do-	30.6-32.8	10.2	
(c) Chrome	-Do-	-Do-	21.5-22.0	9.55	
	-Do-	-Do-	30.6-32.8	6.9	
7. Textile					
(a) Kier	-Do-	-Do-	21.5-22.0	3.9	
	-Do-	-Do-	30.6-32.8	5.0	
(b) Dye	-Do-	-Do-	21.5-22.0	10.45	
	-Do-	-Do-	30.6-32.8	11.75	
(c) Bleach	-Do-	-Do-	21.5-22.0	2.55	
	-Do-	-Do-	30.6-32.8	0.07	
	M. vittatus	-Do-	-Do-	0.11	
Combined	*Macrobrachium* sp.	96 hr	26-30	6.5	Ray Ghosh, Bagchi and Mazumder (MS)
8. Oil Refinery	*Daphinia Lumholtzi*	24 hr	28.0	10	Ghosh and Bagchi (1979)

Table 20. Effect of Industrial Wastes Upon the Plankton Population at the Outfall Region in Different Seasons

Industry	*Location*	*Season* November-February	March-June
1. India Pulp & Paper (Sulphite process)	Above the outfall	339 (P-259 ; Z-80)	59 (P-37 ; Z-22)
	At the outfall	77 (P-41 ; Z-36)	7 (P-3 ; Z-4)
	Percentage reduction	77 (P-84 ; Z-55)	88 (P-92 ; Z-82)
2. Tribeni Tissue (Soda process)	Above the outfall	657 (P-55 ; Z-102)	171 (P-133 ; Z-38)
	At the outfall	20 (P-7 ; Z-13)	87 (P-54 ; Z-33)
	Percentage reduction	97 (P-99 ; Z-87)	49 (P-59 ; Z-23)
3. Distillery & Yeast	Above the outfall	89 (P-48 ; Z-41)	70 (P-45 ; Z-25)
	At the outfall	57 (P-32 ; Z-25)	55 (P-30 ; Z-25)
	Percentage reduction	36 (P-33 ; Z-39)	21 (P-33 ; Z-0)
4. Rubber Industry	Above the outfall	123 (P-70 ; Z-53)	78 (P-44 ; Z-34)
	At the outfall	80 (P-57 ; Z-23)	55 (P-40 ; Z-15)
	Percentage reduction	35 (P-18.2 ; Z-56)	26.6 (P-9 ; Z-55)
5. Kesoram Rayon	Above the outfall	236 (P-180 ; Z-56)	195 (P-140 ; Z-55)
	At the outfall	130 (P-96 ; Z-34)	49 (P-37 ; Z-12)
	Percentage reduction	42 (P-46 ; Z-39)	74 (P-73 ; Z-78)

P = Phytoplankton Z = Zooplankton

Source : Ray, P. (1981).

Value of primary production (gross and net) was estimated by light and dark bottle techniques are given in Table 21A (Ghost *et al.*, 1976) and subsequently by Babu Lal (1987, 1990), Table 21B and C by C^{14} technique. During the summer and monsoon seasons productivity was low around the outfall of pulp and paper mill (Soda process). It however, improved appreciably during winter Ghosh *et al.* (1977) have further shown that primary production was significantly affected at the outfall of pulp and papers mills Sulphite and Soda process (Table 23A & B). In case of tannery outfall gross and net production (Ghosh *et al.*, 1980a) was 71 and 81 per cent, 87 and 74 per cent and 43 and 41 per cent respectively during the summer, monsoon and winter seasons. Considerable oil was discharged into the the esturay along with other effluents mainly from ICI industries (709 m^3/d) at Rishra and from tannery (18,800 m^3/d) Batanagar. Petroleum refinery also disposed 6600 m^3/d or oil on an average ; about 4491 m^3/day of oil is disposed into the estuary, (Ghosh and Bagchi, 1979). The impact of oil refinery effluent on the biota around the outfall have not been evaluated. However, the value obtained from toxicity tests (LC_{50}) for 24-h using *Daphnia lumholtzi* was 10 per cent by volume (Ghosh and Bagchi, 1979). It shows that the effluents are toxic to microorganisms. The impact of oil refinery effluent on the biota around the outfall have not been evaluated.

Mineral wastes contain some amount of organic matter (synthetic rayon waste) and are highly acidic (pH 2.0-4.9). Plankton species were adversely affected at the outfall where they were reduced by 42 per cent in winter (November-February) and 74 per cent during summer (March-June) A sudden fall in pH (from 8.2 to 6.1), alkalinity (from 191 to 96 mg/l and DO (from 4.9 to 3.5 mg/l) observed on several occasions had a detrimental effect on the

Table 21. A Primary production of the Hooghly estuary.
Ghosh B.B., M.M. Bagchi and P. Ray (1979) in Ray P., (1981)
(by light and dark bottle technique) mg/C/m³/hr

	Seasonal Variation					
	Summer		Monsoon		Winter	
Centres	Gross	Net	Gross	Net	Gross	Net
Kalna	69.0	52.1	37.1	50.2	60.6	54.5
Tribeni	7.3	2.6	31.5	8.3	150.5	92.1
Hazinagar	145.5	58.1	10.4	31.8	56.7	27.1
Kakinada	13.5	0.9	7.0	1.7	23.8	0
Barrackpore	46.9	24.7	29.3	8.4	24.7	20.1
Titagarh	13.8	12.6	5.2	5.2	12.1	12.3
Batanagar	56.2	27.4	11.2	9.8	12.9	14.7
Diamond Harbour	30.9	25.5	25.5	51.6	74.3	51.4
Kakdwip	57.5	15.5	43.0	44.5	79.2	42.2

B. Babu Lal (1987) (by C^{14} Technique)

Before Farakka Discharge		Ranged	0.375 to 0.562 mg/c/m^{-3}/d^{-1}
Post Farakka discharge			
Nabadwip	–	38.00 to 95.64	(av. 64.85)
Barrackpore	–	22.63 to 95.16	(av. 55.35)
Diamond Harbour	–	37.66 to 75.05	(av. 53.89)
Fresherganj	–	44.51 to 57.40	(av. 45.55)
Kolaghat	–	29.16 to 53.01	(av. 43.15)

Minimum values of Primary production were recorded during monsoon and maximum during post monsoon period.

C. Annual variation in the range and average rates of carbon assimilation in the Hooghly estuary before and after the construction of Farakka Barrage (1965-86)

Year	Rates of Carbon assimilation Range	mg C m^{-3}/d^{-1} Average	
1965	0.375 - 0.562	0.468	By Light and dark botle metho
1969	Nil - 292.0	–	
1975	24.0 - 106.8	65.4	
1977	80.4 - 117.6	99.4	
1983	78.12 - 163.3	120.71	
1984	40.80 - 187.2	119.0	
1985	37.44 - 150.72	94.08	
1986	29.16 - 95.64	62.50 (By C^{14} technique)	

Source : Babu Lal (1990).

Table 22A. Effect of Pulp and Paper Mill Effluent (Soda Process) on the Primary Productivity

Location	Summer		Monsoon		Winter	
	Gross	Net	Gross	Net	Gross	Net
Above Outfall						
Neap tide	0-14.6 (7.3)	0-52 (2.6)	0-52.5 (31.5)	0-33.3 (8.3)	138.5-162.5 (150.5)	65.6-118.7 (92.1)
Spring/Bore tide	10.2-35.9 (22.6)	6.2-7.9 (7.0)	0-10.7 (9.0)	0-7.3 (2.4)	12.5-91.7 (52.1)	0-12.5 (6.2)
Out fall						
Neap tide	0-18.4 (5.2)	Nil	0-3.1 (2.2)	0-5.2 (1.3)	43.7-65.8 (54.1)	21.9-65.6 (43.7)
Spring/Bore tide	7-12.5 (10.2)	0-21.6 (5.4)	0-14.6 (6.6)	Nil	22.9-25.0 (23.9)	Nil
1.5 Km Below Outfall						
Neap tide	0-8.3 (4.2)	Nil	0-8.5 (4.7)	Nil	55.2-109.3 (82.2)	26.0-72.0 (49.4)
Spring/Bore tide	14.1-28.1 (18.4)	0-198.1 (29.5)	5.2-39.6 (18.3)	0-5.2 (1.7)	12.5-104.2 (60.8)	12.5-47.1 (19.8)

Table 22B. Effect Sulphite Pulp and Paper mill waste on the Primary Production in the Vicinity of the Outfall area

Location	Summer		Monsoon		Winter	
	Gross	Net	Gross	Net	Gross	Net
Above Outfall 750 m	11.8-197.7 (145.5)	33.3-72.9 (58.1)	66.7-187.5 (104.0)	0-80.2 (31.8)	28.2-89.9 (56.7)	0-60.9 (27.1)
Outfall	0-84.4 (41.4)	0-9.4 (3.1)	Nil	Nil	0-36.3 (18.1)	Nil
Below Outfall 1.2 km.	51.8-103.1 (71.6)	7.8-107.3 (54.6)	132.5-154.0 (140.5)	33.5-100 (62.1)	49.9-152.1 (89.3)	0-33.4 (15.0)

(Ray. P. 1981)

ecosystem. Near the outfall the amount of zinc was 2 mg/l in the waste and 3.84 mg/l in the bottom sediment at the outfall region. In the vicinity of the outfall, shrimp *(Macrobrachium roserbergi)* was found to accumulate 0.2 mg/l of zinc. Although some recovery took place up to 250 m below the outfall, it appears to be carried further down in the estuary along with suspensoids. About 3 km below the outfall the littoral zone on the right bank was devoid of any benthic life. The LC_{50} determined for 96 h using *Macrobrachium* sp. and *Daphina similis* ranged between 31-52 per cent and 1.75-75 per cent by volume respectively. These results indicate the need for appropriate treatment or dilution of the effluents before they were discharged into the estuary for protecting the biotic community.

Jhingran and Babulal (1988) have made some interesting observations on the potential resources over the entire region of the Hooghly estuary (600 km from the estuary month to Farakka) and also Rupnarayan river in terms of primary production and energy fixation by following C^{14} techniques. Their finding revealed that the mean value of primary production in the lower stretches was generally higher (1.265 g cm^{-2} d^{-1}) when the intermediate and freshwater zones showed lower values 0.285 and 0.261 g cm^{-2} d^{-1}.

The average annual carbon assimilation in the Hooghly estuary between 1965 and 1986 did not record by signicficant trend. However, immediately after the Farakka discharge had begun, a spurt in production (120.71 mg $cm^{-3} d^{-1}$) was recorded which ultimately stabilised at 62.50 mg $cm^{-3} d^{-1}$ by 1986. The observation indicates that overall production potential did not after much due to Farakka discharge (Babu Lal, 1990).

Babu Lal (*op cit.*) further observed that the carbon assimilation trend at various centres revealed higher production during post monsoon. (41.51 mg $cm^{-3} d^{-1}$) at Frezerganj, while minimum was estimated (22.62 mg $cm^{-3} d^{-1}$) at Barrackpore. On an average the fresh water zone of the Bhagirathi-Hooghly is more productive than saline zone. The primary production of the estuary, studied during different periods are indicated in Table 21A, B & C. Impact of pulp and paper mill wastes on the primary productivity are also shown in Table 22A and B.

4. SPAWNING OF HILSA

In the estuary however, available information reveals that some sporadic work on the potentiality of spawning of hilsa and their seed resources were done by various workers. Hora (1938), Hora and Nair (1940) and Jones and Menon (1951) have indicated that the region of Palta water works near Barrackpore as one of the breeding ground of *Hilsa ilisha* which subsequently Pillay and Rosa (1963) have cofirmed the earlier findings and they extended the freshwater zone up to Medgachi as breeding ground. Bhanot (1973) in the subsequent years made more detailed observation on the spawing behaviour of *H. ilisha* (Ham) based on the observations made during 1960-65. The author recorded breeding grounds between Khusigoli and Uchitpur, a tidal freshwater zone of the 112 km stetch between Nawabganj and Ghusighata. (The spawing took place about a fortnight earlier to the upper non-tidal region). Among the two main breeding seasons, the major one during monsoon with a long duration, while the other comparatively shorter one, which was during winter. Breeding occurred round the year with high collection obtained from Khusigoha and Medgachi. Abundance of Hilsa larvae during 1960-65 are indicated in Table 23.

Table 23. Abundance of H. ilisha larvae from the Hooghly estuary (net/hr)

	1960	*1961*	*1962*	*1963*	*1964*	*1965*
Jan.	0	0	0	90.5	54	10.2
Feb.	0	0	0	1444.3	127.5	26
March	0	0	0	111	261.6	45.2
April	0	0	0	36	55	2.0
May	0	0	32	27	23.7	34.5
June	7.0	30	92	112	110.7	2
July	155.0	33	197.3	166.3	0	32.5
August	773.0	83	49.3	27.5	0	2
Sept.	15.0	43	108	40	0	3
Oct.	487.0	0	44.5	180	0	10.5
Nov.	103.0	0	26.5	182.5	0	20.5
Dec.	1.3	0	16.0	83.5	0	2

Source : Bhanot K.K. (1973).

Further, to the above exploitation De and Sehgal (1989) in the recent past (after Farakka release 1975) tried to make some assessment on the recruitment of hilsa seed (*Tenualosa ilisha* (Ham) at 5 centres Nabadwip, Medgachi, Nawabganj, Uluberia and Diamond Harbour which included nontidal and tidal zone (Table 24). During 1987 and 1988 and 1989 they covered a stretch of about 225 km. Studies were restricted from August to May only, since the post larvae, frog and fingerlings of hilsa were available in the freshwater stretch of the Hooghly estuary from October to May with a peak during October and December. Diamond Harbour and Nabadwip are covering a distance of 75 and 300 km from sea face.

Table 24. Availability of hilsa seed in a 225 km. stretch of the Hooghly estuary during 1987-89

year		*August*	*Sept.*	*Oct.*	*Nov.*	*Dec.*	*Jan.*	*Feb.*	*Mar.*	*Apl.*	*May*
Max. no. of seed per/net/hr.	1987 1988	-	-	37	467	16	36	46	27	71	-
	1988 1991	-	16	130	149	39	45	164	39	93	-
Length range (mm) of seed	1987 1989	-	12-15	9-22	9-25	18-25	8-20	18-20	5-15	10-24	-

Source : De, D.K. and B.N. Sehgal (1989).

The species spawns in the entire freshwater zone of the estuary, while earlier observation inferred that the lower and upper limits of the spawning grounds of hilsa were restricted between Baghbazar (Culcutta) to Medgachi a distance of about 155 and 251 km respectively from the sea face. The probable reason for the extension of spawning grounds as observed and opioned by De and Sahgal (1989) might be due to the increased post Farakka discharge of freshwater into Hooghly estuary. The barrage has changed the entire ecology singnificantly by reducing the salinity and thus converting the gradient zone into almost freshwater, zone which might be conducive for spawning of hilsa. Embryoinc and larval development of *T. Ilisha* observed in the Hoogly estuary is indicated in Table 25.

5. FISH LANDINGS

Progressive changes in the fishery composition have been noticed in the estuary in the context of salinity. These may be summarised as :

Hamilton (1822) : Purely freshwater fish except hilsa : Calcutta and in the lower stretches.

Hora (1943) : Considerable changes in the fishery composition was noticed. Estuarine groups in and around Barrackpore.

Hora and T.V.R. Pilley (1953) : Mainly brackishwater in an around Barrackpore.

David (1954) : Mainly brackishwater fishes (Salinity range 0.204-1.156 ppt)

Ray (1981) : Mainly brackishwater forms in an around Barrackpore (Highest Salinity recorded-4. Oppt).

Table 25. Embryonic and larval development of T. ilisha as observed in the Hooghly estuary.

		Jones & Menon (1951)
1.	Locality	Hooghly river, about 190 km from sea within tidal limits
2.	Season	February, 1950
3.	Meterial obtained for the study	Developing eggs and larvae from natural collection
4.	Size of mature egg	0.70-0.75 mm
5.	Nature of egg membrane	Double layered in preserved eggs
6.	Yolk sac	Segmented
7.	Optic vesicle	Developed during embryonic stage after 11-13 hrs of fertilization
8.	Auditory capsules	Developed during embryonic stage, 19 hr after fertilization
9.	Twitching movement	After 19-23 hr of fertilization
10.	Incubation time	25-28 hr at 23°C
11.	Hreart	Pulsation observed 4-7 hr before hatching
12.	Size of newly hatched larva	2.30 mm
13.	Mouth	Formed on the 3rd day of hatching, that is, two day old larva.
14.	Demarcation of vent of anal opening	Observed in newly hatched larva
15.	Pigmentation	Appeared on the yolk sac 12 hr old larva
16.	Myotome	More than 40 in newly hatched larva

Source : Modified from De, D.K. and P.R. Sen (1986).

"Fishery resources of the Hoogly-Matla Esturaine System, Bull 19, Dec. 1973" a document published by Central Inland Fisheries Research Institute, Barrackpore is perhaps the first of its kind published a comprehensive account on fish fauna covering the period 1960-61 to 1970-71. Although five zones are considered for the Hoogly-Matla estuary and the zone-wise annual landings (Species-wise composition) in terms of kilogram of fish during 1966-67 to 1970-71 have been reported alongwith some other relevant informations specially Gears. From this account it is difficult to pin point the actual species-wise landings of the Bhagirathi-Hooghly estuary. Subseqently, Ray (1981) based on the salinty cited by Dutta *et al.*, (1973) delineated the estuary into three zones Fig. 12.

Zone-I — Nabadwip to Baranagar near Calcutta where traces of salinity (less than 0.1 per cent is found throughout the year).

Zone-II — Middlezone, extends from Barrackpore to Diamond Harbour where salinity ranged between traces and 10 per cent.

Zone-III — Lower zone includes the entire Sunderbans and a tract below Diamond Harbour on the main channel having salinity from 10 per cent to nearly neritic values. From the above, Ray (1981) has processed the data to show Yearwise landing in three different zones during 1966-71 in Table 26.

In another communication Ray and Laha (1981) divided the estuary into two zones. Unpolluted Nabadwip to Bandel, a stretch of 86 km. having approximate water spread area, 4509.84 ha and polluted zone, Bandel to Baj Baj, a stretch of 72 km. having approximate water spread area 4782.24 ha is shown in Table 27A & B.

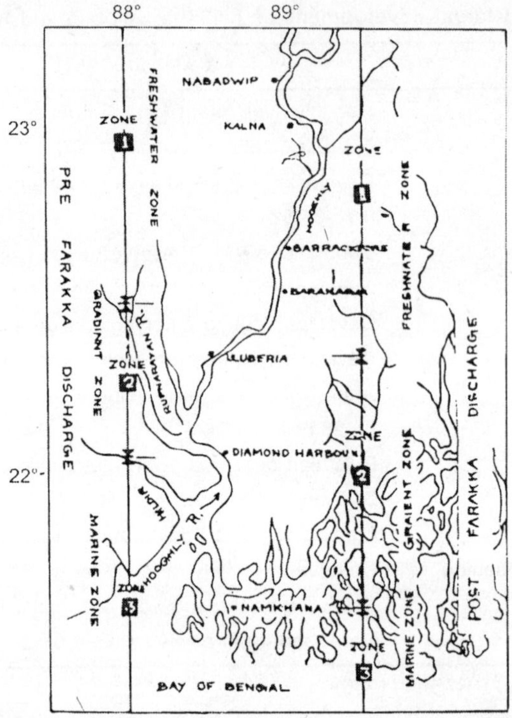

Fig. 12. Showing different zones in the Hooghly estuary during pre and post Farakka discharges.
Source : Nandy, A.C. M.M. Bagchi, and S.K. Mazumdar, (1983).

Table 26. Yearwise landing (tonnes) in three different zones during 1966-71

	Zone-I		*Zone-II*	*Zone-III*
	Unpolluted	*Polluted*		
1966-67	720	57	136	4442
1967-68	742	89	415	598
1968-69	735	88	286	5585
1969-70	910	106	289	5751
1970-71	913	120	383	8526

* Polluted zone means where maximum number of industries are located and pollution load is very high.

@ Personal communication with Laha G. C. CIFRI.

The Table 27A shows that total yield of fish in a period of 16 years at the two zones between Nababganj and Baj Baj of the Hooghly estuary system. Average annual fish yield in the respective zones are 719.25 and 124.94 t and catch/ha are 159 and 26 kg which shows that ecological condition in Zone-I is more favourable for biota than Zone-II. The table further shows that there is a year to year upward tendency of fish yield except a year or two which may be due to some non-sampling error during the pre-Farakka period in the unpolluted zone. On the the contary, polluted zone does not follow the same pattern and shows a decreasing trend mostly in 1960s and also in early '70s except a period when it recorded a bumper catch. In late '70s catch shows an increasing tendency.

Considerable changes in the ecology of Bhagirathi-Hooghly has taken place due to release of freshwater from Farakka barrage. Salinity has gone down remarkably. With the increase of freshwater flow, the salinity is lowered, pushed the gradient zone towards sea. The earlier recognised gradient zone largely become freshwater zone, while the marine zone becomes largely gradient zone (Nandy *et al.*, 1983 of Fig. 12). In general, biotic profile and therir occurrence considerably changed, some of the euryhaline species had shown a diclined trend in the upper and gradient zone (Table 28). The total fish landings as well as species composition in the marine zone of the estuary did not show any significant changes.

Table 27A. Total annual landings (in tonnes) Both in the Unpolluted and Polluted Stretches of the Hooghly Estuary (1962-1978)

Year	*Umpolluted Zone-I*	*Polluted Zone-II*
1962-63	562	297
1963-64	484	142
1964-65	338	103
1965-66	536	68
1966-67	720	57
1967-68	742	89
1968-69	735	88
1969-70	910	105
1970-71	913	120
1971-72	1064	120
1972-73	1064	101
1973-74	1029	106
1974-75	779	154
1975-76	681	170
1976-77	551	161
1977-78	400	118

Note : Zone-I Unpolluted Zone (Nabadwip to Bandel near Jubilee Bridge), a stretch of about 86 km. having water spread area 4509.84 ha.

Zone-II Polluted (Bandel to Baj Baj near Batanagar), a stretch of about 72 km having water spread area 4782.24 ha.

Source : Ray P. and G.C. Laha (1981).

Table 27B. Test of significance for diference in mean values of tudidity, DO and salinity in Zones I and II

Ecological factors	*Mean differences*	*S.E. of mean differences*	*'t'*	*D. F*	*t 0.05.28*	*Inference*
Tubidity	4.3467	5.55391	2.1209	28	2.048	significant
Do	0.9540	1.11355	2.2709	28	2.048	significant
Salinity	1.8392	1.3013	3.8199	28	2.048	significant

Table 28. Fish Landing in the Hooghly estuary before and after Farakka Barrage discharge

Species	Average landing (tons)		Landing after the Commission of Farakka Barrage					
	1972-73	1974-75	1975-76		1976-77		1977-78	
	Zone-I	Zone-II	Zone-I	Zone-II	Zone-I	Zone-II	Zone-I	Zone-II
Pama pama	7.86	13.50	11.00	21.21	8.57	11.18	1.62	5.73
Liza parsia	0.03	4.61	-	2.71	0.01	2.30	0.01	2.51
Setipina phasa & S. taty	31.53	3.17	38.44	3.29	28.65	0.59	22.75	1.48
Harpodon-nehereus	-	57.83	-	44.27	-	0.07	-	2.17
Trichiurus spp.	-	0.95	-	0.36	-	-	-	0.18
Prawns	403.64	67.81	308.99	77.45	229.31	42.00	160.00	28.52
Total Catch	1077.71	328.71	851.18	421.84	711.78	175.00	518.06	110.35

Source : Nancy, A.C., M.M. Bagchi and S.K. Majumder (1983).

Zone-I-Kalna and Barrackpore.

Zone-II-Uluberia and Nurpur.

Based on C^{14} assimilation technique the maximum and minimum computed values for energy fixation were 2.9×10^4 and 1.5×10^3 Cal $m^{-2}d^{-1}$ respectively. The mean values of C^{14} assimilation and energy fixation for the whole estuary works out to about 2.3×10^6 tonnes of carbon and 2.25×10^8 tonnes of calories of energy respectively. Based on this, the potential yield of fish as ratio of carbon production and energy fixation is 5.5×10^4 tonnes. The mean value of actual fish catch in the estuary is about 2.4×10^4 tonnes. Efficiencies of fish : Carbon; fish to dry plankton : Fish Carbon produced and fish energy : energy trapped, works out to 1.05; 0.46; 0.11 and 1.04 per cent respectively.

Thus potential resources over the entire Hooghly estuary (including Rupnarayan river) would be of order of the 2.5 to 3 times the present yield of estuarine fishery (Jhingran & Babulal, 1988).

6. BHAGIRATHI-HOOGHLY, ITS PAST AND PRESENT

The river which once used to carry the main flow of the Ganga had almost similar conditions as was in the main Ganga in the past. In the Ganga even around thirties the volume of water flow was 9,72,000 and in its important tributaries Damodar and Roopnarayan joining the Hooghly had 1,62,000 and 6,21,000 mf^3 of water (Anon, 1919; Oag, 1939) even though Damodar shifted its course of flow around 1770. The flow of the Hooghly thus maintained considerably high volume of water.

Further, from available discharge records for 5 years as stated by Majumdar (1953) the average discharges at Farakka on monthly basis had varied between 2,99,000 and 79,000 cusecs during dry season (November-May) and between 16,15,000 and 1,75,000 cusecs during monsoon (June-October). Some hydrographical conditions of the Hooghly during 1969 as obtained from the record of Sarkar (1988) are as follows. Physico-chemical and biological data are available only from fifties and fishery data from sixty onwards.

Hydrographical status of Bhagirathi-Hooghly during 1869

Discharge of water at Jalangi	:	1,40,160 cusecs
Discharge of water at Baharampur	:	1,40,762 cusecs
Prior to fifties the number of bore days as recorded by Bhattacharyee (1986)	:	20-22/year

Quoting Oag (1939) and Anon (1919), Bose (1956) indicated some hydrological conditions of the Hooghly estuary of the past as indicated below :

Approximate depth and breadth of the Hooghly estuary at different points (Anon, 1919; Oag, 1939)

Centre	*Breadth (miles)*	*Maximum depth at HT during monsoon (in feet)*	*Maximum depth at LT during Winter (in feet)*
Barrackpore	0.34	60	40
Calcutta (Khidderpur)	0.47	51	30
Uluberia	0.47	71	50
Falta	0.76	68	46
Diamond Harbour	2.91	91	69
Kulpi	2.87	56	34

Source : Bose B.B. (1956).

Physico-chemical Status

The range of physico-chemical status are available only from fifties and were recorded by Bose (1956) between Dumurdaha and Birlapur during 1953-55 and are indicated below:

Water temperature	:	18 - 33°C
Maximum difference between surface and bottom temperature	:	1.5°C
Dissolved Oxygen (ppm)	:	4.4 - 8.4
pH	:	7.9 - 8.4
Total alkalinity (ppm)	:	100 - 350
Salinity	:	Freshwater - 30% (Calcutta & above) (Uluberia-Kakdwip)

Pollution Status

In view of high water flow in the Bhagirathi-Hooghly in the past in addition to high tidal influence the delution ratio of wastes (municipal and industrial) was very high as was in the main river. The pollution problems was obviously negligible.

Biological Status

Data pertaining to biological status are also available from fifties. Phyto and Zooplankton concentration (μ/1) with their dominant groups recorded by Dutta *et al.* (1954) are indicated below:

Year	Phytoplankton	Zooplankton
1949 - 50	3,781 - 77,194 (31,956)	10,494 - 17,485 (14,007)
1950 - 51	4,378 - 1,58,613 (56,603)	7,120 - 12,304 (10,002)

Note : Values in the parenthesis are the averages.

Owing to the mingling of waters in the Hooghly system from a larger number of tributaries and to the banking of water during the flood tides from Bay of Bengal the Phytoplankton in the areas surveyed was varied and composite in character.

Relative abundance of phytoplankton were :

Diatoms, Green algae, Blue green algae and

Residual (mainly Englenineal and Dinoflagellates)

The Zoo-plankton in the Hooghly estuary consisted mostly of fresh and brackish water species. The freshwater species were washed down from the upper reaches. A few marine forms were also observed between Falta and Diamond-Harbour. The dominant forms were : Protozoa, Rotifer, Cladocera and Copepoda Nauplic.

Fishery Status

Fish landing data of old days i.e. prior to twenties or around the period when Bhagirathi-Hooghly used to carry enormous volume of water, are not available except some stray reference. It is from sixty onwards regular data became available when considerable damage to the physiographical condition was effected in the river. The progressive changes in the fishery composition have been summaried as follow :

Hamilton (1822)	:	Purely freshwater fish except Hilsa : Calcutta and below.
Hora (1943)	:	Considerable changes in fishery composition noticed. Estuarine groups were available aroundd Barrackpore.
Hora (1953)	:	Mainly brackishwater fishes around Barrackpore
David (1954)	:	Mainly brackishwater fishes around Barrackpore (Salinity 0.204 - 1.156 ppt)

Source : Nandy et al. (1983).

McDowell (1995) in a review stated that the situation existed in 1964 was that the distributary rivers which once fed the western delta from the Ganga were decaying. As a result those feeding the Hooghly had been cutoff from the Ganges during much of each dry season since 1915. The extensive construction of embankments intended for land reclamation and control of flooding had confined tidal flow to the main estuary channels causing tidal range to increase in the upper reaches of all the estuaries in the western delta. Besides, a series of dams constructed in the Damodar Valley affected greatly by reducing peak flows and mass of sediments brought down to the Damodar and Rupnarayan rivers and into the Hooghly at Hooghly point. The piers for the bridge (constructed across the Rupnarayan) would have appreciable effect on the local flow resistance and tidal propagation. Finally, the construction of Farakka barrage should have ensured free flow of freshwater down the Bhagirathi during dry season. But the dry season abstraction of water from the Ganges upstream of Farakka has reduced the quantity available at the barrage for this purpose.

Indication of deterioration of the Bhagirathi-Hooghly started by the end of 19th or earlier twentieth (Bose, 1972) after the change of main flow into the Padma and shifting the flow of Damodar further down. The bed of the Hooghly river was raised to 7.3 m higher (due to formation of a sand bar) than the main bed of the Ganga (Bera, 1984). The seriousness of Bhagirathi-Hooghly has never been felt as acutely as the present time, more so since the partition of the state in 1947. Hooghly has lost its maximum discharge capacity by 40 pc in 1959 compared to the discharge figure of 1954. (Bhattacharjee, 1986). Duration and intensity of the upland supply from the main stream Ganga into Bhagirathi Canal was limited to the months of mid-June to mid-November. The dry weather flow is negligible. Volume of supply diminished from 75-80 thousand cusecs to nil (Bose, 1972). Further, at the time of construction of Farakka Barrage, the flow of the Ganga-Padma did not even reach 40,000 cusecs during summer. Average flow of Bhagirathi at Jangipur being 61 cusecs only for seven months against 1,40,160 cusecs (during 1869). Bhagirathi-Hooghly received strong tidal inflow from the Bay of Bengal throughout the year and during 1986 the number of bore days were more than 180/year against 20-22 days/year prior to fifties (Bhattacharjee, 1986). In the stretch Farakka to Nabadwip (230 km) the slope is in 23,000 when rate of flow was variable due to flood. (Das Gupta, 1984). The slope is 3" inches per mile (Bera, 1984).

Pollution Status

Alongwith the pesticide (DDT and BHC-Y) heavy metals were also estimated not only in the water and soil but also in fish which includes fish muscle and different organs and crustaceans like crab, shrimp and molluscs.

Pesticides :	In sediment	DDT	(17 - 89 ppb);	BHC	(21 - 73 ppb)	
	- Fish	-	(91 - 460 ppb);	-	(46 - 210 ppb)	Joshi, H.C.
	- Molluscs	-	(65 - 963 ppb);	-	(40 - 850 ppb_	(1985)
	- Plankton	-	(15 - 130 ppb);	-	N.D.	
Heavy metals	In sediment,	Zinc 381 ppb	upper level			
	-	Chromium 9.6				
	In Fish -	Zinc	(2.5 - 120.4 ppm);	Copper	(2 - 20 ppm)	
	Shrump	-	(14.8 - 112.2 ppm);	-	(2 - 38 ppm)	Ghosh, B.B.
	Crab	-	24.4 ppm	-	14.0 ppm	*et al.*
	Molluscs	-	83.0 ppm	-	20.0 ppm	(1983)

Their concentrations are appreciably high. In the Hooghly estuary, the estimated zinc was found to be six times higher than the highest value obtained around Kanpur.

Biological Status

Plankton estimated (μ/1) by Nandy *et al.* (1983) in Zone I (Kalna and Barrackpore) and Zone-II (Uluberia and Nurpur) during pre 1960-61 and post (1975-77) Farakka discharges is indicated below.

	Pre Farakka discharge (1960-61)		Post Farakka discharge (1975-77)	
	Zone I	Zone II	Zone I	Zone II
Phytoplankton	68.04	32.18	219.38	162.16
	(Average 50.11)		(Average 190.77)	
Zooplankton	26.88	7.97	84.99	60.62
	(Average 17.42)		(Average 72.80)	

The above table indicates that there is considerable increase in the average value of phytoplankton (280.77 p.c.) and Zooplankton (317.9 p.c.) in both the zones (I and II) during the post Farakka discharge and is dominated mainly by fresh water forms. The excess water released from Farakka barrage has pushed the brackishwater zone further down.

The dominant forms were diatoms (Bacillarocophyceae and Chlorophyceae in phytoplankton and copepod (Calanoid and cyclopid) among Zooplankton.

However, these values are definitely much lower than the values estimated during 1949-50.

Fishery Status

The estimated total yield of fish in a period of 16 years at the two zones, I (unpolluted : Nabadwip-Bandel, 86 km (4509.84 ha) and zone II (polluted : Bandel to Batanagar, 72 km (4782 ha) between Nabadwip and Baj Baj of the Hooghly estuarine system covering the periods 1962-63 to 1977-78 has been estimated which states that average annual yield in the respective zones were 719.25 and 124.94 tonnes and the catch/ha were 159 and 26 kg, which shows that ecological condition in zone I is more favourable for biotic life than zone II. Further, there is a year to year increasing tendency of fish landing in zone I and on the contrary in zone II (polluted zone) it does not follow the same pattern and shows a decreasing trend (Ray and Laha, 1981).

Mainly brackish water forms around Barrackpore was recorded (Ray, 1981). According to Nandy *et al.* (1983) the immediate impact on fisheries resources of the estuary after the commssioning of the Farakka barrage has not been studied in details. So far, although some trends have been recorded in the upper estuary (Kalna to Nurpur). Some of the typically euryhaline species had been showing a declining trend.

Fish landings before and after Farakka Barrage discharge in a stretch Kalna to Nurpur

Annual average landing (tonnes)		Landings after the commissioning of the Farakka Barrage					
1972 - 73 to 1974 - 75		1975 - 1976		1976 - 1977		1977 - 1978	
Zone-I	Zone-II	Zone-I	Zone-II	Zone-I	Zone-II	Zone-I	Zone-II
107.71	328.71	851.18	421.84	711.78	175.00	518.06	110.35

Source : Nandy, *et al.* (1983).

Zone I : Kalna and Barrackpore, Zone II : Uluberia and Nurpur.

Bhagirathi-Hooghly is also facing similar problems. In the past when the main flow of water was through this system, purely freshwater fish except Hilsa was recorded around

Calcutta (Hamilton, 1922), has now been converted into its tributary with minimum flow during non-monsoon months. A gradual change in the fishery composition *i.e.* estuarine group around Barrackpore was distinctly predominant with the increase in salinity (Hora, 1943), which later changed mainly into brackishwater fishes Hora and Pillay (1953); David (1954).

Added to it the release of more toxic metallic wastes which caused metals to accumulate in different organs of fish possibly has affected the growth and reproduction and ultimately fishery as a whole.

After commissing of Farakka barrage from 1970 entire biotic life indicated gradual changes due to freshwater influx, therby the fauna and flora have been shifted to zone II down stream. Nandy *et al.* (1983).

Besides natural calamities like draught or late monsoon etc., which effects the growth and Farakka Barrage constructed across the Ganga has affected free migration of commercial important fish like Hilsa in the upstream. This has resulted dwindling of Hilsa fishery appreciably. The estuarine fishery is also considerably affected on account of their restricted migration.

Addendum 2
A. (December 1996)

The recent 30 year. Water Pact on Ganga water at Farakka between India and Bangladesh has invited the following comments from different experts. (*Statesmen*, 13 December, Calcutta, 1996).

1. During the two cycles of 10 days period in the crucial month of April, Calcutta will get from Farakka barrage between 25,000 and 28,000 cusecs, which according to C.P.T. experts is "too low a head water flow to flush silt into the sea."

   The agreement in no way reverses the process of deterioration of the Hooghly. Heavy siltation will increase the intensity and frequency of tidal bores which will seriously impede navigation.

2. C.P.T. experts and Mr. Debesh Mukherjee, Former General Manager of Farakka Barrage seriously questioned the sharing formula worked out on the basis of average total flow at Farakka between 1949 and 1988.

   "The average data on water flow does reflect the ground reality. Flow of water varies from day to day. Under the agreement the average flow in April has been shown to be between 60,992 cusecs and 63,180 cusecs whereas the *actual flow* during the month for the *past decade* has been about *54,000 cusecs. It would have been somewhat proper if the average flow had been arrived at on the basis of data of the past one decade rather than the past 40 years.* Naturally, the basis for the sharing formula is flowed as also the quantum," they said.

3. A senior port official said "*Calcutta will get much less than what has been stated in the agreement* as the *offtake in the upstream, especially in Uttar Pradesh and Bihar, has been growing fast in recent years.*

The above statements of hydraulic experts indicate the fate of the Bhagirathi-Hooghly is heading in future and its resultant impact on biological life.

B. (2001, March, The Statesman) 23.03.2001

Calcutta (now Kolkata) has a net work of ferry services meant to cater the public between the city and the Hooghly. But this is greatly affected on account of heavy silting to such an extent that it has affected the normal ferry service around Calcutta and is evident from the fact reported by Maity (2001) in his interesting article "Siltation could ferry Services Closure". According to him several ferry services between Howrah and the city are on the verge of closure. Several vessels got stuck up in front of Ramkrishnapur during low tide. This might have led to an accident. Bally and Babu Ghat had to be stopped several months ago. Ramkrishnapur-Chapdani Ghat and Ghusuri – Babu Ghat might have to be stopped in the comming month. The "Hooghly Nadi Jalpath Paribahan Samity" has informed commuters about disruption in services for more than five hours every day between Ramkrishnapur and Babu Ghat during low tide. They apprehended that other jetties like Shobhabazar - Ghusuri could face similar problem in the summer month. The State Government is incurring a heavy loss due to disruption in ferry services between Ramkrishnapur and Chandpal Ghat during peak hours.

A huge pillar of Vidyasagar Setu was also obstraciting the tidal wave. The above facts in which recent heavy flood (September, 2000) might have contributed significantly by showing the alarming stage of the river Hooghly has already attained. This was predicted by the hydraulic experts.

Sen (2005) has also presented a list of common freshwater endangered species of Bhagirathi-Hooghly which is shown below in Addendum 3.

Addendum 3

Common freshwater endangered/species of Bhagirathi-Hooghly (Sen P.R., 2005) (Sr. Scientist - Retd. ICAR).

			Local Names
A.	1.	*Hilsa ilisha* (Hem)	Ilish
	2.	*H. toli* (Valve)	
	3.	*Gudusia chapra* (Ham)	Khoyra, Chapila
B.	1.	*Oxygaster bacaila* (Ham.)	Chela
	2.	*O. Phulo* (Ham.)	
	3.	*Esomus danricus* (Ham.) - Carpminow	
	4.	*Rasbora rasbara* (Ham.)	Rasbora.
	5.	*Cirrhinus reba* (Ham.)	Reba, Tatkini
	6.	*Labeo bata* (Ham.)	Bata
	7.	*L. Calbasu* (Ham.)	Kalobos
	8.	*L. fimbriatus* (Bloch)	
	9.	*L. Gonius* (Ham.)	Ghania
	10.	*L. nandia* (Ham.)	Nandini

11.	*Osteobrama belangeri* (cuvier and valenciennes)	
12.	*Puntius sarana* (Ham.)	Sarna Puti
13.	*P. Sophore* (Ham.)	Puti
14.	*P. ticto* (Ham.)	Titputi
15.	*Tor. putitora* (Ham.)	Mohasol.
16.	*T. tor* (Ham.)	
17.	*Ompok bimaculatus* (Bloch)	Kalopapda
18.	*O. pabo* (Ham.)	Papda
19.	*O. pabda* (Ham.)	
20.	*Mystus cavasica* (Ham.)	
21.	*M. Vittatus* (horai, Jayaram) *M. Vittatus Vittatus* (Bloch)	Misti Joler tengra
22.	*Rasbora, rasbora* (Ham.)	Rosbora.
23.	*Gambusia affinis patruelis* (Baird and Girard)	
24.	*Rhinomugil Corsula* (Ham.)	Khorsula.
25.	*Anabas testudineus* (Bhoch)	Koi mach
26.	*Ailia coila* (Ham.)	Kajali Mach
27.	*Heteropneustes fossilis* (Bloch)	Singh mach
28.	*Clarias batrachus* (Linnacus)	Magur Mach
29.	*Amblypharyngdon mola* - (Ham)	Mourola Mach
30.	*Colisa fascita* (Bloch)	Kholsa Mach
31.	*Notopterus Chitala* (Hamilton)	Chital Mach
32.	*N. notopterus* (Pallas)	Falui Mach.
33.	*Colisa chuna* (Hamilton)	Kholse Mach
34.	*C. fasciata* (Bloch)	
35.	*C. Pectoralis* (Regan)	
36.	*C. lalin* (Hamilton)	
37.	*Osphronemus goramy* (Lacepede)	Gourami

7. SUGGESTIONS FOR FUTURE INVESTIGATION

Detailed investigations on the seasonal dynamies of the estuarine biota are urgently needed. Data on various aspects of the Bhagirathi-Hooghly would be extremely useful in formulating effluent standards with due consideration to seasonal variations.

Systematic monitoring of the estuarine zone must be carried out to collect the baseline data on physical, biological and chemical characteristics of the system. In view of the fact that plankton, fish and other benthos are affected by metals like, Zinc, Copper, Chromium, Lead and also by pesticides, through their bioaccumulation, studies on heamatological, serological, histopathological and cytochemical tests are essential to find out any changes in their physiological aspects which affect healthy growth, survival, maturation and reproduction. The treatment such pollutants may be given top priority. A systematic monitoring programme is necessary for a proper assessment of man's impact on the estuary.

The effect of pesticides and metals on the biota (fish and spawn of different sizes, eggs etc.) needs through investigation. Synergistic and antagonistic effects, and chelation of metal ions with organic compound in relation to their effect on aquatic life remain to be studied. Bioaccumulation tests should be carried out in fish samples collected from different outfall regions to determine index of heavy metals accumulated in living system.

In view of the above it is worthwhile to undertake a separate study on this aspect specially with commercial important species at various stretches of the estuary as this will reveal valuable information on breeding success, since the matter is related to production potentiality of the ecosystem.

Studies on avoidance behaviour of fish to toxicants on the lines suggested by Stott (1975), Sprague (1964a, 1968; Sprague and Drury (1969) and Hoglund (1975) or modified, under natural ecosystem may reveal valuable information.

Critical studies are needed to determine the retention period of effluents in the estuary. This would help to determine the pollution levels and self purification capacity of the estuary. Environmental impact on assessment of heat-waste disposal by six thermal power plants on the estuarine ecosystem deserves urgent attention.

Technology and industrialisation must come but they should not effect the critical balance of nature endangering the very basic needs of life.

8. A WORD ON GAP – A SCIENTIFIC STUDY

The GAP (Ganga Action Plan) was initiated in 1986 by the Central Environment and Forest Department to combact the problem of pollution of the river Ganga.

In West Bengal the report of the findings conducted jointly by the Central Pollution Control Board and State Pollution Control Board after a considerable period of time (18 years) were as follows (*Statesman,* News Service 3.2.2004).

1. Most households either have poorly maintained sewer line at all. As a result, pollutants end up in the Ganga inspite of STPS.
2. All the sewage of the Command area is not being intercepted, which is why, the sewage flowing down to STPS are comming from open drains or surface runoffs.
3. In case of system failure of STPS untreated sewerage flows straight into the river. There is no arrangement to hold it back.
4. Some fully-equipped STPS (Sewage Treatment Plant) are not functioning as they are not receiving adequate sewage. If the waste water quantity is below average the plant can not function.
5. Almost all STPS have been installed without the permission of WBSPCB.

In West Bengal 15 important towns in an around Kolkata from which maximum pollutants flow into the Ganga were earmarked for inclusion in the Ganga Action Plan. About 527 million litres of polluted waste water from these towns flow into the river daily. The flow of pollutants into the river was arrested by diverting waste water into sewerage treatment plants to lower the pollution by keeping the BOD below 3 milligram per litre and suspended solids less than 50 milligram per litre. Thus it is evident that more stress is given to reduce the BOD values as indicated in the article "Rejuvinating Ganga" by Sharma (1994).

Nevertheless, it may be remembered that BOD itself is not a pollutant and exercises no direct harm. Only by depressing the DO content to levels that are inimical to fish life and other beneficial uses, does BOD exert an indirect effect. Where reaeration, dilution and or photosynthetic action offset or minimise their depletion, BOD does not interface to a reasonable extent for uses of its water.

BOD is important only insofar it produces septicity or decreased DO or subsequent growth of saprobic bacteria which increases the turbidity or other undesirable characteristics of the streams. In a slow sluggish stream a 5-day BOD of 5 mg/l might be significant to produce deoxygenation resulting in anaerobic conditions, where as a swift mountain stream can easily handle 50 mg/l of 5 day BOD without appreciable depletion of DO. For this the study of physiographical and limnological conditions are important, as thes will provide information on slope, discharge figure, velocity of water, dilution, shape factor etc. These are the most important parameters which contribute in the self purification of a stream.

The stream characteristics along different sections of the Ganga as indicated by Das Gupta (1984) in his document, "The Ganga Basin II" is a clear example of such situation indicated below which are quite alarming.

Stretch	*Section*	*Length (km)*	*Av. Slope of land*	*Mean annual rate of flow m^3/s*	*Stream velocity m/sec*
Source Rishikesh	Mountaineous	250	1 in 67	850	May : 0.09-0.6
Rishikesh-Allahabad	Upper plain	800	1 in 4100	850-1700	July : 0.22-00.9
Allahabad-Farakka	Middle plain	1005	1 in 13800	4000-10200	
Farakka-Nabadwip	Deltaic non-tidal plain	230	1 in 23000	1000-1300	
Nabadwip-Outfall	Deltaic tidal plain	240	1 in 24000	Variable due to tides	

Natural calamities like heavy storm rainfall, soil erosion from deforestation at associated with high abstraction of water may have far reaching effect on the ecosystem by bringing ecological imbalance.

Hence, instead of giving adequate attention on these aspects if only treatment of municipal wastes is given importance a time may come when the river, during non-monsoon months may be converted into several disconnected large water spread areas or a channel as happened in the case of Bhagirathi-Hooghly due to continuous silt precipitation and large scale use of river water and we will have to remain satisfied with the sewage treatment only when simultaneously on the other side, a grim picture will gradually appear in the river for which we are spending so much of fund for its improvement. We are to see that the river also holds sufficient water with adequate flow. May be that, three fourth of the pollution on the Ganga is from untreated municipal wastes (as opined by CPWC) but the river has its own self purification capacity and by that not only the wastes will be purified, but simultaneously such wastes may cause eutrofication to the river water that may result good growth of flora which is the ultimate goal of our study.

Bhargava (1981b) in his article "Ganga, the most self purifying river" reported that the rate of exocellular polymers excreated during endogenous growth phase of various species of bacteria is considerably reducing BOD and turbidity. The interaction of exocellular polymers present in the river with colloidal matters present in the river with colloidal matters present in seepage results in a very rapid and significant reduction of the BOD. The author further states that the data analysis has shown that Ganga has the highest BOD rate constant value and the reaeration rate constant value, both of the which are higher by an order of magnitude than the so far accepted values for any river. The coliform group of organisms are also found to reduce significantly in very short time.

It is worthwhile to give more attention for the treatment of metals and pesticides and simultaneously see that the river gets adequate depth, flow and discharges even during the lean months.

Industrial waste containing metallic toxicants and pesticides which are not easily biodegradable. These are the main problems.

In the Bhagirathi-Hooghly (lower stretch of the Ganga) then concentrations are more and fishes and crustaceans have accumulated much more in their tissues.

Fujaya (1977) has observed that if the normal requirement of DO in water for fish falls, the fish has to gulp more oxygenated water to replenish their DO deficiencies and by doing so more of water containing metal toxins or pesticides are likely to enter the system to cause damage. The problem is more in the estuary (where the effluents spread in a wider area due to tidal influence) than the other stretch of the Ganga where the water is flowing in the littoral Zone. If adequate flow is there the pollution will be restricted to the marginal area only for a short stretch. The optimum requirement of DO in warm water fish is believed to be 5 ppm. (Doudoruff and Shumway, 1967) which would require at least 65-75% saturation of water with oxygen. Around outfall especially where the flow is minimum and in the Hooghly estuary (Lower stretch of Ganga) where DO falls below 5 ppm/similar situation may arise as observed by Fujiya (1977). Hence, bioaccumulation of metals/ pesticides in fish in their different organs should be studied to see the impact of pollutants on growth and reproduction of fish and entire biotic population.

Although BOD (out of 42 parameters used to assess the water quality) has come down to 3.0 mg/l and slight improvement in DO has become possible by treating the sewage (as claimed by Virendra Vats, 1994) it is not possible to find out the difference between the past and present condition of the estuary to assess the overall extent of recovery of the estuarine water. Further, the sludge which settles down in the treatment plant is dried and compacted into a compost like manure. Such manure prepared from dried sludge, in all likelihood contains adequate metallic pollutants which may be harmful for any agriculture use directly.

Hence, under the present degrading physiographical and limnological condition of the estuary which is heading from bad to worse, how far only such treatment (sewage) will help the river condition to improve where biomass has to survive, grow and reproduce, as they did in the past, are the aspects, which are yet to be seen in future.

Besides industrial, municipal, agricultural (Pesticides etc.) and other sources of pollution, shipwrecks which takes place at Sand heads between Jamboo and Mausomi islands is another source of pollution as they carry various polluting materials which can affect biotic life (Das, 2003). Details of shipwrecks which occurred during 2003 at Sand heads thrice in a month's time and their probable impact on fishes are given below :

Table 29. Ship wrecks carrying pollutants - their probable impact on aquatic life.

	Name of ship and country	*Date of Ship wrecks*	*Particulars of material carrying*	*Remarks on aquatic life*
1.	M.V. Sijitikabiru. (Indonesia)	May, 15, 2003	5000 tons of sodaash, 20 tons of diesel oil, 100 tons of furnace fuel oil (Reported by West Bengal Coast Guard (R. K. Wadwa).	Officials from Dept. of Fisheries Govt. of West Bengal reported : 1. Tiger prawn, bhetki, pomfret and hilsa which normally lay an estimated 34000 crores of eggs between March-July are likely to stay away from the sand heads where the water is supposed to be polluted with diesel and soda-ash. 2. The annual yield of fish from the sand heads is 82000 metric tons, a figure they fear will dwindle significantly. 3. The sandheads is normally favourable zone for fish for laying eggs. 4. The water level here maintains a depth between 9-12 feet.
2.	M.V. Fortune carrier (Belize)	June, 11. 2003.	100 tons of furnace-fuel oil, 20 tons of gasolin, around 150 wooden logs. (Reported by R.K. Wadwa).	Initial report suggested by the Fisheries Dept.-Fish catch was poor. 45000 and odd, fishermen from Kakdwip, Namkhana Diamond Harbour and areas adjoining Haldia who did venture out on June 6 are now trying their luck else where to meet their demands.
	Ship wrecks also occurred at Sandheads during the earlier years but their details are not available. These were :			
3.	M.V. Luknam (Vietnam)	July 10 (2001)		It is interesting to note that ship wrecks occurred mostly during May to July when fishes especially hilsa are almost mature and ready for freshwater search for the purpose of breeding.
4.	M.V. Jubilient	October 16, (2000)		
5.	M.V. Prime value.	July 19 (2000)		
6.	M.V. Orient Greece	May 28 (1998)		
7.	M.V. Greece opal.	June 19 (1997)		

Source : Rahul Das (Kolkata) - *The Statesman,* 13.6.2003.

The above facts clearly indicate that shipwrecks at Sand heads (a favourable zone for fish breeding) normally occur during May-July, the most important period when breeding activities of some commercial important fishes takes place and it is quite likely that the materials the ships carry can cause heavy pollution in the surrounding areas of Sand heads. Such pollutants may adversely affect the marine aquatic life and their breedings. Hence, a seperate full fledged bio-chemical and avoidance reaction study is needed around the sand heads keeping in view of the ship wrecks carrying pollutants and based on their findings appropriate measures should be suggested to save the aquatic life-their breeding and finally the production of commercial prized fishes like Hilsa prawn etc. alongwith other fishes.

REFERENCES

Anon, 1919, The report on the Hooghly river and its headwaters, 1 and 2. The Bengal Secretarist Book Department Calcutta. In : Observations on the Hydrology of the Hooghly estuary-Benoy B. Bose (1956), Vol. 3, No. 1.

Babu Lal (B) 1987, Impact of Farakka Barrage on The Hydrological changes and Productivity Potential of Hooghly Estuary. *Symposium on the Impact of current land use pattern and Water Resources Development on Riverine Fisheries.* April 25-27, 1987 (abstract). Central Inland Capture Fisheries Research Institute, Barrackpore, 743601, West Bengal.

Babu Lal, Impact of Farakka Barrage on the Hydrological Changes and Productivity Potential of Hooghly Estuary. J. *Inland Fish Sec. India 22* (1 & 2) : 28 - 42.

Banerjee, B. N. 1972, Navigation in a tidal river with particular reference to river Hooghly. *The interdisciplinary cympotium Organised by the Department of Geography.* University of Calcutta in collaboration with other organisations. (ed.) Kanan Gopal Bagohi, December.

Basu, A.K. 1966, Studies on effluents from Puls and Paper mills and its role in bringing the physico-chemical changes around several discharge points of the Hooghly River Estuary. *India. J. Instt.* of *Engres.* India, Vol. *XIVI* No. 10 pt. PH. 3

Basu, A.K. and B.B. Ghosh 1970, Observations on diurnal variations in some selected stretch of the Hooghly estuary (India) *J. Hydrol. 32* (1), 271-283

Basu, A.K. B.B. Ghosh and R.N. Pal 1970, Comparison of the polluted Hooghly estuary with Unpolluted Matla estuary. *India. J. Wat. Poll. Contr. Fed.* 42 (10): 1771-1781

Basu, A.K., C. S.G. Rao, R.S. Dhaneswar, G. Mukherjee, M.K.J. Kutty D. Ray and R.C. Trivedi 1973, Studies on Hooghly estuarine Pollution in the Vicinity of Tissue Mill. *Proc. Symp. on Environmental Pollution,* pp. 26-40. National Environmental Engineering Research Institute. Nagpur.

Bera, S., 1984, *Nadibigyaner Katla* (The science of rivers) 10, Raja Ram Krishna Street, Calcutta-700006.

Bhargava, D.S. 1981b, Ganga the most self purifiying river *International Symposium Water Resources Conservation pollution and Abatement.* December 11-13, Roorkee, India.

Bhanot, K.K. 1973, Observation on the spawning of *Hilsa ilisha* (Ham) in the Hooghly estuary. *J. Inland Fish Sec. India* Vol. V. December.

Bhattacharyya, R. and De Sankar, 1972, Some aspects of Agriculture in Damodar-Hooghly inter fluve. "The Bhagirathi-Hooghly Basin" *Proceedings of the interdisciplinary symposium Organised by the Department of Geography. University of Calcutta.* 1967 Rep. FAO/UNDP (TA) (2449) Reported by E.W. Surber.

Bhattacharya, Kapil, 1986, *Sadhin Bharate Nadnadi Parikaipana* (Planning of river system in independent India) Kalam, 52 B, Braja Nath Lane, Calcutta-700012.

Bhattacharya, Ravik (Kolkata), 2005, Sundarbans Endangered-Trouble in Tide Country. In : The Statesman, South Bengal Phs. April 2, 2005.

Bose, B.B. 1956, Observations on the Hydrology of the Hooghly Estuary *Indian J. Fish*, 3 (1), pp. 101-118.

Bose, N.K., 1972, The Bhagirathi-Hooghly - A few remarks. In : The Bhagirathi-Hooghly Basin *Proceeding of the interdisciplinary Symposium* Organised by the Department of Geography, University of Calcutta, in collaboration with the Geographical Society of India, India Association of Geohydrologists, Indian Statistical Institute, Regional Science Association (India). The Commissioners for the port of Calcutta. The Institute of Engineers (India) and River Research Institute, West Bengal, Edited by Kanan Gopal Bagohi in collaboration with Sunil Kumar Nandi, December.

Chakrabartry, R.D. and M.K. Banerjee 1966, On the depth conditions in the Hooghly estuary as recorded by echo sounding. *Proc. Nat. Acad. Sci, India* Vol XXXVI. Sec. B : 35. pt. IV, 401-416.

Chatterjee, S.P. 1972, The Bhagirathi-Hooghly Basin Procudings of the interdisciplinery symposium. Organised by Dept. of Geography, University of Calcutta in collaboration with– The Geographical Society of Indis, Indian Association of Geohydrologists, ISI, Regional Science Association, the commissioners for the port of Calcutta. The Institute of Engineers (India) and River Research Institute, West Bengal. (ed) Kanan Gopal Bagohi.

Das Gupta, S.P., 1984, *The Ganga Basin Part II* Centre for study of man and environment for the Central Board for the prevention and Control of Water Pollution. In Assesment and Development study of River Basin Series ADSORBS/7/1982-83.

David, A., 1954, A preliminary survey of the fish and fisheries of a five mile stretch of the Hooghly river near Barrackpore. *India J. Fish.* 1 (1 & 2) : 231-255.

De, D.K. 1980, Fishing Gear and Fishing methods of Hooghly-Matla Estuarine System. *J. Inland Fish, Soc.* India *12* (1) : 54-63

De, D.K. and P.R. Sen, 19'6, Observations on the embryonic and early larval development of Indian shad, *Tenualosa ilisha* (Hamilton) *J. Inland Fish. Soc. India.* 18 (2): 1-12

De, D.K. and B.N. Sehgal, 1989, Spawning of Hilsa, *Tenualosa ilisha* (Ham) in the Hooghly estuary, *J. Inland Fish Soc. India 21* (2) : 46-48.

Doudoruff, P., and D.K. Shumway, 1967, Dissolved Oxygen Criteria for protection of Fish. *Am. Fish. Soc.* Spec Publ. 4 : 13-19.

Dutta, N., J.C. Malhotra, and B.B. Bose, 1954, Hydrology and seasonal fluctuations of the plankton in the Hooghly estuary. *Symp. on marine and freshwater plankton in the* Indo-Pacific. 1-13.

Dutta, P., G.G. Laha, P.M. Mitra, D.K. De, A. Chaudhury, P.K. Pandit, A.R. Chaudhury, R.N. De, B.K. Saha, H.S. Mazumder, N.D. Sarkar, N.C. Mondal, G.P. Bhattacharya, A.K. Namasudra, S.P. Ghosh and A.R. Paul 1973. Fishery Resources of the Hooghly Metla Estuarine System. *Bull No. 19* December. Central Inland Fisheries Research Institute (ICAR). Berrackpore. West Bengal.

Forstner, U, and G.T.W. Wittmann 1979, Metal pollution in the Aquatic environment. Springer-Verlag. Berlin. Heidelberg, New York.

Fujiya, M., 1977, Pollution indused fish diseases and their effects. *Fresh water Fisheries Research Laboratory* Hino-Shi, Tokyo, Japan.

Ghosh, B.B., P. Ray and V. Gopalakrishnan, 1973, Survey and characterization of waste waters discharged into the Hooghly estuary. *J. Inland Fish. Soc. India* 4 Dec. 82-101.

Ghosh, B.B. and A.K. Basu, 1968, Observations on estuarine pollution of the Hooghly by the effluents from a chemical factory complex at Rishra, West Bengal *Envir. Hlth.* 10 (3) : 204-218.

Ghosh, B.B., P. Ray, D.D. Halder and M.M. Bachi, 1976, Some observations on the pollution of the Hooghly estuary in the vicinity of a Rayon industry with a note on the toxicity of effluent. In ACMRR/IASO. Espert consultation on Bio-assay with aquatic organisms in relation to pollution problems. Dubrovnik Yugoslavia 16-19 Nov. FAO publication FI : ACMRR/BE/7, 18 October.

Ghosh, B.B., P. Ray, M.M. Bagchi and D.D. Halder, 1977, Effects on the Biotic condition of the Hooghly estury due to disposal of effluent from sulphite pulp and paper Industry. *Proc. of the Seminar on Industrial Wastes.* December 8-9, pp 167 - 201, Metropolitan Development Authority and National Environmental Engineering Research Institute, Calcutta.

Ghosh, B.B., and M.M. Bagchi, 1979, Some observations on oil pollution in the Hooghly Estuary with special reference to oil refer refinery complex at Haldia. *National seminar on Protection of Marine Environment.* November 13-15. National Institute of Oceanógraphy, Dona Paula, Goa (sponsored by the Department of Science and Technology. Government of India).

Ghosh, B.B. and M.M. Bagchi 1981, Observations on Environmental pollution caused by zinc-borne wastes disposed from Rayon Industry into the Hooghly estuary. In *Proc. Internal. Symp wat. Resour Conser. Poll & Abatement, Dept. of Civil engineering.* University of Roorkee. 11-13 December, pp. 131-140.

Ghosh, B.B., P. Ray, M.M. Bagchi, D.D. Halder and R.K. Banerjee, 1979 a, Biotic and abiotic condition of the Hooghly estuary in the vicinity of a Rubber Industrial Comples. *Proc. Symp. Environ Biol,* pp. 139-156 Academy of Environmental Biology. India.

Ghosh, B.B., M.M. Bagchi and P. Ray, 1979b, Stuies on organic production in the Hooghly estuary polluted by industrial and domestic wastes. *National Seminar on Protection of Marine* Environment. November, 13-15 National Institute of Oceanography, Dona Paula. Goa, Sponsored by the Department of Science and Technology, Govt. of India.

Ghosh, B.B., M.M. Bagchi and S.B. Saha, 1980a, Characterization of Tannery wastes and their effect on the Primar Productivity of the Hooghly estuary near Calcutta (India) *3rd. International Congress on Industrial waste water and wastes.* Stockholm, February, 6-8, pp. 1-9.

. 1980 b, Effect of Industrial pollution by Soda process pulp and paper Mill waste on the Primary Productivity of the Hooghly estuary near Tribeni. *Symposium on Coostal Aquaculture,* Cochin, India, January, pp. 12-18

Ghosh, B.B., M.M. Bagchi and D.K. De, 1983, Some observations on the status of pollution in the Hooghly estuary (West Bengal) with reference to heavy metal disposed through Industrial wastes. *J. Inland Fish Soc. India* 15 (1-2) : 44-53.

Ghosh, J.J., 1993, In "Ganga Dusaner Bhayabaha Kahini" (The alarming news of the Ganga pollution) by Roop Kumar Basu. Barteman, 9th November.

Ghosh, Somendra and Sudipta Bhattacharya, 2005 : In : Bhasaner Por Fulo Annya Upokarane Gangar Dusan Cramoso Barche, UDBEG. (After imersion, flowers with other ingradients gradually increase The pollution of The Ganga, Worries), Bartaman (19.10.2005).

Gopala Krishnan, V., P. Ray and B.B. Ghosh, 1978, Problem of estuarine pollution with special reference to the Hooghly estuary. *Proc. of the seminar on Pollution and Human environment,* pp. 313-324 Bhabha atomic Research Centre. Trombay.

Gopala Krishnan, V., P. Ray and B.B. Ghosh, 1976, Present status of pollution in the Hooghly estuary with special reference to the adverse effects observed on the fishery resources. *Proc. of the symposium on environmental pollution,* pp. 1-8 Central Pubic Health Engineering Research Institute, Nagpur.

Hamilton, Francis, *An account of the fishes found in the river Ganges and its branches,* Edbinburg : Archibald constable and company, pp. 405.

Hoglund, L.B., 1975, Avoidance reactions of the cod (Gadus morhua) to certain surfactants. In Document FIR/TPLR/75 : 2/19 FAO Rome.

Hora, S.L. 1938, A preliminary note on the spawning grounds and bionomics of the so-called Indian shad, *Hilsa ilisha* (Ham) in the river Ganges, *Rec. Indian Mas.* 40 (2), pp. 147-148.

Hora, S.L. and Nair, 1940, Further observations on the bionomics and fishery of the Indian shad, *Hilsa ilisha* (Ham.) in Bengal waters. *Rec. Indian Mas,* 42 (1) : 35-50

Hora, S.L., 1943, Evidence of distribution of fishes regarding rise in salinity of the river Hooghly. *Current Science,* 12 : 189-90.

Hora, S.L. and T.V. R. Piliay, 1953, Further evidence from fish distribution of the rise in salinity of the river Hooghly. *Current Science,* 22 : 49-50.

Jhingra, A.G. and Babulal, 1988, Primary production Energy fixation and potential fishery of the Hooghly-Matla estuarine system. (Abstract) *In National seminar on management of fisheries in Inland open water system of India.* Central Inland Capture Fisheries Research Institute, Barrackpore (West Bengal).

Jones, S., and P.M.G. Menon 1951, Observations on the life-history of the Indian shad *Hilsa ilisha* (Ham.) *Proc. Indian Acad. Sci,* 31 (3) : 101-125.

Josh, H.C. 1985, Pesticide residue monitoring in the Bhagirathi-Hooghly stretch of the Ganga river system. *Proc. National seminar on pollution control and environmental management,* NEERI, Nagpur, March 17-19, 1985, pp. 80-86.

Joshi, H.C., 1986, DDT residues in fish in river Hooghly. *Proc. National symposium on pesticide residue and environmental pollution.* Musaffarnagar, (October, 2-4, 1985).

Joshi, H.C., 1988, Impact of manmade environmental modifications on the riverine ecology and productivity-A case study of the river Ganga. Lecture delivered at the summer institute on Resource management and conservation of Inland Capture Fisheries of India, 4-23 July, 1988. CICFR, Barrackpore.

Maity, 2001, Siltation could ferry services closure In : *The Statesman,* 23.3.2001.

Mc Dowell, D.M., 1995, The Development of the river Hooghly for navigation *Port of Calcutta, 125 years (1870-1995)* Calcutta Port Trust, October 15, Strand Road, Calcutta.

Mazumdar, S.C., 1959, Ganga Barrage and the Bhagirathi-Hooghly Problems. *Indian Journal of Power and River Valley Development.* Vol III, No. 3.

Mukherjee, M., 1993, Silting of Hooghly may inundata Calcutta, *The Statesman,* June 5.

Nandy, A.C. M.M. Bagchi, P.M. Mitra and S.K. Majumder, 1980, The impact of Farakka Berrage on fisheries of the Hooghly estuary. Lecture delivered at the *Summer Institute on Brackishwater Capture and Culture Fisheries,* held at the Central Inland Fisheries Research Institute (3rd July to 2nd August) Barackpore.

Nandy, A.C., M.M. Bagchi and S.K. Majumder, 1983, Ecological changes in the Hooghly estuary due to water release from Farakka Barrage. Mahasagar-*Bulletin of the National Institute of Oceanography* 16 (2) : 209-220.

Neeri, 1978, National Environmental Enginering Research Institute water quality survey of the Hooghly estuary (1974-76), Sponsored by CMDA, Calcutta.

OAG, T.M. 1939, Report on the river Hooghly and its headwaters. The commissioners for the port of Calcutta. In : Observations on the Hydrology of the *Indian J. Fish. 3,* 101-118 Hooghly estuary. - Benoy B. Bose (1956).

Pillay, S.R. and H. Rosa (Jr.) 1963, Synopsis of biological data on Hilsa ilisha (Ham) 1822, *FAO Fisheries Biology aynopsis,* No. 25

Ray, P., 1961, Evaluation of toxicity of some Industrial effluents to fish by bio-assays, *Indian J. Fish.* 8 : 233-40.

Ray, P., and B.B. Ghosh, 1976, Some ecological considerations of the Hooghly estuary polluted by domestic and inudstrial wastes under tropical environment. ACMRP/IABO, Expert Consultations on Bio-assays with aquatic organisers in relation to pollution problems. Dubrovnik Yugoslavia, November 16-19, FAO publication FI : ACMRR/EE/8, 29th September.

Ray, P., B.B. Ghosh and M.M. Bagchi, 1977, The effects of variations in physico-chemical parameters on plankton bio-mass around a distillery out fall in the Hooghly estuary. *Proc. of the Seminar on Industrial wastes.* pp. 141-153, Dec. 8-9. Convened by National Environmental Engineering Research Institute and Calcutta Metropoliton Development Authority.

Ray, P., B.B. Ghosh and M.M. Bagchi, 1979, Effects of Pulp and Paper mill waste (Soda Process) around the outfall in the Hooghly estuary. *Proc. Symp. environ. Biol.,* pp. 453-464. Academy of Environmental Biology, India.

Ray, P. and E. Mitra, 1980, Effects of suspensoids on biotic life around the out fall area of a Sulphite Pulp and Paper Mill waste in the Hooghly estuary. *Symposium on Coastal aquaculture, Cochin.* India January, pp. 12-18.

Ray, P. 1980, A pollution survey of the Hooghly estuary as influenced by industrial and domestic wastes under tropical environment. *3rd International Congress on industrial waste water and estates.* Stookholm, February, 6-8, pp. 1-11.

Ray, P., and G. Laha, 1981, Effects of Industrical and Municipal wastes on the trend of Fisheries in the Hooghly Estuary. *International Pollution and Abatement,* December, 11-13, 1981 Roorkee, India.

Ray, P., 1981, *Impact of man on the Hooghly estuary.* In Water Pollution and Management Reviews. *South Asian Publication Pvt.* Ltd., Madras.

Ray, P., 1993, *Aquaculture in Sundarban Delta - its perspective* (An assessment) International Books & Periodicals Supply Service, Delhi.

Roy, H.K., 1949, Some potamological aspects of the river Hooghly in relation to Calcutta water supply. *Sci and Cult.* 14 (8) : 318-324.

———————— 1955, Plankton ecology of the river Hooghly at Palta, West Bengal. *Ecology,* 36 (2) : 169-175.

Saha, S.B., B.B. Ghosh and V. Gopalakrishnan, 1975, Plankton of the Hooghly Estuary with special reference of salinity and temperature, *J. Var. Biol. Ass. India,* 17 (1) : 107-120.

Saha, S.B., 1980, Studies on the Biotic characteristics in and around the outfall areas of paper mill at Tribeni, Hazinagar and Titagarh, West Bengal. *In Symp, Coast, Aquacult mar. Biol. Ass.* Cochin, 12-18 January.

Sarkar, Birendra Nath, 1988, "Gangadhara Lupta Hote Choleche" (The Ganga flow is on the way of extinct) The "AAJKAL" 6th November, Calcutta.

Seth, G.K. and T.R. Bhaskaran, 1950, Effects of Industrial Wastes disposal on the sanitary condition of the Hooghly river in and around Calcutta *Indian J. Med. Res.* 38 (4) : 341.

Sharma, Kallegala, 1994, Rejuvinating Ganga *Science Reporter,* February.

Shetty, H.P.C., S.B. Saha and B.B. Ghosh, 1961, Observations on the distribtution and fluctuations of plankton in the Hooghly-Matla estuarine system, with notes on their relation to commercial fish landing *Indian J. Fish.*, 8 (2) : 326-363.

Sprague, J.B., 1964a, Avoidance of copper-Zinc solutions by young salmon in the Laboratory, *J. Water Pollut. Control Fed*, 36 : 990-1004.

Sprague, J.B., 1968, Avoidance reactions of Rainbow tront to Zinc-Sulphate Solutions, *Water Research.* Pergamon Press. Vol. 2, pp. 367-372 (Great Britain).

Sprague, J.B. and D.E. Druruy, 1969, Avoidance reaction of Salmonid fish to representative pollutants. Reprinted from Advances in water pollution Research. In Proceedings of the 4th International Congress held in Prague.

Stott, B., 1975, Developments toards a new method for investigating avoidance behaviour in fish *In* : Fourth FAO/SIDA training Course on Aquatic Pollution in relation to protecting of living resources. FIR/TPLR/75 : 2/26. FAO, Rome.

Vats, Virendra, 1994. In : Rejuvinating Ganga *Science Reporter,* February.

World Health Organisation, 1976, Report on Pesticide residues in food. *Tech. Rep. Ser.*, 592, p. 28.

APPENDIX I

Gear used in the estuarine Zone

Source : Dutta, *et al.*, 1973

Gear with local name	*Area of Operation*	*Season of Operation*	*Species Caught*
1. Trawl nets (Moi, Kelta, Buro, Kantni and Kachhi)	Upper and Middle Stretch of Zone I	Nov./June	Prawns, also miscellaneous.
2. Sine nets (a)Large : (Jungla, Kachal and Jagatber) (b)Medium & Small (1) Ber (2) Charghers, (3) Kachal and (4) Chatber	Coastal region of Zone III (1) Zone III (2) III also upper area of Zone I (3) Upper area of Zone I (4) Upper area of Zone I	Nov.-Feb. (Occasionally in Monsoon (1) All seasons, (2) All seasons, (3) March-July. (4) Nov.-August	*H. ilisha, T. jella, P. pama* *Setipinna* spp. & S. biauritus (1) Mullets, Perches, Poly- nemids and sciaemids (*P. pama*) (2) Mullets, perches *(S. panijus)*, Polynemids and sciaenids (*P. pama*) (3) *H. ilisha,* (young) and Miscellaneous fishes. (4) *H. ilisha* (young). *S.* *Phasa,* Prawns and Miscellaneous fishes. *H. ilisha* only
3. Purse nets : (sanglo, Dar and Khorke)	Zone - I	July-Oct. and Jan. - March.	
4. Drift nets* (1)Chhandi (2)Dholi (3)Kona (Bhasa-Kona Ghai-Kona, Ghalo-Kona (4)Shele and Bhasa-larang,	(1) Zone III, I, II (2) Zone I (3) Zone I (4) Zone III	(1) Monsoon and Winter (June to February) (intensity greater in Monsoon) (2) Nov. - May. (3) Feb. - Sept. (4) Nov. - May.	(1) *H. ilisha,* other (*I. elong-* *ata, H. toli, Setipinna* spp.) and *S. Cinereus* (2) *H. ilisha.* & *S. phasa* (3) *H. ilisha* (4) *P. indicus* and *S. biauritus*
5. *Lift nets :* Bhesal, Gara-bhesal, Nauka-bhesal, Pata, Prawns. Thela, Seitki, Thapa, Dhain, Tana, Char and Koli	Zone I & III	All seasons	Mullets, *L. elongata,* *Setipinna* spp. and
6. *Cast nets :* Bachari & kepla	Zone I & III	All seasons	*P. paradiseus, P. pama,* *P. pangasius and Prawns*
7. *Bag nets :* (Stationary) *nehe-* (1) Been *Trichiu-* (2) Behundi (3) Thor	(1) & (2) All Zones (3) zone II, also, main channel of Zone III (4) Zone III	All seasons	(1) (2) and (3) Almost all all species, especially *H.* *reus,* prawns, *P. pama,* *rus* Spp. *Setipinna* spp. *Setipinna* spp. *I. elongata,*

contd....

Appendix I – *contd...*

Gear with local name	*Area of Operation*	*Season of Operation*	*Species Caught*
(4) Atone and (5) Gopa			mullets & *S. Paningus* (4) and (5) *L. Calearifer* and *S. biauritus.*
8. *Set gill nets :* (1) Barang (Med-barang, Bhetki-barang) (2) Nangar (3) Khota (4) Khuti, (5) Patang (6) Geba-Chandi	(1) Zone III (2) (3) (4) Zone III (4) & (5) Zone I	(1) Nearly all seasons (2) - (5) Monsoon & & winter.	(1) *L. calcarifer* and *S. biauriturs* (2) - (5) *H. ilisha* & other clupeids.
9. *Set-Barrier nets* (1) Pata, (2) Char-para (3) Khat pata *tetradactylum* (4) Khal ghera (5) Kumor.	Zone III	All seasons	All species, particularly Prawns, *M. parsia, Setipinna* spp., *P. pama, E.* *L. calcarifer,* and *I. elongata.* Prawns.
10. *Traps :* Bitti, Dwar-bitti, Chaibitto and Bhasa-pata	Zone I	Monsoon May-October	
11. *Hooks & Lines* (Barsi)	Zone I, II, III	All seasons	Catfishes, Perches, Polynemids, Scienids and *Setipinna* spp.

* Except (3) all are gill nets. (3) is a labyrinth form of net which drifts.

Zones I, II and III are indicated in Fig. 24.

Zone I Nabadwip - Khusigoli.

Zone II Khusigoli - Diamond Harbour.

Zone III Tract below Diamond Harbour on the main channel.

APPENDIX II

Gill nets used in the Hooghly Matla estuarine System

(Figs. 1-6)

Source : Dutta, S.N., 1973.

Gill Nets

- Without foot rope
  - Without sinker
    1. Shele jal
    2. Chandana jal
    3. Ilish-dholi jal
  - With sinker
    1. Phase-dholi jal
    2. Khoira-dholi jal
- With foot rope
  - Without anchor but with sinker
    - Chhandi jal
  - With anchor
    - Without sinker
      1. Barang jal
      2. Khota jal
    - With sinker
      1. Nongar jal

A. Drift gill net

(a) Gill nets without foot rope

1. *Shele Jal (Fig. 1) :* Made of either cotton or nylon, a fleet is formed by 150 to 200 shots. A single shot has 70 meshes across and 13-½ meshes in depth with a mesh size of 20 cm. The head rope of each piece measures 6.65 m. The main webbing is directly hung on the head rope by the reeving method. At the joining of 2 shots, a bamboo float of 1 m long is attached vertically to the head rope of the net. Sinkers are totally absent. The hanging coefficient of the net is 0.48 (or 48%). The net is operated as bottom drift by 7-8 men with a larger sized boat, near coastal area of the middle

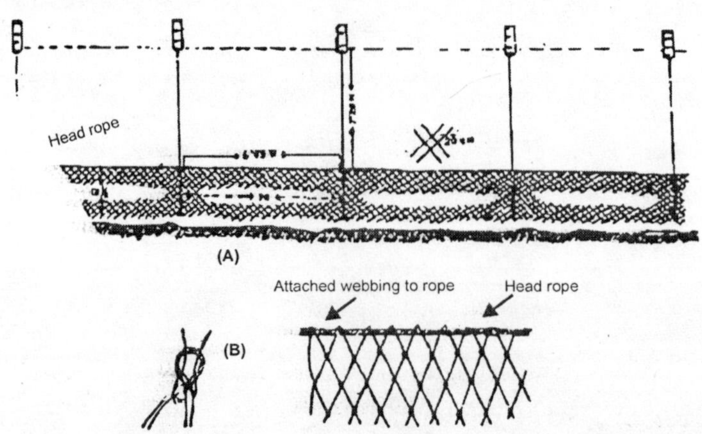

Fig. 1 : Shela Jal : (a) Way of setting net; (b) Type of knot used; (c) Attachment of webbing with the head rope

zone of the Hooghly. The operation is restricted between October to middle or March, when the water remains comparatively calm. Large sized *Polydactylus indicus* and *Otolithodes biauritus, Epinephelus* spp., *Tachysurus* spp. *Pampus* spp. and *Pristis* spp. are largely landed with this net. Other local names are bhethi jal, bhasapatan jal, karal jal and bhasa-barang jal.

2. *Chandana Jal :* Made of cotton and very rarely nylon, is formed by 50 to 100 shots. A single shot has 320 meshes across and 120-½ meshes in depth, with a mesh size of 8.70 cm. The length of head rope in each shot measures 14.3 m. Attachment of webbing with the head rope and arrangement of bamboo floats are same as in the case of shele jal. The hanging coefficient of the net is 0.51. The net is operated as surface drift. Number of men type of boat, area of operation and season of operation are more or less same as in the case of shele jal. *Hilsa sinensis, Tachysurus* spp., *Osteogeniosus militaris, Chirocentrus dorab* and *Pampus* spp., constitute the major species in the landings. Another local name is kajal jal.

3. *Ilish-dholi Jal :* Made of either cotton or nylon, a fleet of net is formed by 4 to 6 shots. A single shot has 720 meshes across and 51 meshes in depth, with a mesh size of 8 m. Either reeving or stapling method is followed in the attachment of webbing to the head rope. Triangular shola floats (Eschynomene aspera), 7 in number, are attached to the head rope of each piece at equal distances. Sinkers are absent. Hanging coefficient is 0.35. The net is operated as surface drift. One to two men with smaller boat operate this net in the upper stretches of the Hooghly estuary during December to June. Medium sized *H. ilisha* forms the major catch. Another local name is dholi jal.

4. *Phasa-Dholi Jal (Fig. 2) :* Made of cotton, a fleet of this net is formed by 4 to 6 shots. A single shot has 1080 meshes across and 81 meshes in depth, with uniform mesh size of 6 cm. The head rope of each shot measures 24 m followed by stapling type of attachment with the webbing. Oval shola floats, 4 in numbers are attached to the head rope of each shot, similarly 4 burnt clay sinkers are attached to the (bottom)

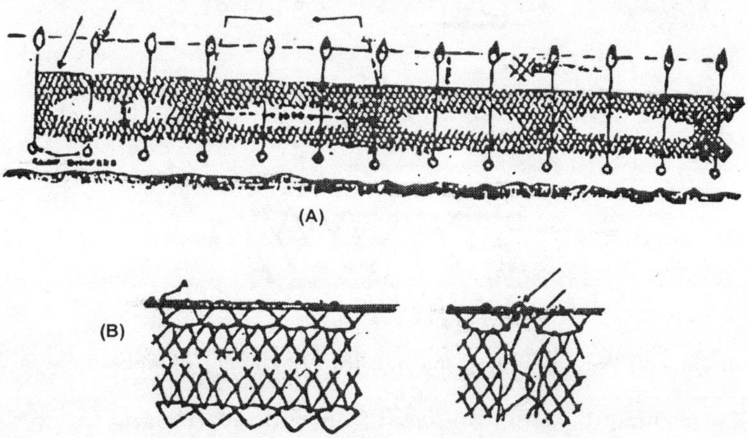

Fig. 2 : Phase-Dholi Jal : (a) Way of setting the net; (b) Attachment of webbing with the head rope; (c) Joining the two pieces.

side. The hanging coefficient is 0.37. The net is operated as column drift. Number of men, type of boat and area of operation are same as in the case of ilish-dholi jal. Season of operation is from October to February. *Setipinna phase, Eutropiichthys vacha, Chupisoma garua, Pama pama* and small sized *H. ilisha* are caught with this net. Another local name is phase jal.

5. *Khoira-dholi Jal :* This net, constituting 4 to 6 shots, is almost similar to phase-dholi jal, except for the measurements. A single shot comprises 1080 meshes across and 101 in depth, with uniform mesh size of 4.1 cm. The head rope of each shot measures 17.85 m. The hanging coefficient is 0.40. The net is operated as surface drift. Season of operation is from January to July. Small sized *H. ilisha, Gudusia chapra, E. vacha. C. garua* and *S. phasa* are caught.

(b) Gill Nets with Foot Rope

6. *Chhandi Jal (Fig. 3) :* These days the net is mostly made of nylon and is one of the very important gear of this estuary. A fleet is formed by 25 to 150 shots and a single shot has 260 meshes across and 35 meshes in depth, with a regular mesh size of 11 cm. The length of the head rope in each shot is 8.85 m. followed by reeving type of attachment with the webbing. At the joining of 2 shots a bamboo float and in between 2 bamboo floats a fin float are attached to the head rope. Similarly sinkers are attached to the foot rope. Hanging coefficient is 0.31. The net is operated as surface drift by 4 to 8 men, with a chhandi naoka, almost throughout the entire. Hooghly and Rupnarayan estuaries. The net is operated from June to February to cover both monsoon and winter runs of *H. ilisha, H. ilisha, H. sinensis, Pampus* spp., *C. dorab* and *Tachysurus* spp. are caught in this net. Other local names are bhasa jal, bhasa chandi jal, ilish jal and dora jal.*

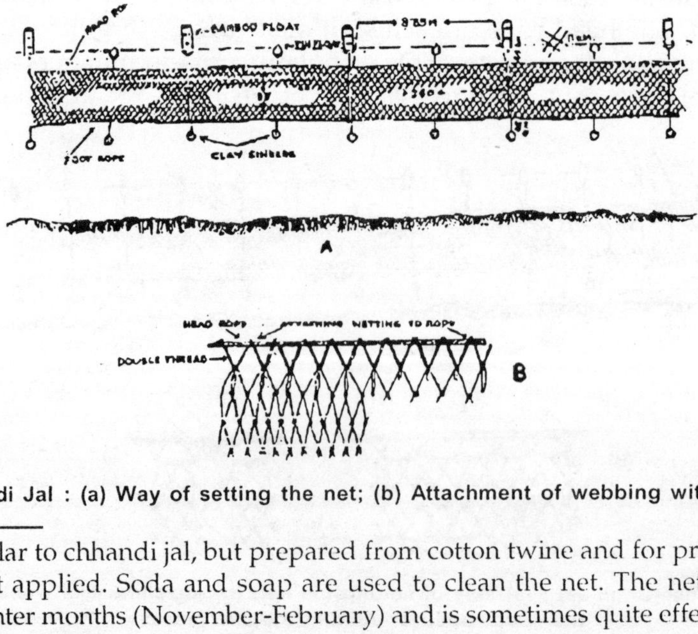

Fig. 3 : Chhandi Jal : (a) Way of setting the net; (b) Attachment of webbing with the head rope

---

* (Dora jal is similar to chhandi jal, but prepared from cotton twine and for preservation of the net decoction is not applied. Soda and soap are used to clean the net. The net is used to catch *H. ilisha* during winter months (November-February) and is sometimes quite effective because during this period water remains clean and as the net is also white, it is not easily noticeable in the water. Sometimes, even nylon made gill nets are called dora jal.

(c) Set Gill Nets with Foot Rope

These nets are kept stationary with the help of anchors.

1. *Barang Jal (Fig. 4) :* Made of synthetic material or cotton, a fleet of net is constructed by joining 4 to 8 shots. A single shot has 280 meshes across and 29 meshes in depth with a mesh size of 215 cm. The head rope of each shot measures 29.4 m. Reeving method is followed in the attachment of webbing with the head rope. 6 bamboo floats, are attached to the head rope of each shot at equal distances. Sinkers are absent, but iron anchors are tied to head rope and foot rope with the help of anchor line in order to keep the net fixed at one place. Empty tins or drum floats are used to join the anchors as shown in Fig. 4. One or two bamboo poles, at an interval of 2 shots, join the head rope and foot rope to support the net while in operation. A hanging ratio of 0.49 is always maintained. Eight to twelve men with a larger size boat operate this net in the lower zone of the Hooghly estuary between October to March as bottom set net and *lates calcarifer P. indicus, O. biauritus, Eleutheronema tetradactylum, Tachysurus* spp. and *C. dorab* are generally captured. Other local names are shele barang jal, bhetki barang jal and daura jal.

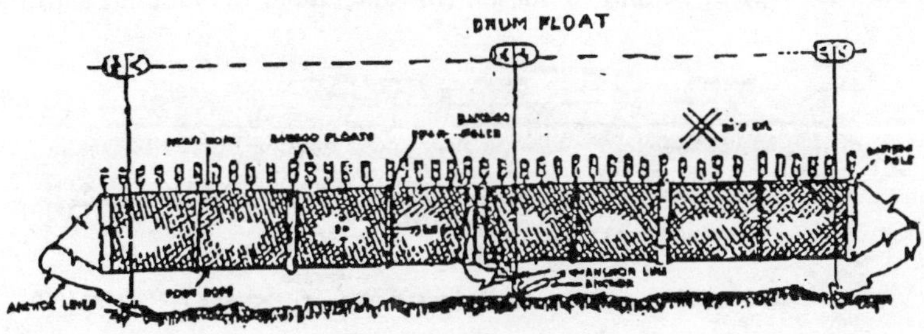

Fig. 4 : Barang Jal

2. *Khota Jal (Fig. 5) :* Mostly made of nylon, 4 to 7 shots are joined to make a fleet of gill net. A single shot has 500 meshes across and 76 meshes in depth with a mesh size of 10 cm. The length of float line in each shot measures 17.2 m and is made of wire. The attachment of webbing with the float line is of reeving type. Empty drum floats are attached to the float line at an interval of 2 shots. Sinkers are absent. Two special type of anchors are used at both the ends of the net, joining the float line and lead line through anchor line. The lead line and anchor lines are also made of Wire. These anchors are bigger and heavier than ordinary iron anchors and are made of hollow bamboo poles filled with sand and with openings closed with wooden pieces. Bamboo poles, at an interval of 2 pieces, join the float line and lead line in order to support the net. The hanging coefficient is 0.34. The net is operated as column set. Two to three men with a medium size boat operate this net in upper and lower zones of the Hooghly estuary. Seasons of operation extend from January to March and June to October. *H. ilisha, Pangasius pangasius* and *S. phasa* form the bulk of the catches. Another local name is khuti jal.

3. *Nongar Jal (Fig. 6) :* The chhandi jal when operated as a set type of net with the help of anchors is termed as nongar jal. Iron anchors, at an interval of 4 pieces, join the

head rope and foot rope through anchor lines. The net is operated as surface set and is operated in the upper zone of the Hooghly estuary for catching *H. ilisha.* Other local names are longar jal, nongar-chhandi jal and patan jal.

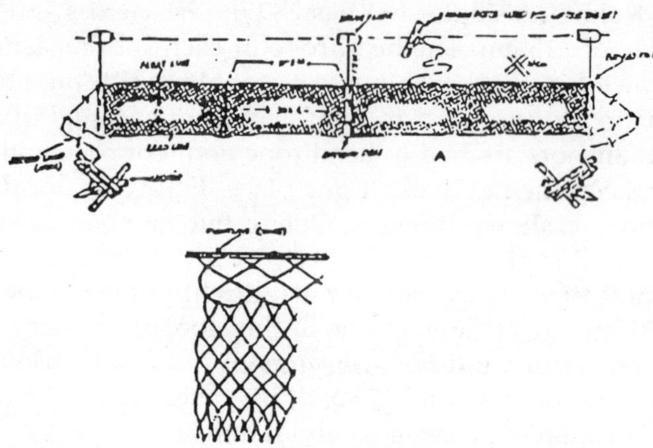

Fig. 5 : Khota Jal : (a) Way of setting Khota jal; (b) Attachment of webbing with the head rope

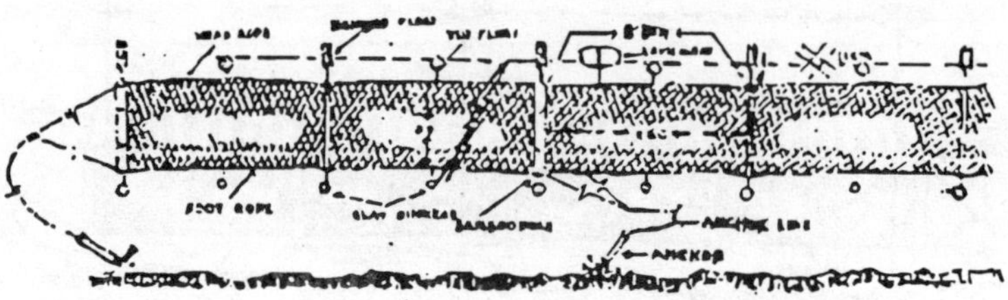

Fig. 6 : Nongar Jal

APPENDIX III

List of gears used in the Hooghly estuary
Source : De, D.K., 1980

De (1980) further enlighted the fishing gear and their use in the estuarine system. Based on the mode of operation De (1980) has classified the fishing gear into nine groups. The classification followed here is that of von Brandt (1958 and 1964). A list of gears with their local name together with the method of operation are presented here except Gill net which is already discussed by Dutta (1973).

Type of gear		*Local name*
1. Long lines	-	Barsi
2. Traps & Barrier	-	Bitti Charpata, Khal pata
3. Bag net	-	Been, Thor
4. Dragged net	-	Buro Jal
5. Seine net	-	Kochal jal, Gaiber jar or Tanaber jal, Berjal, Jagetberjal, Chatberjal, Kona jal.
6. Dip or lift net	-	Gara besal, Nanka besal
7. Falling net	-	Khepla jal
8. Gill net	-	*Discussed earlier by Dutta* (1973) in detail and hence excluded here
9. Clap or Purse net	-	Kharki jal, Sangla jal.

Description of Fishing Gear and Method of Operation

1. Longlines

The principle used in line fishing is to offer the fish a bait to entice it to bite.

Borsi

Long lines, known as barsi, are employed extensively throughout the estuary. It is stout line made up of cotton. The length of the main line ranges upto 1000 m, armed with a larger number of 450 to 500 short snoods, set at frequent regular intervals. To each snood a hook is tied at the free end. Small fishes, dried fishes and flour balls are attached to the hooks as a bait. Lines ar fished in different ways, they can be anchored or left adrift. They may be fixed in any position from the surface to the bottom. The size of the line, snoods and hooks depend on the fishing depth and species of fish caught. Longlines are operated throughout the year. Catfishes, perches, polynemids, scienids and anchovies are generally caught. This gear contributes 3.4% of the total catch of the estuary.

2. Trap and Barrier

The catching principle is based on allowing the fish to enter the trap.

Basket

They attain a diversity of shapes and sizes. Generally they are rectangular, cylindrical or sub-cylindrical in shape having only a silt like mouth through which fish can enter but can not come out. The traps are generally made of fine bamboo splinters. Sometimes a bait is also kept inside the trap. The type of fish caught in basket trap are chiefly prawn in the upper zone and mullets, *Eleutheronoma tetradactylum, Lates calcarifer* and different prawns spp., are in lower zones of the estuarine system.

Charpata

In tidal regions of lower estuary, large areas may be surrounded by a long length of netting forming a wall at the time of high tide. At the time of ebb tide, all the fish present are stranded at the bottom, from where they are caught. Charpata has a number of pieces, each about 7 m long and 3 to 4 m wide. The mesh size at upper area is 5 cm and at lower zone it is 1.3 cm. The total length of the net varies from area to area.

Khalpata Jal

The mouths of shallow tidal creeds that completely drain out at low tide may similarly be cut off during flood tide by a wall of bamboo screens called Khalpata.

Both the traps are operated throughout the year and fishes caught are prawns, *Lize parsia, Setipinna* sp., *Pama pama, E. tetradactylum., L. calcarifer* and *ilisha elongata.*

3. Bag Net

A bag net has a totally or partially framed mouth. The fixed form of bagnet is kept open by the force of the water flowing through it. The fish are actually filtered from the water.

Beenjal or Behund Jal

Bag net with a wide mouth is narrowing towards the cod end. Bag nets have wings extending from the mouth. There is a flap inside near the cod-end to prevent the fish from swimming out. The size of the net varies in different parts of the estuary. Generally, the size of the bag portion is 20 m along, mouth 6 m wide and wings 9 m long. Mesh size near the mouth is 4.0 cm and narrowing to 0.5 cm at cod-end. Number of meshes around the periphery of the net varies between 700 and 1000. The nets are operated with the help of heavy anchors and floats. The mouth of the net, facing the current of water, remains open due to force of water. In Hooghly-Matlah Estuarine System about 4000 bag nets are in operation and constitute more than 60% of the total catch of the estuary. The bag nets are used throughout the year. More than half the catches of the bagnets consists of *Harpodon nehereus* and different prawn spp. Remaining catches are dominated by *P. pama, Trichiurus* spp., *I elongata, mullets* and *Silago panigius.*

Thor Jal

It is similar to been jal but smaller in size. The number of mesh around the periphery ranges between 300 and 700. These nets are operated in pairs from boats with the help of wooden framework.

4. Dragged net

This group contains all nets and gear which are towed through the water.

Buro Jal

The net has a bag and a pair of short wings. Its shape is identical to that of a bag net and operation is similar to that of a typical boat seine. This net is mainly operated in the upper stretches of estuary during November to June for catching prawns.

5. Seine Net

In seining, one end of the net is short from a fixed point, a certain area is surrounded and the other end of gear returned to the starting point; after which the gear is then hauled.

Kochal Jal

It consists of 22-24 pieces and each piece has a length of about 12 m and depth of 30 m. Overall length of a net is about 280 m. Mesh size varies from 8-11 cm. The head rope is provided with floats. On sighting a shoal of hilsa the net is laid out from two boats encircling the shoal. It is mostly operated between the November and February in the lower estuary.

Gaibor Jal or Tanaber Jal

This is a very long rectangular seine net measuring 300 metres or more with puckerings in the lower portion. The depth of the net ranges from 3 to 9 m. Floats are attached to the head rope. The net is operated as encircling net, either in the middle of the river, or from the shore enclosing a large area of water. It is used primarily for hilsa.

Ber Jal

The simple type of seine net without puckerings is known as berjal. It is operated during October and May for catching mullets, perches, polynemids and scienids.

Jagat Ber Jal

Long ber jals are known as jagat ber jal measuring over 1600 m in length. The operation lasts for several days and large variety of fish are caught.

Chatber

Fine meshed seine nets that are operated in the upper stretches of the estuary for catching fry and fingerlings of hilsa, anchovies, prawns and miscellaneous fishes during November to August.

Kona Jal

This is a boat seine with a series of bags, each with a funnel shaped pouch placed at intervals of 9 to 12 m., provided with a valve to prevent the escape of trapped fish. It measures about 30 to 300 m in length and 9 m in breadth. The number of bags varies according to the length of net. Floats are attached to the head line but sinkers are absent. The mesh size of the net is about 5 cm.

The mouth of the bag is rectangular in shape, 6 to 9 m high and 4 to 5 m wide and the mesh is about 2.5 cm. The mouths of each bag is kept distended by two diagonally placed poles. The net could be operated as a drag net from shore or fixed to stakes in the river where there is a strong current. The catches are mainly hilsa and the season of operation is from May to October.

6. Dip or Lift Net

This method is to submerge a hanging net, then pull it rapidly out of the water so as capture any fish which happen to be over it.

Gara Besal

This is a fixed lever net. It is roughly triangular and is tied to a "V" shaped frame of bamboo. The size of the net varies from about 11 to 14 m in length and 8 to 11 m across the cord. The mesh is about 4 cm towards the middle. The nets are fixed in the river bed close to the bank.

Nauka Besal

This is similar to gara besal but smaller is size and is operated from a boat.

Both the nets are operated throughout the year and fishes caught are mullots, *I. elongata*, *Setipinna* spp. and prawns.

7. Falling Nets

This method is to cover the fish with a cone-shaped net with a stiff opening.

Khepla Jal

It is conical shaped net, widely used in almost all parts of the estuary. These nets occur in different sizes. A standard kepla jal measures about 4 m in length and peripheral circumference is about 20 m. The height of the pocket measures 0.5 m. The fishes get caught by the pockets provided at the periphery of the nets. The peripheral pockets are provided with sinkers. Nets are operated throughout the year and catches consist of *Polynemus paradiseus, P. pama, Pangasius pangasius* and prawns.

8. Clap or Purse Net

The principle is to purse fishes in bag shaped nets. The shape of this net is like a purse and the upper and lower lips of the mouth are framed with two long flexible split bamboo. Since there is every chance of the fish escaping, the closing mechanism of these nets is very well developed with the help of two lips which can be shut instantly as and when it is desired.

Kharki and shangla jal are commonly used in the upper stretches of the estuary during the period from July to December and January to March for catching migratory hilsa.

Kharki Jal

It is rectangular in shape and is operated from a boat which is drifted with the current. When fish enters the mouth of net is closed by pulling up the bamboo pole which is vertically attached to lower lip and passing through the upper.

Sangla Jal

It is an improved modification of kharki jal. The bag portion is rounded in shape and slightly larger than Kharki jal. In outline the mouth is semicircular and lips are often over 25 feet in length. The size of net is about 100 feet from front to back. Net can be set at any desired depth from a boat. Mouth is kept open by weight tied to lower lip and can be closed by pulling rope attached to middle of lower lip which passing through the upper.

Materials Used for Fishing Gear

Nets are made of either cotton or hamp twines. Presently synthetic twine is in extensive use for all kind of gears. Pieces of Bamboo (*Bamboosa* spp), Sola (*Cechynomena aspera*), various kind of light wood, earthen pots, scaled tin cans of various sizes are used as floats. Pieces of stones and brick, burnt clay, lead and iron of various shapes and size are also common.

Preservatives applied for net made of either cotton or hemp are extracts of 'Gab' fruits (*Diospyros embryopteris*), Coal tar, bark extract of Babul (*Accacia arabica*) are also used as preservatives.

III
The Sundarban Delta

1
Introduction

The Sundarbans, once the largest mangrove forest in the world maintained its ecological balance by receiving upland freshwater resources, maintained normal tidal flow in the estuaries and creeks adequately deep, lying crisis cross the Sundarban delta was enriched by biodiversity reserve. Large areas of wet lands were created. The morphoecological conditions of the entire Sundarbans thus maintained an ideal habitat for healthy growth and reproduction of aquatic life. Prawns and hilsa were the dominant fishery along with other commercial important fishes. But such healthy natural ecosystem of Sundarbans could not prolong for long, as the head water supply through major sources like R. Jamuna, Ichhamati and Bhagirathi – Hooghly along with other tributaries and network of channels were virtually in a state of moribond condition due to shifting of the Ganga flow to Padma and Damodar shifted further down. Change in the morphoecological conditions thus gradually appeared. Since 1582 upto date there were 100 cyclones crossed over Sundarbans. This was followed by high surge of sea waters and some time associated with earth quake. During 1737 high cyclone of 250 km/hr crossed over the Sundarbans associated with high surge of sea waters whose height was about 41 feet flushed the entire Sundarban forest. All these resulted silting of river basin and thus the important rivers and tributaries starting deteriorating fast one after another. Such conditions induced people to rehabilitate.

Thus the growing human population demanded progressive course of reclamation, resettlement, colonisation, agricultural development, construction of roads etc. The reserve forest area was gradually reclaimed to a considerable extent. Subsequently population influx especially during partition of Bengal (1947) into East Pakistan which later emerged as free Bangladesh (1971) and the continued cross-border infiltration of the population has further complicated the situation.

The banks and even beds of rivers were occupied by encroachers, constructed bheries, brick kilns etc. thus prevented the normal tidal-water flow. The surging affect of tidal water further encouraged silting by eroding the banks. Silting gradually increased which further prevented normal growth of forest. The vast wetlands located in the east of Kolkata that once used to purify city's waste water by their self purifying capacity could not continue any longer. The hectic construction of houses and housing complex, industrial complex, heaping up of garbage rapidly increased and resulted an increased volume of complex waste water. To dispose such waste water several canals were excavated. The wet lands thus gradually headed towards an unbalanced ecological condition resulting its fast deterioration.

Even the mighty river Bidyadhari–Hooghly has changed into a narrow channel. Various problems are appearing one after another. Even banks of canals are occupied by encrochers to lead life of shanty township, thereby prevent normal wastes flow. Some where canal beds are being used for agriculture purposes. Untreated complex wastes generated from various sources which are being used in bheries directly for fish culture with low fish production are likely to be affected by various heavy metals. Thus the Sundarbans delta which was once proud of its rich bio-diversity reserve, flora and fauna is now loosing its ecological balance very fast, while the area of the delta is gradually increasing. Some important aspects of the ecosystem of the Sundarbans and their rapid changes are presented through some plates under the caption "The Sundarbans at a glance."

The important sequential changes in the Sundarbans are :-

1. With the gradual suspension of head water, rivers, tributaries, estuaries and network of channel's are convered into moribund condition.
2. Silting gradually increased from eroded banks by tidal waves.
3. The growing human population demanded progressive course of forest reclamation, resettlement, colonisation, agricultural development, bheries for fish culture, construction of roads etc. and thus morphoecological conditions of the Sundarbans gradually altered.
4. The vast wetlands thus showed signs of deterioration and gradually converted into waste land.
5. Fishing effort rapidly increased, so much so, that even people occupied river banks, converted the river beds into bheries and even constructed brick kilns for their livelihood.
6. Bhagirathi-Hooghly, once a mighty river has gradually changed into a narrow channel due to high silt precipitation resulting low water supply after the construction of Farakka barrage.
7. Prawn and hilsa alongwith other commercial important fishery shows a decline while fishing effort on the other hand shows an increase.

2
Biological Characteristics of the Main Sundarbans

In the Sundarbans mangrove swamps, as reported by Naskar (1987), diatoms form a predominant group; the common diatoms identified from the Sundarban's fishery water are *Navicula* spp., *Pleurosigma* sp., *Gyrosigma* spp., *Cymbella* sp., *Coscinodiscus* sp., *Cyclotella* sp., *Fragilaria* sp., and *Amphipleura* sp. The Cyanophycean plankton are *Oscillatoria* sp., *Lyngbya* sp., *Spirulina* sp., *Anobaena* spp., *Microcoleus* sp., *Nodularia* sp., and *Rivularia* sp. and the Chlorophycean plankton are *Protococcus* sp., *Pediastrum* sp., *Hydrodictyon* sp., and *Ankistrodesmus* sp., while the common desmid is *Cosmarium* sp. The Cyanophyceae dominates the ill-consolidated soils whereas the Chlorophyceae the consolidated one.

The common benthic flora of this backishwater fisheries of the Sundarbans are *Oscillatoria* spp., *Gleocapsa* sp., *Symploca* sp., *Protococcus* sp., *Enteromorpha* sp., *Chaetomorpha* sp., *Polysiphonia* sp., *Gyrosigma* sp. and *Spirulina* spp.

The pneumatophores, prop-roots, stilt-roots and the lower trunk regions of *Avicennia* spp., *Sonneratia* spp., and the members of Rhizophoraceae hold the dense cover of Bostrychielum like *Bostrychiam* sp., *Cologlobossa* sp., *Contella* sp. and *Murrayella* sp. These periphyton have also much value as fish food in the mangrove forests (Schuster, 1952).

From the brackishwater paddy-cum-fish culture plots at Kulti-Minakhan area of the Sundarbans, the *Enteromorpha tubulosa* has been identified and studied in details. It is found that this littoral benthic algae is very much relished by the mullets, tilapia and prawns (Naskar and Santra, 1986).

Sewell (1934), Pillay (1954), Chakraborty (1984) and others have pointed out that in the Sundarbans waterways, copepods are dominant. The important species among the copepods are *Diaptromus* sp., *Pseudodiaptomus* sp., *Acartia* sp., *Cyclopina* sp. and *Cyclopsis* sp. The Cladocerans are predominant : Zoea and Megalopa larvae, mysids are also present. During monsoon Flagellates, Heminthes and Rotifers are also found.

For agriculture and certain other purposes, the original forest lands were cleared and numerous river banks were built up to check the tidal flow which resulted new types of eco-climatic conditions in the lower Ganga delta regions. Presently the oiginal halophytic tidal mangrove forests under the Sundarbans are only restricted in the southern and south-eastern parts of the 24-Parganas district. No or a little sign of the haloplytic flora except some fossil records is now traceable from the northern clearings of the Sundarbans delta. As the ecology and the soil-water salinity changed remarkably, the mesophytic flora have now occupied the places of the halophytes in the norther parts of the northern clearings of the Sundarbans delta are now highly populated, about 700 inhabitants per square Km. (Blasco, 1975). For this higher biotic pressure and interaction, the nbatural vegetations like halophytes and mangroves are gradually declining. All the mesophytic flora appeared here during the initiation of cultivation of the crop plants and other types of human interactions. Along with the mesophytes, a large number of aquatic and semi-aquatic plants are also distributed in the marshy places of the nothern clearings. Naskar (1981) has collected and described some 1175 angiosperm species under 154 angiosperm families from this part of lower Ganga. The ecology of this area is typically tropical dominated by mesophytes, halophytes, xerophytes, aquatic and semiaquatic. The ten dominant families under angiospermae are Fabaceae (135 spp), Poaceae (127 spp), Asteraceae (64 spp), Cyperaceae (62 spp), Euphorbiaceae (47 spp), Malvaceae (31 spp), Rubiaceae (27 spp), Acanthaceae (27 spp), Convolvulaceae (26 spp) and

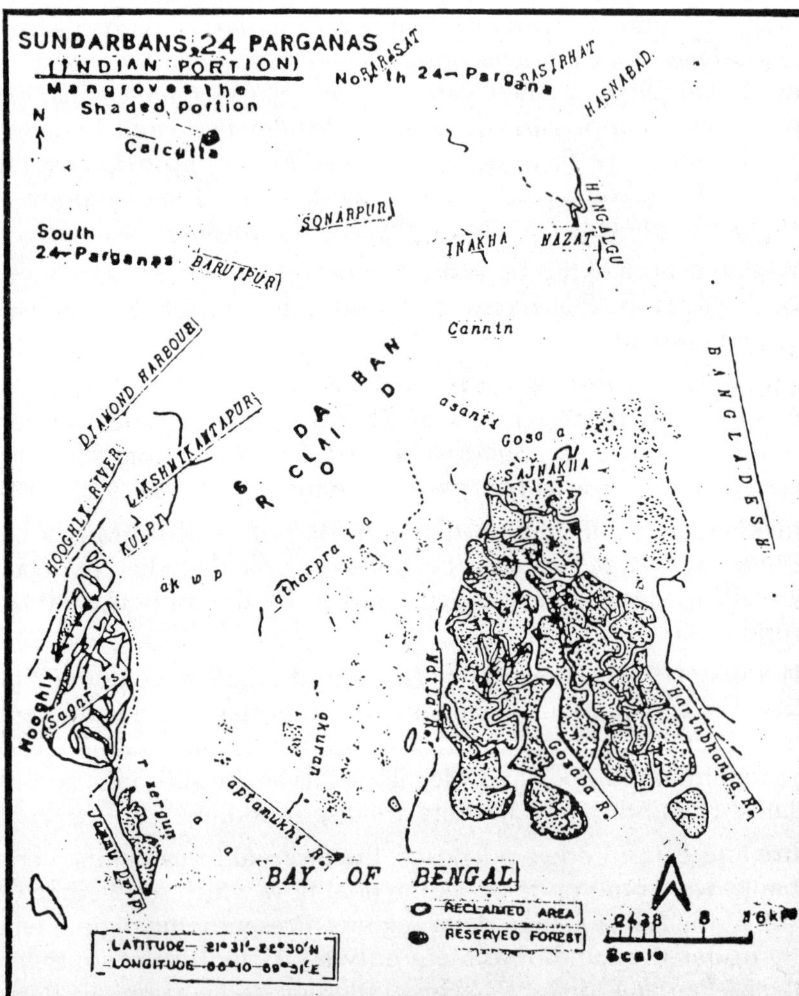

Fig. 1 : Mangroves of the Sundarbans

Scrophulariaceae (24 spp). These ten families comprise about the 50% of the total plant species of the Sundarbans.

Frequency of occurrance of common mangrove species and their associates in the Sundarbans are indicated in Fig. 1 and Table 1.

Table 1. Common Mangrove Species and their Associates in the Sundarbans Delta (Naskar, 1981)

Sl. No.	*Family and Species*	*Habit*	*Habitat*	*Frequency of Occurrence*		
				C	*O*	*R*
1.	*ACANTHACEAE*					
	*1.1 *Acanthus ilicifolius* L.	Shrub	Mud-flat	•		
	+ 1.2 *Acanthus volubilis* Wall.	Herb	Epiphytic		•	
2.	*AIZOACEAE*					
	o.2.1 *Sesuvium portulacastrum* L.	Herb	River-slopes	•		
3.	*AMARANTHACEAE*					
	o3.1 *Aerva (Aerua) lanata* (L.) Juss.	Herb	River-slopes		•	
4.	*APOCYNACEAE*					
	+4.1 *Cerbera odollam* Gaertn.	Tree	Flat lands		•	
	+4.2 *Nerium indicum* Mill.	Shrub	Flat lands	•		
	+4.3 *Thevetia peruviana* (Pers.) Sch.	Shrub	Flat lands	•		
5.	*ARECACEAE*					
	*5.1 *Nipa fruticans* Wurmb.	Tree	River-banks			•
	*5.2 *Phoenix paludosa* Roxb.	Shrub	River-flats	•		
6.	*ASCLEPIADACEAE*					
	+6.1 *Hoya parasitica* Wall.	Herb	Epiphytic	•		
	o6.2 *Sarcolobus carinatus* Wall.	Shrub	River-flats		•	
7.	*AVICENNIACEAE*					
	*7.1 *Avicennia alba* Bl.	Tree	River-slopes	•		
	*7.2 *Avicennia marina* (For.) Vierh.	Tree	River-slopes		•	
	*7.3 *Avicennia officinalis* L.	Tree	River-slopes	•		
8.	*BIGNONIACEAE*					
	+8.1 *Dolichandrone spathaceae* Sch.	*Tree*	*River-banks*		•	
9.	*BORAGINACEAE*					
	+9.1 *Heliotropium curassavicum* L.	*Herb*	*River-slopes*	•		
10.	*CHENOPODIACEAE*					
	o10.1 *Salicornia brachiata* Roxb.	*Herb*	*River-slopes*	•		
	o10.2 *Suaeda maritima* Dumort.	*Herb*	*River-slopes*	•		
	o10.3 *Suaeda nudiflora* Moq.	Herb	River-slopes	•		

contd...

Table 1 - *contd...*

Sl. No.	*Family and Species*	*Habit*	*Habitat*	*Frequency of Occurrence* C	O	R
11.	*COMBRETACEAE*					
*11.1	*Lumnitzera racemosa* Willd.	*Tree*	*River-* banks		•	
12.	*CONVOLVULACEAE*					
+12.1	*Ipomoea pescaprae* (L.) Sweet.	*Herb*	*Sandy* beach		•	
13.	*CYPERACEAE*					
+13.1	*Cyperus exaltatus* Retz.	Sedge	River-flats	•		
+13.2	*Fimbristylis halophila* L.	Sedge	River-flats	•		
+13.3	*Fimbristylis sub-bispicata* Fees and Mey.	*Sedge*	*River-* flats	•		
+13.4	*Scirpus triquetra* L.	Sedge	River-flats	•		
14.	EUPHORBIACEAE					
*14.1	*Excoecaria agallocha* L.	Tree	All habitats	•		
*14.2	Excoecaria bicolor Muell.	Tree	River-slopes		•	
15.	*FABACEAE (including CAESALPINIACEAE)*					
+15.1	*Caesalpinia crista* L.	Shrub	River-slopes	•		
+15.2	*Caesalpinia nuga* (L.) Ait.	Shrub	River-slopes	•		
+15.3	*Canavalia cathartica* Thw.	Twinner	Hedges		•	
*15.4	*Cynometra raniflora* L.	Tree	River-slopes	•		
+15.5	*Dalbergia spinosa* Roxb.	Tree	River-banks	•		
+15.6	*Derris indica* (Lam.) Benn.	Tree	River-banks	•		
*15.7	*Derris scandens* Benth.	Shrub	River-flats	•		
*15.8	*Derris trifoliata* Lour.	Shrub	River-flats	•		
*15.9	*Derris umbellatum* DC.	Shrub	River-flats	•		
16.	*LORANTHACEAE*					
+16.1	*Viscum orientale* Willd.	Herb	Epiphytic	•		
17.	*LYTHRACEAE*					
o17.1	*Ammannia baccifera* L.	Herb	River-flats	•		
18.	*MALVACEAE*					
+18.1	*Hibiscus tiliaceus* L.	Tree	River-banks	•		
*18.2	*Hibiscus tortussus* Wall.	Tree	River-banks	•		

contd...

Table 1 - *contd...*

Sl. No.	Family and Species	Habit	Habitat	Frequency of Occurrence C	O	R
*18.3	*Thespesia lampus* (Cav.) Dalz. & Gibs	Tree	River-banks		•	
+18.4	*Thespesia populnea* (L.) Sol. ex. Corr.	Tree	River-banks	•		
19.	*MELIACEAE*					
*19.1	*Amoora cucullata* Roxb.	Tree	Swamp forests		•	
*19.2	*Xylocarpus granatum* Koenig.	Tree	Tidal forests		•	
*19.3	*Xylocarpus molucensis* (L.) Roem.	Tree	Tidal forests	•		
20.	*MYRSINACEAE*					
*20.1	*Aegiceras corniculatum* (L.)	Shrub	Swamp forests	•		
21.	*PANDANACEAE*					
+21.1	*Pandanus fascicularis* Lamk.	Shrub	Outer-side		•	
22.	*PLUMBAGINACEAE*					
*22.1	*Aegialitis rotundifolia* Roxb.	Shrub	Tidal zone		•	
23.	*POACEAE*					
+23.1	*Aeluropus lagopoides* (L.) Trin.	Grass	River-flats		•	
+23.2	*Leersia hexandra* Sw.	Grass	River-slopes	•		
+23.3	*Phragmites kakra* Trin ex. Steud.	Grass	River-flats	•		
*23.4	*Porteresia coarctata* (Roxb.) Takeoka.	Grass	River-beds	•		
24.	*RHIZOPHORACEAE*					
*24.1	*Bruguiera cylindrica* Bl.	Tree	Tidal forests	•		
*24.2	*Bruguiera gymnorhiza* (L.) Lam.	Tree	Tidal forests	•		
*24.3	*Bruguiera parviflora* W. & A.	Tree	Tidal forests	•		
*24.4	*Bruguiera sexangula* (Lour.) Poir.	Tree	Tidal forests	•		
*24.5	*Ceriops decandra* (Griff.) Ding Hou.	Shrub	Tidal forests	•		
*24.6	*Ceriops tagal* (Perr.) Robins	Shrub	Tidal forests	•		
*24.7	*Kandelia condel* (L.) Druce	Shrub	Tidal forests	•		
*24.8	*Rhizophora apiculata* Bl.	Tree	Tidal forests	•		
*24.9	*Rhizophora mucronata* Lam.	Tree	Tidal forests	•		
25.	*SONNERATIACEAE*					
*25.1	*Sonneratia apetala* Ham.	Tree	Tidal forests	•		

contd...

Table 1 - *contd...*

Sl. No.	Family and Species	Habit	Habitat	Frequency of Occurrence C	O	R
	*25.2 *Sonneratia caseolaris* (L.) Engl	Tree	Tidal forests	•		
26.	*STERCULIACEAE*					
	*26.1 *Heritiera fomes* Buch. Ham.	Tree	Tidal zones		•	
	*26.2 *Heritiera littoralis* Dry.	Tree	Outside tidal zones		•	
27.	*TAMARICACEAE*					
	+27.1 *Tamarix dioica* Roxb.	Shrub	River-banks	•		
	+27.2 *Tamarix gallica* L.	Shrub	River-banks	•		
28,	*TILIACEAE*					
	*28.1 *Brownlowia Kosterm* L.	Tree	Tidal forests		•	
29,	*VERBINACEAE*					
	+29.1 *Clerodendrum inerme* (L.)	Shrub	River-banks	•		
	+29.2 *Premna intergrifolia* L.	Shrub	River-banks	•		

NB. C-Common; O-Occasional; R-Rare.

Symbol against each species indicates the ecological condition for that plant :

The symbols : '*' denotes True Mangrove, '+' denotes Mangrove Associate, 'o' denotes Obligate Mangrove.

Total No. of Tree Species	: 30	Total No. of Families	:	29	True Mangroves	:	35
Total No. of Shrub Species	: 20	Total No. of Genera	:	49	Mangrove Associates	:	28
Total No. of Herb Species	: 20	Total No. of Species	:	70	Obligate Mangroves	:	7
Toal No. of Species	: 70						

3
Common Fish and Crustaceans of the Mangrove

In the Sundarbans water, over 120 euryhaline fish species are caught in the commercial catches.

Among the various fishes. Clupeides like *Hilsa ilisha* constitutes an impotant fishing in the Sundarbans delta; the next important are *Hilsa tolley, Harpodon nehereus, Liza* spp., *Late calcarifer*, catfishes of the genera *Tachysurus, Pangasius* and *Mystus*. Smaller clupeids and scianiedes are abundant and landed in considerable quantities, which have good demand in the local markets. Among them the *Penaeus monodon* occupies the attention of most of the fish traders for its high export potentialities. Jumbo shrimp or *P. monodon* has provided an all round economic development to the Sundarbans delta in the recent years to a great extent.

All those fishes are harboured in these mangrove swamps of the Sundarbans because these halophytic adaptation and eco-climatic configuration provided ideal dwelling places, food and ideal breeding ground. The fallen leaves from the mangrove trees and shrubs provide food by decay and forming peryphyton and other algal vegetation (Ghosh, 1980). Jana and Chatterjee (1974) has stated that with the change in ecological condition in the Sundarbans regions, a number of euryhaline fishes are gradually diminishing from these areas. Frequency of occurrance of common fish and crustaceans are indicated in Table 2.

Table 2. List of the Common Fish and Crustaceans of the Mangrove Regions in the Sundarbans

Family and Species	*Very common*	*Common*	*Occasional*	*Rare*
A. FISHES :				
1. CLUPEIDAE				
1.1.	*Hilsa ilisha* (Hamilton)		x	
1.2.	*H. toli* (Valenciennes)		x	
1.3.	*H. elongata* (Bennett)			x
1.4.	*Raconda russeliana* (Gray)		x	
1.5.	*Nematolosa nasus* (Bloch)		x	

contd...

Table 2 – *contd...*

Family and Species	*Very common*	*Common*	*Occasional*	*Rare*	
1.6.	*Corica soborna* (Hamilton)		x		
1.7.	*Gudusia chapra* (Hamilton)				
1.8.	*Pellona elongata* (Benett)		x	x	
1.9.	*Andontostoma chacunda* (Hamilton)			x	
2. ENGRAULIDAE					
2.1	*Anchovialli indica* (Van Hass.)		x		
2.2	*Setipinna phasa* (Hamilton)		x		
2.3	*S. taty* (Cuvier & Valen.)		x		
2.4	*Coilia borennsis* (Hooker)			x	
2.5	*C. dussumieri* (Cuvier & Valen.)			x	
2.6	*C. ramcarati* (Hamilton)				
2.7	*Trissocles hamiltonii* (Gray)			x	
2.8	*T. purava* (Hamilton)			x	x
3. CHANNIDAE					
3.1	*Chanos chanos* (Forskaal)			x	
4. CHIROCENTRAIDAE					
4.1	*Chirocentrus dorab* (Forskaal)			x	
5. SYNODIDAE					
5.1	*Hapodon nehereus* (Hamilton)		x		
5.2	*H. tumbil* (Bloch)			x	
6. SIRORIDAE					
6.1	*Baoarius baoarius* (Hamilton)		x		
7. TACHYSURIDAE					
7.1	*Tachysurus iella* (Day)			x	
7.2	*T. sons* (Hamilton)			x	
7.3	*Oatsogeniosus militaris* (L.)			x	
8. PLOTOSIDAE					
8.1	*Plotosus canius* (Hamilton)		x		
9. BAGRIDAE					
9.1	*Mystus gulio* (Hamilton)				

contd...

Table 2 – *contd...*

Family and Species	*Very common*	*Common*	*Occasional*	*Rare*
9.2	*M. vitatus* (Bloch)			
9.3	*M. cavasius* (Hamilton)			
10. SCHILBEIDAE				
10.1	*Pangasius pangasius* (Hamilton)			
11. ANGUILLIDAE				
11.1	*Anguilla bengalensis* (Gray)		x	
12. MUGILIDAE				
12.1	*Rhinomugil corsula* (Hamilton)		x	
12.2	*Mugil macrolepis* (Smith)			x
12.3	*M. cephalus* (L.)			x
12.4	*M. cunnecius* (Valenciennes)			x
12.5	*M. troacholi* (Bleeker)			x
12.6	*M. speigleri* (Bleeker)		x	
12.7	*Liza parsia* (Hamilton)	x		
12.8	*L. tade* (Forskaal)	x		
13. POLYNEMIDAE				
13.1	*Eleutheranema tetradactylum* (Shwa)		x	
13.2	*Polydactylus indicus* (Shaw)			x
13.3	*Polynemus paradiseus* (L.)	x		
14. CENTROPOMIDAE				
14.1	*Ambassis nama* (Hamilton)			x
14.2	*A. baculis* (Hamilton)			x
14.3	*A. ranga* (Hamilton)			x
14.4	*Lates calcarifer* (Bloch)	x		
15. THERAPONIDAE				
15.1	*Therapon iarbus* (Forskaal)		x	
16. **SILLAGINIDAE**				
16.1	*Sillago sihama* (Forskaal)		x	

contd...

Table 2 – *contd...*

Family and Species	*Very common*	*Common*	*Occasional*	*Rare*
16.2	*S. panidius* (Hamilton)		x	
17. CARANGIDAE				
17.1	*Caranx carangus* (Bloch)			
17.2	*Megalesis cordyla* (L.)			
18. LUTIANIDAE				
18.1	*Lutianus argentimaculatus* (Forskaal)		x	
19. LEIOGNATHIDAE				
19.1	*Leiognathus blochii* (Cuvier & Valen)		x	
19.2	*L. equulus* (Forskaal)		x	
19.3	*L. fasciats* (Lacepede)		x	
19.4	*Gazza minuta* (Bloch)			x
20. GERRIDAE				
20.1	*Garras oblongus* (Cuvier)		x	
20.2	*G. oyena* (Forskaal)			x
20.3	*G. astifer* (Hamilton)		x	
21. SCIAENIDAE				
21.1	*Pseudoscisena soldoldo* (Hamilton)		x	
21.2	*Sciaenoides bieuritus* (Cantor)		x	
21.3	*Pama pama* (Hamilton)		x	
21.4	*Otolithus maculatus* (Cuvier)		x	
22. DREPANIDAE				
22.1	*Drepane punctatus* (L.)			x
23. SCATOPHAGIDAE				
23.1	*Scatophagus argus* (Bloch)		x	
24. CICHLIDAE				
24.1	*Etroplus suratensis* (Bloch)		x	
25. TRICHIUPYDAE				
25.1	*Trichiurus haumala* (Forskaal)			x

contd...

Table 2 – *contd...*

Family and Species	*Very common*	*Common*	*Occasional*	*Rare*
25.2	*T. muticus* (Gray)			x
25.3	*T. savala* (Cuvier)		x	
26. SCOMBEROMORIDAE				
26.1	*Cybium commersonii* (Lacepede) Rare			x
27. STROMATEIDAE				
27.1	*Pampus argenteus* (Euphrasen)		x	
27.2	*P. chinensis* (Euphrasen)		x	
27.3	*Stromoteus sinensis* (Day)			x
28. SOBIDAE				
28.1	*Glossogobius giuris* (Hamilton)		x	
28.2	*Boleophthalmus boddarti* (Pallas)		x	
29. TAENIOIDIDAE				
29.1	*Odontomblyopus rubrcundus* (Hamilton)		x	
30. PERIOPHTHALMIDAE				
30.1	*Periophthalmus Koetreuteri* (Pallas)		x	
31. PLATYCEPHALIDAE				
31.1	*Thyeenophyra indicus* (L.)			x
31.2	*Platycephalus insidiator* (Forskaal)			x
32. SOLEIEAE				
32.1	*Colea bleekeri* (Boulenger)			x
33. CYNOGLOSSIDAE				
33.1	*Cynoglossus brevis* (Gunthar)			x
33.2	*C. cynoglossus* (Hamilton)		x	
33.3	*C. lingua* (Hamilton)			x
34. TRIACANTHIDAE				
34.1	*Triacanthus brevirostris* (Temm. & Sch.)			x
35. BATRACHOIODAE				
35.1	*Batrachus oumbiens* (L.)			x
36. MEGALOPIDAE				
36.1	*Megalops cyprinoides* (Broussonet)			x
37. CYPRINODONTIDAE				
37.1	*Oryzias melastigma* (Mc Chelland)			x
38. AMPHIPNOIDAE				
38.1	*Amphipnous cuchia* (Hamilton)			x

contd...

Table 2 – *contd...*

Family and Species	Very common	Common	Occasional	Rare
39. NANDIDAE				
39.1	*Nandus nandus* (Hamilton)			x
40. ELOPIDAE				
40.1	*Elops saurus* (L.)			x
41. PLOTOSIDAE				
41.1	*Plotossus anguillaris* (Bloch)	x		
41.2	P. *Canius* (Hamilton)	x		
B. PRAWNS :				
42.	**PENAEIDAE**			
42.1	*Penaeus monodon* (Fabricius)	x		
42.2	*P. indicus* (Edwards)	x		
42.3	*Metapenaeus monoceras* (Fabricius)	x		
42.4	*M. brevicornis* (Milne-Edwards)	x		
42.5	*Paravenaeopsis sculptilis* (Milne-Edwards)	x		
42.6	*P stylifera* (Milne-Edwards)	x		
43. PALAEMONIDAE				
43.1	*Palaemon rude* (Heller)	x		
43.2	*P. styliferus* (Milne-Edward)	x		
43.3	*Macrobachium rosenburghii* (Fabricius)		x	

Plenty of crabs belonging to different genera and species are available in the Sundarbans region, but economically most important are the *Genus-Scylla* and *Neptunus Scylla serrats* (Forskaal) and *Neptunus pelagieus* (L.) are the two common edible and economically important crabs of the Sundarbans water systems.

Cultural Potentialities of Some Important Fishes and Prawns

Among the culturable fishes, mullets and tiger shrimps are the most important members in the Sundarbans estuary. These fishes and prawns are cultured in about 2,500 small to large-sized brackishwater fisheries (Ghosh, 1980). Beside culturable species, some predator fishes enter into these brackishwater culture systems. These predator species directly kill and feed on the mullets and shrimps (Anonymous, 1978). Jana *et. al.* (1974) mentioned that several important fishes and prawns are now declining from the Sundarbans estuaries due to deterioration of the mangrove vegetation.

Some of the most impotant cultural practices are as follows:

A. Mullet Culture

Several species of grey mullets belonging to the family Mugiladae are distributed throughout the mangrove swamps of the Sundarbans. The juveniles of mullets enter these estuaries along with the tide water but adult mullets migrate towards the bay for spawning (Jhingran, 1982). The tropical mullet species belonging to the genera *Liza, Mugil* and *Rhinomugil* are also suitable for culture in the mangrove associated areas of the Sundarbans (Chakraborty *et al.*, 1981). These mullet species feed on the benthic algae, mangrove plant detritus and are associated with the aquatic micro-flora and fauna of the magrove ecosystem. In juvenile stage these grey mullets migrate to the mangrove water in a massive scale for food and shelter (Luther, 1973). Pillay (1954) have mentioned that fishes of *Liza* group are the most suitable for culture in the brakishwater fisheries, because of their intimate association with the mangroves. Possibilities of potential yield from intensive mullet farming in the Sundarbans mangrove associated brackishwater fisheries were worked out by several cutural trials with *Liza* spp. and *Mugil* spp. (Anonymous, 1978).

B. Prawn/Shrimp Culture

The common prawn and/or shrimp of the Sundarbans are mentioned by Naskar and Chakraborty (1984). Hall (1962) attempted to relate marine prawn fisheries with the condition of the mangroves. Macnae (1974) and Namin (1977) explained that higher yield of prawns were derived from mangrove associated tropical coastline. Destruction of mangroves is one of the major constraints for declining trends of prawn catches (Daugherty, 1975). The genera, *Penaeus* and *Metapenaeus* are commercially, important, prawns in the Sundarbans. Adult penaeid prawns migrate towards the sea for maturation and breeding and on their post-larval stage they return to these mangrove forests of the Sundarbans estuaries, which serve as nursery bed (Chakraborty, *et al.*, 1981). Post-larvae of prawns occur almost throughout the year in the mangrove estuaries, though the abundance fluctuate seasonally.

4

The Estuarine Fishery of Sundarban Delta : Its Trend

With the diversion of major flow of the Ganga into the Padma and shifting of the Damodar further down, the major sources of fresh water resources at the head water that used to feed the Bidyadhari, The Matla, The Ichhamati, The Kulti and other estuaries through their major distributary the Jamuna and redistributories like Nawee, Sunti and other innumerable channels criss crossed the delta completely, suspended due to silting through tidal waters. Deprived of the upland water supply all estuaries of Sundarban delta had to depend entirely on other resources i.e. local drainage and spill area to prolong their life. At this stage another characteristics peculiar to tidal rivers carrying large portion of silt especially during August and September every year during spring tide hasten their decay. This fact is revealed from the Ganga in the west and Kulti in the east of Sundarbans from their high shape factor value, thereby indicating gradual rise in the river beds. By this process the deeper water area becomes shallow while the shallower zones get gradually swampy or exposed. The mangrove forests show greater role in arresting the silt to form/ extend the islands. This is how extension of delta ara by 33.7 percent in 40 years have become possible. Thus it is clear that flood water is essential to get a higher catch when increased effort may not be a factor. The facts thus emerges are -

1. The gradual silting of the estuaries is leading fast towards their deterioration. Mangrove has considerable role in this process.
2. Formation of shallow areas in the river bed and also large spill areas on both sides of the estuary and in the cognomination of 54 islands interconnected through tidal estuaries/creeks etc. has every possibility of increasing the deltaic area of Sundarbans. The increase of population will also similarly increase the reclaimed area.
3. Deeper zone gradually reduces its depth and thereby increases the shape factor and gradually advances towards the confluence.

   By this process the swampy areas and the spill areas infested with mangrove forest creat another ecosystem which are possibly used as grazing grounds and nersury beds for estuarine or offshore fish and other threatened endangered and endamic fauna, prawns and other Crustaceans. Mac Nae (1968) states that mangroves are the nature's own aquaculture system which means "no mangrove, no prawn."

4. Lastly, the Cyclonic weather when persists for a longer duration, the effects would be more pronounced. The surging effect is felt not only in the area of the tract of the cyclone but also in the surrounding regions. The upsurge of the water level in the tract area will cause water at higher levels to move towards the surrounding regions with lower levels and with this set up a long wave motion which would penetrate into the areas outside the storm region. When this wave propagates into the shallower region of the estuarine and river system its height may increase importantly. As a result surging effects may be experienced in areas far beyond the limits of the storm affected region. Such surging effect may influence the fishery to a great extent.

Information on the annual fish landings during 1966-67 to 1970-71 of the Hooghly-Matla estuarine system of five zones* (Zone-I-V, Fig. 2) Dutta *et al.*, 1973 stated below and subsequent landings during 1990-91 although shows a steady rise, a fall (13.7%) in the total fishery was noticed in the following year (1991-92), while hilsa fishery registered a sharp decline (42.9%) in 1991-92 compared to the previous year (1990-91).

Table 3. Shows annual fish landing (in tonnes) in the Hooghly-Matla estuarine system during 1966-67 to 1970-71 and subsequently from 1990-91 and 1991-92

Year	*Fish landing (in tonnes)*	*Remarks*
1966-67	5900.8	
1967-68	8376.7	
1968-69	7592.2	
1969-70	7978.1	
1970-71	10754.9	
1990-91	409811.0	During these 2 year's period
1991-92	35353.9	Hilsa fishery T. *ilisha* dwindled by 42.9 percent and the total fishery by 13.7 percent.

* Zone-I = Nabadwip - Khusigoli ; Zone-II - Khusigoli - Diamond Harbour. Zone-III - Lower zone including entire estuarine area of Sundarbans and tract below Diamond Harbour ; Zone-IV - Rupnarayan estuary and Zone-V - Small zones in the Matla estuary.

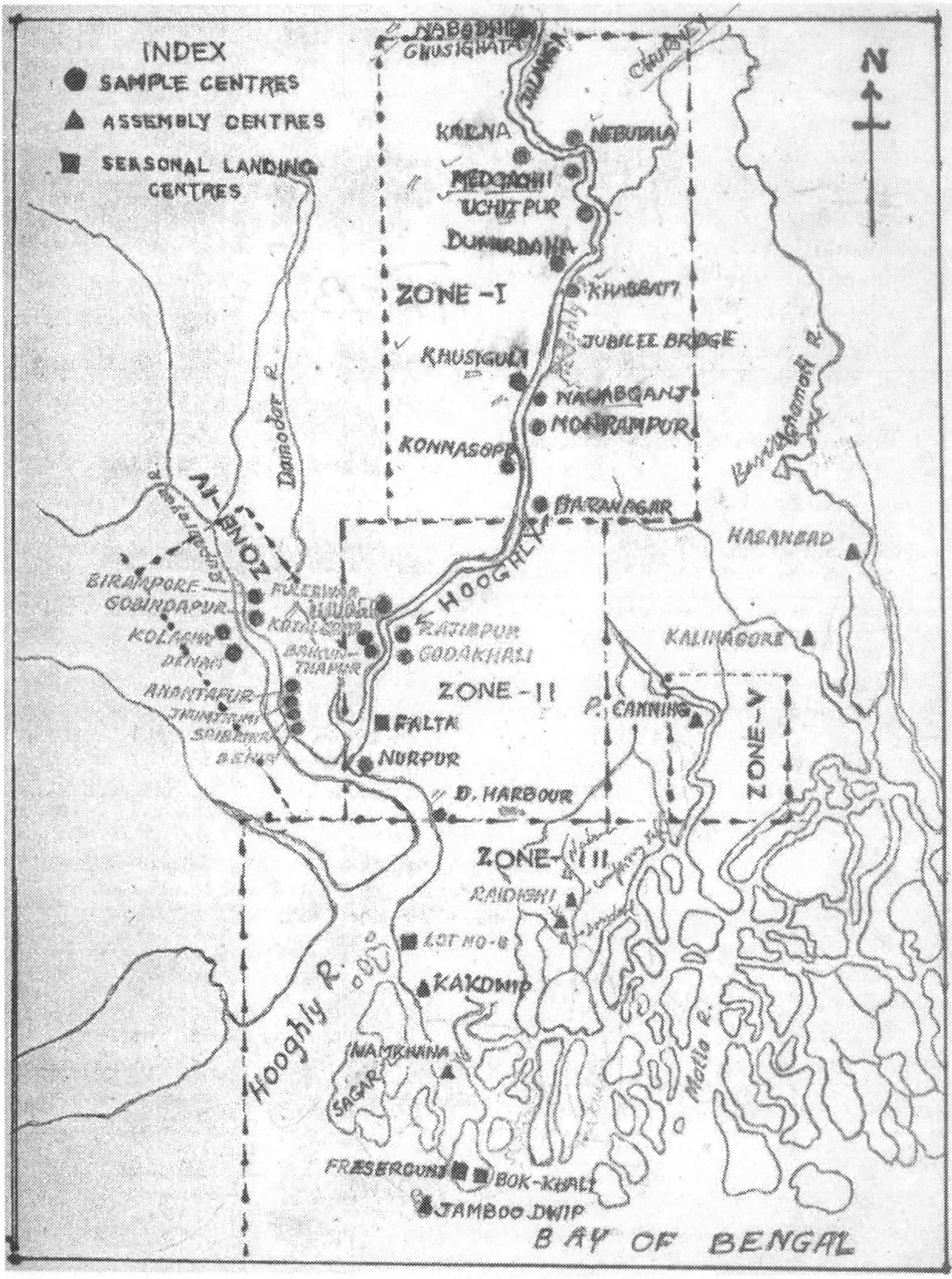

Fig. 2 : Hooghly-Matlah Estuary and its Various Fishing Zones

5

Analysis of Fish Landing from the Hooghly-Matla Estuarine System of Sundarban Delta during 1966-67 to 1970-71*

Fisherman operate their nets in the remote areas of the lower region of the estuary and catches are brought by them or sometimes by other parties to the landing centres. The main landing centres are - Namkhana, Diamond Harbour, Kakdwip, Raidighi, Port Canning, Hasnabad and Kalinagar (Fig. 2, Table 4).

The assembly centres like Port Canning, Raidighi, Kalinagar and Hasnabad also received good amount of catcher from the bheries (Brackishwater impoundment). The catches from the main channel of the Hooghly, Saptamukhi and Muriganga rivers mostly come to Namkhana, Kakdwip and Diamond Harbour. The catches from Thakuran river and adjoining canals come to Raidighi. Port Canning receives the catches made in the Matla estuary. The catches made east of Matla river come to Hasnabad and Kalinagar. Namkhana and Diamond Harbour appears to receive more catches than other centres. During winter season fishing activities increases because of favourable conditions in the South of Sundarbans.

Various aspects of fish landings-average monthly landings, species composition of annual catches, annual species wise landings in different centres, zonewise landings during different years, specieswise composition of catches in various zones etc. are indicated in different Tables 5 (I-VII), 6, 7, 8 (I-V) and 9) for the period 1966-67 to 1970-71 and (1961-62 to 1970-71) only of which some important aspects are discussed here.

Species composition and their relative abundance depends largely upon the morphoecological changes of ecosystem associated with fishing effort. It is interesting to note that composition for dominance was mainly among five species during three major seasons.

Fish landings are normally recorded at seven centres spread over the entire brackish water zones of Sundarban delta. These centres are Namkhana, Kakdwip, Diamond Harbour, Hasnabad, Posh Canning, Kalinagar and Raidighi (Table 4).

The overall pictures of landings (in kg.) in Namkhana, Kakdwip and Diamond harbour dominated all through the year 1966-67 to 1970-71 with some changes in their position viz.

* The entire analysis of fish landings is based on the data obtained from Dutta *et al.* (1973).

Namkhana dominated in all the four years 909067, 895707, 962757 and 2270030 kg. except during 1968-69 (760801 kg.) ; while Diamond Harbour dominated only during 1969-70 and 1970-71 (1403807 and 1396354 kg.). This was followed by Raidighi and Port Canning. Raidighi dominated duing 1967-68 (96624 kg.), 1969-70 (133079 kg.) and 1970-71 (148876 kg.); while Port Canning dominated only during 1966-67 (76733 kg.) and in 1968-69 (77534 kg.); Lastly the two centers Hasnabad and Kalinagar were showing the least value. Hasnabad was showing the lower value (70905 kg.) than Kalingar (97397 kg.) only during 1969-70 (Table 4).

It will be interesting to reveal further information on some commerial important species landed at different centres during these five years (Table 5. (I-VII).

At Diamond Harbour during 1966-67 the dominated species were Hilsa, *P. indicus, P. pama* and *S. biauritus.*

During 1967-68 the order was Hilsa, P. indicus, *H. toli* and *T. jella.* In 1968-69 the important species were Hilsa. *T. jella, P. indicus* and *P. pama.* In 1969-70, Hilsa, *T. Jella, P. indicus* and finally in 1970-71 the species were *P. indicus, T. jella* and *P. pama.*

The order of abundance of species at Namkhana centre were *H. neharius,* Prawn, Setipinna and *T. salalyar* and *T. Pantulu* (1966-67). In 1967-68 they were Hilsa, *H. neharius,* Prawn and setipinna. *H. neharius,* Prawn, Setipinna and *P. pama.* were the order during 1968-69. In 1969-70 the order was Prawn, *H. neharius, P. setipinna* and Hilsa, and finally in 1970-71, the order of landing was *H. neharius,* Prawn, Setipinna and *P. phasa.*

Species dominated at the Kakdwip centre were *T. jella,* Prawn, *H. neharius* and Hilsa during 1966-67. In 1967-68, Hilsa, Prawn, *P. indicus* and *T. jella* dominated, in 1968-69, the order was *H. neharius,* Prawn, *P. indicus* and *S. biauritus,* in 1969-70 they were Prawn *H. neharius, P. indicus* and *T. jella.* Lastly, the order was Hilsa, Prawn. *T. jella* and *H. neharius* in 1970-71.

At Raidighi centre the order was Prawn, Hilsa, *H. naharius* and *L. calcarifer* in 1966-67; the order was however different in 1967-78. They were Hilsa, *P. pangasius* and Prawn ; In 1968-69 the order was *P. paugasius, I. elangata, S. senarius* and Setipinna ; *I. elougata, P. paugasius,* setipinna and Hilsa in 1969-70 and in 1970-71 they were *I. elongata, P. pangasius,* Setipinna and *P. canius.*

At port Canning, the species recorded were Prawn, setipinna, *M. Parsia* and *L. calcarifar* in 1966-67 ; Prawn, Hilsa and Setipinna and *L. calcarifer* in 1967-68 ; Prawn, Satipinna, *P. parsia* and *T. jella* in 1968-69 ; Prawn, *L. Calcarifer, L. parsia* and *P. carrius* in 1969-70 and lastly the abundance of species were Prawn, calcarifer, *T. jelly* and setipinna in 1970-71.

The order of species dominance were prawn, *H. neharius,* Setipinna and *L. calcarifer* at Hasnabad during 1966-67; Prawn *H. neharius* setipinna and *P. Pangasius* in 1967-68 ; Prawn, *P. pangasius, H. neharius* and *P. pama* in 1968-69 ; Prawn, *H. neharius, P. pangasius* and *P. pama* in 1969-70 and lastly in 1970-71 the order was prawn, *P. pangasius, S. biauritus* and *L. Calcarifer.*

At Kalinagar the order was prawn, Setipinna, *L. calcarifer, P. panna* during 1966-67; Prawn, Setipinna, *L. Calcarifer* and *E. tetradactylum* in 1967-68; Prawn, *L. Calcarifer, E. tetradactylum* and Setipinna in 1968-69; Hilsa, Prawn, *H. parsia, E. tetradactylum* in 1969-70 and lastly the order was prawn, setipinna, *P. indicus* and *E. tetradactylum* in 1970-71.

The average monthly landings (in kg.) of different species for 5 years in the Hooghly-Natla estuarine system indicates an interesting picture (Table 6). Initially, prawn, *H. neharius* and *S. Phasa* and *S. tatty* combined dominated from March to May when in June *S. phasa* and *S. tataty* group was replaced by Hilsa. From July onwards till December, Hilsa replaced prawn and in tern prawn replaced *H. neharius* further down (July-September) and from October, *H.. neharius* regained its second position pushing prawn to third position till December. In January *H. neharius* althouh occupied the first position followed by prawn and hilsa, in February, prawn regained its first position followed by *H. neharius* and Hilsa. *S. phasa* and *S. tataty* combined mostly occupied third position from March-May, from June to August fourth position, and from September and October-fifth position and lastly in November and December regained the fourth position.

It is interesting to note that in all the 4 years 1966-67, 1968-69, 1969-70 and 1970-71 all the three zones III, I and IV dominated due to high landings of which zone III was leading followed by zone I and IV. Only during 1967-68 zone IV and I followed zone III (Table 7).

Zonewise number of occurrence of commercial important species with their order of dominance during the period of 5 years (1966-67 to 1970-71) indicate interesting picture. In zone I the dominant species were Prawn, Hilsa, Satipinna. This was followed by *P. pangasius, P. pama* and *P. paradiseus.* In zone II besides hilsa and prawn, *H. nehereus* came in the picture in place of satipinna and satipinna occupied the fourth place followed by *P. pangasius, P. Pama* and *S. biauritus.* In zone III prawn and satipinna dominated followed by Hilsa, *H. neherus* and Trichirus. In zone IV Prawn, Hilsa, and *H. neherus* dominated followed by *P. pama C. borneensis* and *P. paradiseus.* Finally in zone V, *H. nehereus* and prawn topped the list followed by Setipinna, *H. parsia,* Trichiurus and *P. Pangasius.*

From the above it is clear that Prawn, Hilsa, setipinna, *H. nehereus* dominated interchanging their position but interestingly in all the zones Prawn maintained its supremacy followed by *H. nereneus* and Hilsa Table 8 (I-V).

Yearwise species composition covering the period March to February of the Hooghly-Matla estuarine system of Sundarban during the 10 year period (1961-62 to 1970-71) indicate an interesting picture of composition among dominant species.

The yearwise catch of dominant species can be divided into two groups. In the first group the species dominated were Hilsa, prawn and *H. nehereus* and a competition to dominate was noticed among them. During 1961-62 and 1962-63 Hilsa as usual maintained the spremacy, followed by prawn and *H. nehereus.* In 1963-64, Hilsa as usual maintained the supremacy and was followed by *H. nehereus* and prawn. During 1964-65 to 1968-69, *H. nehereus* maintained its supremacy followed by Hilsa and prawn by interchanging their position except in 1968-69 when setipinna entered in place of Hilsa which lost its position from the dominant group. Again in 1969-70 the cycle of dominance of species-Hilsa, prawn and *H. nehereus* regained their old position noticed in 1961-62 and 1962-63. In 1970-71 *H. nehereus* again regained its first position followed by Hilsa and prawn.

In the second group no systimatic dominance of any particular species was noticed except interchanging their position among them. These species were *P. pama, T. jella, P. pangasius, S. Phasa, S. taty, T. pantului, T. savala,* Trichiurus, *S. biauritus,* setipinna and *P. indicus.* (Table 9).

Table 4. Volume of landings (in kg) at Assembly Centres on The Hooghly-Matlah Estuary

Centres	*1970-71*	*1969-70*	*1968-69*	*1967-68*	*1966-67*
Diamond Harbour	1396354	1403807	577653	579148	375636
Namkhana	2270030	962757	760801	895707	909067
Kakdwip	736993	841155	791511	785515	773031
Raidighi	148876	133079	58017	96624	44356
Canning	123777	72086	77534	73273	76733
Hasnabad	49796	70905	59807	63966	79139
Kalinagore	35201	94397	47895	55176	59877
Total for the above centres	4761027	3578186	2373218	2549409	2317839
Total for zone III	8526676	5751612	5585694	5984215	4442530

Table 5(I). Annual Specieswise Landings at Diamond Harbour Assembling

	Kilogrammes				
Species	*1970-71*	*1969-70*	*1968-69*	*1967-68*	*1966-67*
M. tade	629	2192	1023	417	121
M. parsia	389	154	131	303	83
L. calcarifer	172	166	670	711	149
S. panijus	3055	2305	3198	2879	1138
P. paradiseus	1102	3928	1071	532	752
P. indicus	81102	139567	101585	108964	13165
E. tetradactylum	5538	1806	936	3582	359
B. biauritus	8254	47618	9281	215	3095
S. miles	3	-	-	-	-
P. pama	24545	9142	45235	4989	6931
H. ilisha	1040870	991911	255329	363414	335290
H. toli	11244	5149	6640	13716	1708
I. elongate	3697	2500	12416	2120	1443
C. ramcarati	246	-	75	71	310
C. borneensis	-	-	-	-	-
S. phasa *S. taty*	1218	1454	5965	3155	313
P. pangasius	5180	2401	1283	1023	15
T. jella	36236	171038	106430	6944	2406
O. militaris	81	3	17	742	-
P. canius	27	4	-	-	-
T. savala *T. pantului*	33	-	187	91	7
H. nehereus	59	20	394	895	209
S. cinereus	573	3805	5415	4505	8
Prawns	5089	2806	1228	2720	2362
Miscellaneous	166412	15838	19144	57160	5772
Total	1396354	1403807	577653	579148	375636

Table 5(II). Annual Specieswise Landings at Namkhana Assembling Centre

	Kilogrammes				
Species	*1970-71*	*1969-70*	*1968-69*	*1967-68*	*1966-67*
M. tade	12608	1051	1408	569	472
M. parsia	33772	23576	24662	26003	23352
L. calcarifer	7342	3804	4319	2316	3077
S. panijus	16002	1621	1681	1098	1669
P. paradiseus	12489	4398	3251	1724	4027
P. indicus	8914	3689	10147	2000	2780
E. tetradactylum	5760	2365	3015	2804	1090
B. biauritus	32279	43284	10308	27532	3838
S. miles	-	11	242	99	-
P. pama	66165	26450	26986	30076	24526
H. ilisha	9468	56762	3473	158705	8384
H. toli	4738	7283	6206	11578	2296
I. elongata	42689	17699	19452	9824	24814
C. ramcarati	55445	6197	13796	17893	28174
C. borneensis	-	-	120	1	97
S. phasa *S. taty*	243313	88352	68982	68812	97943
P. pangasius	36830	17309	6221	2324	2990
T. jella	12658	27825	17939	31542	26830
O. militaris	7856	20754	6729	769	1505
P. canius	588	50	70	68	1081
T. savala *T. pantului*	113387	23213	24978	25094	55882
H. nehereus	603515	178461	172408	150020	231042
S. cinereus	36162	16749	24149	34616	3845
Prawns	500436	184125	137118	137096	194667
Miscellaneous	405814	207729	173141	153144	164686
Total	2270030	962757	760801	895707	909067

Table 5(III). Annual Specieswise Landings at Kakdwip Assembling Centre

	Kilogrammes				
Species	*1970-71*	*1969-70*	*1968-69*	*1967-68*	*1966-67*
M. tade	759	409	359	628	542
M. parsia	552	386	437	602	405
L. calcarifer	1751	808	272	599	487
S. panijus	2067	1333	613	160	794
P. paradiseus	996	948	915	1017	2105
P. indicus	49320	62779	77965	95912	54554
E. tetradactylum	15790	4629	3673	2651	4534

contd...

Table 5(III) - *contd...*

Species	Kilogrammes				
	1970-71	*1969-70*	*1968-69*	*1967-68*	*1966-67*
B. biauritus	17143	9951	75836	11225	3126
S. miles		-	-	20	50
P. pama	9873	17626	31771	23089	37846
H. ilisha	91879	40192	10971	134590	65895
H. toli	1310	1741	1201	229	1452
I. elongate	18599	16819	5340	16570	7489
C. ramcarati	5325	6237	2281	12389	13028
C. borneensis	10	246	61	458	118
S. phasa *S. taty*	31583	18735	20405	27070	13185
P. pangasius	800	959	370	95	1356
T. jella	69971	52922	63306	68061	111743
O. militaris	631	68	2768	3	46
P. canius	-	-	-	-	35
T. savala *T. pantului*	3765	4904	3351	3740	3090
H. nehereus	57010	93341	122392	74458	98560
S. cinereus	6936	3642	21565	6144	3198
Prawns	91497	147242	100803	102064	81946
Miscellaneous	259426	355238	244883	203741	267447
Total	736993	841155	791511	785515	773031

Table 5(IV). Annual Specieswise Landings at Raidighi Assembling Centre

Species	Kilogrammes				
	1970-71	*1969-70*	*1968-69*	*1967-68*	*1966-67*
M. tade	2339	3292	1141	1971	2177
M. parsia	5093	6553	2447	3451	3411
L. calcarifer	5070	5220	1746	2887	2495
S. panijus	1261	1714	739	702	626
P. paradiseus	171	892	229	462	402
P. indicus	3100	3795	1345	1992	1811
E. tetradactylum	3438	4041	1376	2197	1585
B. biauritus	4902	3737	101	822	1117
S. miles		1			
P. pama	3890	4902	1794	2563	2135
H. ilisha	2980	9792	1187	37294	3920
H. toli		2		7	4
I. elongate	27600	19345	8067	5066	2593
C. ramcarati	4382	4608	1603	2538	1709
C. borneensis			1		2
S. phasa	12586	10222	3359	2867	1839

contd...

Table 5(IV) - *contd...*

Species	Kilogrammes				
	1970-71	*1969-70*	*1968-69*	*1967-68*	*1966-67*
S. taty					
P. pangasius	17628	13474	9384	8835	2220
T. jella		8			235
O. militaris	4				
P. canius	10467	5975	316	509	185
T. savala	912	1081	316	394	503
T. pantului					
H. nehereus	135	52	169	43	16
S. cinereus	7961	3226	5135	3507	860
Prawns	5809	7073	2652	3611	5568
Miscellaneous	29148	24074	14906	14910	9143
Total	148876	133079	58017	96624	44356

Table 5(V). Annual Specieswise Landings at Canning Assembling Centre

Species	Kilogrammes				
	1970-71	*1969-70*	*1968-69*	*1967-68*	*1966-67*
M. tade	147	54	100	358	366
M. parsia	1241	1895	2656	2690	2885
L. calcarifer	4810	3750	2399	4099	2251
S. panijus	7	5	-	13	53
P. paradiseus	32	38	186	54	80
P. indicus	1276	788	926	1109	1403
E. tetradactylum	896	632	923	1670	2160
B. biauritus	946	78	766	1996	1174
S. miles	-	-	-	-	-
P. pama	753	1038	1090	1596	1874
H. ilisha	38	261	101	8047	191
H. toli	-	-	-	-	-
I. elongata	-	-	-	-	-
C. ramcarati	87	87	223	527	527
C. borneensis	-	-	-	-	-
S. phasa	2414	1625	4050	4307	3136
S. taty					
P. pangasius	376	774	1115	656	974
T. jella	4069	1390	2535	651	1173
O. militaris	-	-	-	-	-
P. canius	1138	1804	1244	1656	1928
T. savala	769	548	735	944	40
T. pantului					
H. nehereus	583	563	842	327	97
S. cinereus	-	-	-	-	-
Prawns	90186	37808	42386	20246	19708
Miscellaneous	14009	18243	15257	22327	36713
Total	123777	72086	77534	73273	76733

Table 5(VI). Annual Specieswise Landings at Hasnabad Assembling Centre

	Kilogrammes				
Species	*1970-71*	*1969-70*	*1968-69*	*1967-68*	*1966-67*
M. tade	81	20	74	120	212
M. parsia	169	450	264	333	369
L. calcarifer	1154	1095	1580	1134	2727
S. panijus	60	163	56	63	77
P. paradiseus	425	667	893	343	642
P. indicus	44	388	404	301	895
E. tetradactylum	110	76	85	115	261
B. biauritus	2577	1421	630	309	202
S. miles		-	-	-	
P. pama	153	2203	2223	1982	1717
H. ilisha	14	1	2	-	2
H. toli		-	-	-	
I. elongata		21	105	95	92
C. ramcarati	948	803	588	350	225
C. borneensis		5	5	14	12
S. phasa	987	1738	1664	2819	2987
S. taty					
P. pangasius	4194	5230	3903	2092	2160
T. jella		831	1451	1402	1878
O. militaris	4	-	2	-	26
P. canius	167	63	225	306	758
T. savala	10	-	-	-	119
T. pantului					
H. nehereus	180	6783	3455	3107	7679
S. cinereus			3		529
Prawns	31447	38362	31697	42083	41042
Miscellaneous	7072	10585	10498	7028	14518
Total	49796	70905	59807	63966	79139

Table 5(VII). Annual Specieswise Landings at Kalinagar Assembling Centre

Species	Kilogrammes				
	1970-71	1969-70	1968-69	1967-68	1966-67
M. tade	958	2274	2101	1528	815
M. parsia	2328	5438	2447	2516	2968
L. calcarifer	1274	3562	3576	3463	3561
S. panijus	201	589	387	513	235
P. paradiseus	120	446	419	244	282
P. indicus	1419	2643	1559	1797	1527
E. tetradactylum	1412	4273	3563	3193	1405
B. biauritus	500	1299	927	1162	885
S. miles	97	476	334	247	201
P. pama	1213	4757	2715	2332	2780
H. ilisha	2	11929	5		2
H. toli		60	8	21	
I. elongata	1173	2760	1156	1625	1067
C. ramcarati	469	716	300	440	640
C. borneensis	1	16	23	28	2
S. phasa *S. taty*	3121	7517	3007	4572	4168
P. pangasius	1059	2280	1248	1371	694
T. jella	670	2406	469	693	215
O. militaris	551	1619	1308	659	553
P. canius	873	2252	2154	2595	2454
T. savala *T. pantului*	31	145	102	179	255
H. nehereus	142	1040	1065	1214	2376
S. cinereus	3	10		49	
Prawns	6814	10740	6342	10017	23797
Miscellaneous	10764	25150	12680	14718	8495
Total	35201	94397	47895	55176	859877

Table 6. Average Monthly Landings of Different Species at The Hooghly-Matlah Estuary During 1966-67 to 1970-71

Species/Months	*March*	*April*	*May*	*June*	*July*	*Aug.*	*Sept.*	*Oct.*	*Nov.*	*Dec.*	*Jan.*	*Feb.*
M. tade	971	855	852	809	641	858	797	1930	1613	994	835	1549
M. parsia	3180	2746	3253	3398	3431	6701	5166	5404	7655	4141	5129	4968
L. calcarifer	2093	2318	1917	1851	1395	1736	2095	3942	2961	1582	1416	1741
S. panijus	2052	2201	2646	3152	879	1100	747	2983	5620	3985	6766	2591
P. paradiseus	758	902	3905	9861	2471	917	608	3452	14388	8831	4397	2265
P. indicus	5012	1266	1259	1232	977	1326	1345	1374	12780	65300	5704	43643
E. tetradactylum	1537	1532	1321	1654	1160	1341	1462	1858	2991	4044	3508	2799
B. biauritus	7848	9077	1596	1457	1001	972	682	1947	49559	63228	36523	15127
S. miles	324	131	337	110	112	236	217	265	1695	3209	2960	537
P. pama	9191	7032	7258	6865	4914	7766	8626	14760	20276	34311	21545	14369
H. ilisha	11749	6379	7173	26911	160977	164447	236358	94098	133176	296605	69997	43205
H. toli	592	160	3	106	1435	1137	1191	938	3523	4253	3215	1267
I. elongata	3156	4369	3931	5077	3922	4338	4870	6494	18907	27720	20072	10529
C. ramcarati	4734	2474	2363	3581	1730	4737	5319	13462	14833	14412	12748	10746
C. borneensis	447	116	188	175	265	13	37	98	3637	3741	1801	507
S. phasa *S. taty*	21850	11515	11848	10551	9665	13507	17403	21612	123315	146327	104241	34590
P. pangasius	8914	6722	6792	9554	4556	7381	9750	13539	11764	10259	11633	8080
T. jella	1437	270	76	186	1537	11153	30769	24202	28909	27470	37407	9996
O. militaris	689	211	130	850	70	96	130	94	3060	4086	4972	3456
P. canius	1102	1213	1065	1123	1183	1303	1602	1759	1639	1272	1568	1552
T. savala *T. pantului*	1986	1014	941	1124	1413	6537	7794	6494	86236	101390	94281	13166
H. nehereus	35799	28850	42168	41997	20978	43249	31047	81038	600979	483403	297019	59814
S. cinereus	3757	4694	1597	4413	3033	949	55	221	10630	12429	6181	13042
Prawns	70933	54217	66088	61992	41423	53116	50440	61985	271169	238982	177324	80995
Miscellaneous	61054	54865	57499	59755	33299	43568	41832	76494	325749	415962	350515	135103
Total	261161	205129	226207	257783	302468	378486	460345	440457	1757044	1977937	1333098	515636

Table 7. Zonewise Annual Landings (in kg) of Fish during 1966-67 to 1970-71 at the Hooghly-Matlah Estuary

Years	*Zone I*	*Zone II*	*Zone III*	*Zone IV*	*Zone V*
1966-67	777250 (13.17)	136089 (2.31)	4442530 (75.29)	382996 (6.49)	161972 (2.74)
1967-68	831036 (9.92)	415101 (4.95)	5984215 (71.44)	1007486 (12.03)	138942 (1.66)
1968-69	823556 (10.85)	286938 (3.78)	5585694 (73.57)	779262 (10.26)	116808 (1.54)
1969-70	1015576 (12.73)	289242 (3.63)	5751612 (72.09)	801679 (10.05)	120008 (1.50)
1970-71	1033508 (9.61)	383068 (3.56)	8526676 (79.28)	692768 (6.44)	118917 (1.11)
Total					
Average (Year)	896185 (11.04)	302088 (3.72)	6058145 (74.60)	732838 (9.02)	131329 (1.62)

Figures within brackets relate to %

Table 8(I). Zonewise Composition of Catches (in Kg) of the Hooghly-Matlah Estuary

March 1966 to February 1967

Species	*Zone I*	*Zone II*	*Zone III*	*Zone IV*	*Zone V*	*Total*
M. tade	-	-	6285	-	431	6716
M. parsia	-	70	44511	62	2979	47622
L. calcarifer	-	-	20364	33	154	20551
S. panijus	10714	327	8248	2432	482	22203
P. paradiseus	8310	755	14926	15788	328	40107
P. indicus	-	-	84617	-	174	84791
E. tetradactylum	27	174	15120	-	94	15415
B. biauritus	-	-	147461	-	938	148399
S. miles	-	38	5554	-	2306	7898
P. pama	11300	2118	90291	10589	1677	115975
H. ilisha	133170	27491	591871	46605	52	799189
H. toli	7	-	8410	229	17	8663
I. elongata	40	1204	59272	548	202	61266
C. ramcarati	3900	1288	75092	2315	3410	86005
C. borneensis	18	23	434	31070	361	31906
Setipinna						
1. phasa	22616	1556	249214	6873	6243	287576
2. taty	-	896		178		
P. pangasius	19009	12214	16028	9299	3749	52899
T. jella	-	-	147330	5	-	147335
O. militaris	-	385	11240	307	16	11948
P. canius	-	-	10108	121	23	10252
Trichiurus						
1. savala	-	600+171	145849	22	3281	149923
2. pantului				-		
H. nehereus	159	30882	1191549	26173	78119	1326882
S. cinereus	-	-	9187	-	471	9658
Prawns	301053	27451	553510	104431	36399	1022844
Miscellaneous	268927	28446	936059	125916	20066	1379414
Total	777250	136089	4442530	382996	161972	5900837

Table 8(II). Zonewise Composition of Catches (in Kg) in zones of the Hooghly-Matlah Estuary

March 1967 to February 1968

Species	Zone I	Zone II	Zone III	Zone IV	Zone V	Total
M. tade	-	-	8287	-	490	8697
M. parsia	-	2346	42758	1978	4290	51372
L. calcarifer	-	8	18836	2099	320	21263
S. panijus	998	1544	12097	9069	426	24044
P. paradiseus	10253	834	9348	13262	291	33988
P. indicus	-	-	221376	-	259	221635
E. tetradactylum	35	228	19903	317	139	20622
B. biauritus	-	6779	98831	-	1861	107471
S. miles	-	322	5285	-	2045	7652
P. pama	8474	1450	79566	43749	1876	135115
H. ilisha	135561	187842	752174	261907	33	1337517
H. toli	-	-	26177	91	784	27052
I. elongata	-	112	62721	-	122	62955
C. ramcarati	2018	693	55032	24907	2786	85436
C. borneensis	235	-	501	1398	454	2588
Setipinna *1. phasa* *2. taty*	34353	3114	353754	11270	4602	407093
P. pangasius	7123	4070	42566	38692	2283	94734
T. jella	-	-	116363	-	-	116363
O. militaris	9	7	5781	86	19	5902
P. canius	-	59	6371	-	30	6460
Trichiurus *1. savala* *2. pantului*	-	669	345750	821	1016	348256
H. nehereus	4188	70992	2007606	115553	69978	2268317
S. cinereus	-	589	56774	35591	-	92954
Prawns	297521	72704	681590	188214	23753	1263782
Miscellaneous	330268	60739	954938	258482	21085	1625512
Total	831036	415101	5984215	1007486	138942	8376780

The morpho-ecological conditions which changes from season to season and year to year possibly have controlled the changes in the pattern of fishery.

The fishing efforts and their corresponding catch for the years 1960-61 to 1970-71 recorded by Dutta *et al.* (1973) are indicated below in Table 10.

Table 8(III). Zonewise Composition of Catches (in Kg) in various zones of the Hooghly-Matlah Estuary March 1968 to February 1969

Species	*Zone I*	*Zone II*	*Zone III*	*Zone IV*	*Zone V*	*Total*
M. tade	-	-	8906	-	595	9501
M. parsia	-	2471	39155	2563	5442	49631
L. calcarifer	-	167	19179	-	348	19694
S. panijus	1225	1378	21195	13683	246	37727
P. paradiseus	8982	4459	12865	15768	471	42545
P. indicus	-	47	224612	120	45	224824
E. tetradactylum	-	277	16859	1430	69	18635
B. biauritus	-	71	190962	905	709	192647
S. miles	-	11	11285	-	220	11516
P. pama	9824	3493	132649	36184	919	183069
H. ilisha	118576	148544	305367	105551	-	678038
H. toli	150	21	15142	422	184	15919
I. elongata	15	833	101708	664	156	103376
C. ramcarati	850	613	36951	22724	916	62054
C. borneensis	-	-	239	4175	101	4515
Setipinna	41991	7839	576227	18249	4344	648650
1. phasa						
2. taty						
P. pangasius	9166	3393	51813	28041	2092	94505
T. jella	-	-	197038	-	-	197038
O. militaris	-	-	17043	1576	-	18619
P. canius	-	-	5359	1298	89	6746
Trichiurus	-	1073	386030	1537	2922	391562
1. savala						
2. pantului						
H. nehereus	6943	34721	1617535	203732	61319	1924250
S. cinereus	-	6042	75266	4	-	81312
Prawns	332060	34089	476046	192679	25400	1060274
Miscellaneous	293774	37396	1046263	127957	10221	1515611
Total	823556	286938	5585694	779262	116808	7592258

Table 8(IV). Zonewise Composition of Catches (in Kg) in various zones of the Hooghly-Matlah Estuary March 1969 to February 1970

Species	*Zone I*	*Zone II*	*Zone III*	*Zone IV*	*Zone V*	*Total*
M. tade	-	-	13835	-	500	14335
M. parsia	-	1371	50466	3015	2341	57193
L. calcarifer	-	65	27936	-	334	28335
S. panijus	1598	806	18474	10304	303	31485
P. paradiseus	8760	1074	16729	38396	378	65837
P. indicus	-	18	246711	359	53	247141

contd...

Table 8(IV) - *contd...*

Species	*Zone I*	*Zone II*	*Zone III*	*Zone IV*	*Zone V*	*Total*
E. tetradactylum	-	270	24367	2257	293	27187
B. biauritus	628	331	183892	546	921	192318
S. miles	-		8793	189	3219	12201
P. pama	12049	2231	89492	45320	742	149834
H. ilisha	155157	178210	1158404	261050	180	1753001
H. toli	120	-	15510	192	522	16344
I. elongata	277	133	121151	974	418	122953
C. ramcarati	3879	591	48905	35284	1918	90577
C. borneensis	-	51	267	8017	-	8335
Setipinna	52296	3646	399621	24801	4194	484558
1. phasa *2. taty*						
P. pangasius	7201	3542	78997	34037	2400	126177
T. jella	-	-	260381	-	-	260381
O. militaris	28	-	30023	1161	-	31212
P. canius	-	10	21043	6	147	31206
Trichiurus	-	851	250969	753	3676	256249
1. savala *2. pantului*						
H. nehereus	585	33585	951758	58211	59054	1103193
S. cinereus	-	2519	36288	-	-	38807
Prawns	433947	31241	584876	145507	25319	1220890
Miscellaneous	339051	28697	1106724	130800	13096	1618368
Total	1015576	289242	5751612	801679	120008	7978117

Table 8(V). Zonewise Composition of Catches (in Kg) in various zones of the Hooghly-Matlah Estuary March 1970 to February 1971

Species	*Zone I*	*Zone II*	*Zone III*	*Zone IV*	*Zone V*	*Total*
M. tade	-	-	23647	-	248	
M. parsia	-	5215	58184	3051	3355	69805
L. calcarifer	-	-	35069	-	328	35397
S. panijus	976	10259	30264	14365	564	36428
P. paradiseus	2486	3355	27315	47569	575	81300
P. indicus	-	-	183561	471	275	184307
E. tetradactylum	-	33	42600	857	257	43747
B. biauritus	-	32	303390	586	185	304193
S. miles	-	-	9197	1314	881	11392
P. pama	3692	24148	122377	46561	969	197747
H. ilisha	185907	204685	1263678	76828	41	1731139
H. toli	21	-	20572	234	98	20925
I. elongata	541	776	210767	4170	119	216373

contd...

Table 8(V) - *contd...*

Species	*Zone I*	*Zone II*	*Zone III*	*Zone IV*	*Zone V*	*Total*
C. ramcarati	827	1303	93650	33670	2099	131549
C. borneensis	-	50	53	7744	24	7871
Setipinna	22345	1089	762816	15322	2474	804055
1. phasa *2. taty*						
P. pangasius	6618	10093	121884	30014	2393	171002
T. jella	-	-	145598	-	-	21409
O. militaris	18	-	20227	1164	-	37250
P. canius	-	-	37106	4	140	37250
Trichiurus	-	375	459692	1300	3534	464901
1. savala *2. pantului*						
H. nehereus	2045	32356	1993436	113819	67260	2208916
S. cinereus	242	-	79393	3	55	79693
Prawns	364915	45436	979686	162907	22139	1575083
Miscellaneous	442866	43863	1502514	130015	10904	2130962
Total	1033508	383068	8526676	692768	118917	10754937

Considering the area of fishing operation almost same, the above table indicates some interesting findings stated below :

1. When the fishing operation is more the catch increases which is quite natural. For example the catch was more in the years 1962-63 ; 1963-64 ; 1964-65 and 1967-68 when the fishing effort increased.
2. But reverse is also noticed when fall in the catch are noticed inspite of increased effort. For example during 1965-66 and 1969-70 fall in the catch was noticed.
3. During 1970-71 when the effort was more catch also increased. This time severe storm or cycline occurred and for similar reason (heavy storm) in 1962-63, the catch increased alongwith the effort. Similar incidents occurred during 1998 when high tidal water resulted from severe storm during monsoon, high landings of hilsa occurred at the Hooghly confluence near Diamond Harbour. The above findings leads to a conclusion that mere increase in fishing effort may not always enhance the production. Possibly it is supplemented by the meteorological condition (storm, cyclone, heavy rainfall or high flood etc.) associated with the tidal waves of monsoon brings considerable volume of water to push the water deep into the shallow estuarine area where by ordinary traditional gear, fishes are caught. The higher production during 1962-63; 1963-64; 1964-65 may be attributed to reasons stated. In the absence of such meteorological disturbances lower production was also recoded even with higher fishing effort (1965-66; 1969-70).

Shallow river beds/creeks are used for construction of bheries for prawn culture, brick kilns etc. by putting embankments on the river bed and catchment area deforested for the purpose of agriculture etc. bring enormous volume of silt into the river bed. This is again

Table 9. Yearwise Species Composition of Annual Catches (March to February) of The Hooghly-Matla Estuary

		Kilogrammes															Average	
Species	Local Name	1961-62	1962-63	1963-64	1964-65	1965-66	1966-67	%	1967-68	%	1968-69	%	1969-70	%	1970-71	%	(66-67 to 70-71%)	
M. tade	Bhangone	4451	8194	12470	7850	7714	6716	0.11	8697	0.10	9501	0.13	14335	0.18	23895	0.22	12629	0.16
M. parsia	Parse	22765	54491	38399	39197	59065	47622	0.81	51372	0.61	49631	0.63	57193	0.72	69805	0.65	55125	0.68
L. calcarifer	Bhetki	55492	45396	45597	25658	29756	20551	0.35	21263	0.25	19694	0.26	28335	0.36	35397	0.33	25048	0.31
S. panijus	Tuladandi, Tulbele	29757	51005	46408	43938	23570	22203	0.38	24044	0.29	37727	0.50	31485	0.39	56428	0.52	34377	0.42
P. paradiseus	Topse	28632	28731	61965	56619	46389	40107	0.68	33988	0.41	42545	0.56	65837	0.83	81300	0.76	52755	0.65
P. indicus	Sele	62953	67469	134266	137147	91483	84791	1.44	221635	2.65	22824	2.96	247141	3.10	184307	1.71	192540	2.37
E. tetra-dactylum	Gurjaoli	11356	23027	22698	17383	17331	15415	0.26	20622	0.25	18635	0.25	27187	0.34	43747	0.41	25121	0.31
B. biauritus	Norebhola	28818	69455	138841	94052	24902	148599	2.51	107471	1.28	192647	2.54	192318	2.41	304193	2.83	189006	2.33
S. miles	Bhola	-	9243	11763	6377	5965	7898	0.13	7652	0.09	11516	0.15	12201	0.15	11392	0.11	10132	0.12
P. pama	Bhola	121286	170918	142546	138011	175709	115975	1.97	135115	1.61	183069	2.41	149834	1.88	197747	1.84	156348	1.93
H. ilisha	Ilish	1058356	1584260	1413154	1475576	887764	799189	13.54	1337517	15.97	678038	8.93	1753001	21.97	1731139	16.10	1259777	15.51
H. toli	Chandana ilish	26223	33426	17844	21851	7480	8663	0.15	27052	0.32	15919	0.21	16344	0.20	20925	0.19	17781	0.22
I. elongata	Dhala	3766	43622	113853	39010	85145	61266	1.04	62955	0.75	103376	1.36	122953	1.54	216373	2.01	113385	1.40
C. ramcarati	Telchite, Amadi	33558	18266	54603	55040	186649	86005	1.46	85436	1.02	62054	0.82	90577	1.14	131549	1.22	91124	1.12
C. borneensis	Telchite, Amadi	-	1335	1807	1385	12918	31906	0.54	2588	0.03	4515	0.06	8335	0.10	7871	0.07	11043	0.14
S. phasa	Phasa	79704	127864	252488	95834	147024	287575	4.87	407093	4.86	648650	8.54	434558	6.07	804055	7.48	526386	6.48
S. taty	Phasa	51635	126244	64805	264313	289067												
P. pangasius	Pangas	55147	46125	59605	55352	48747	58299	0.99	94734	1.13	94585	1.24	126177	1.58	171002	1.59	108943	1.34
T. jella	Medkanta	121200	234722	394918	183558	244625	147335	2.50	116363	1.39	197058	2.60	260381	3.26	145598	1.53	173343	2.13
O. militaris	Aar-tangra	10540	16328	32792	25282	6690	11948	0.20	5902	0.07	18619	0.25	31212	0.39	21409	0.20	17818	0.22
P. canius	Kan-magur	7845	11993	9093	2737	9385	10252	0.17	6460	0.88	6746	0.09	21206	0.27	37250	0.35	16383	0.20
T. savala	Rupavati	73648	41533	29808	39669	390891	149923	2.54	348256	4.16	391562	5.16	256249	3.21	464981	4.32	322178	3.97
T. pantilui	Rupavati		81304	297634	146005													
H. nehereus	Nehere, Bomla, Lutea	851719	741969	1293197	2053941	1585860	1326882	22.49	2268317	27.08	1924258	25.35	1183193	13.83	2208916	20.54	1766312	21.75
S. cinereus	Complet, Pomfret	-	-	-	5200	11671	9658	0.16	92954	1.11	181312	1.07	38807	0.49	79693	0.74	60485	0.74
Prawns	Chingri	1012592	797577	919829	979693	856383	1022844	17.33	1263782	15.09	1860274	13.97	1220890	15.30	1575083	14.64	1228574	15.13
Miscellaneous		411003	806374	552423	1234393	1261769	1379414	23.38	1625512	19.40	1516611	19.96	1618360	28.29	2130962	19.81	1653973	20.37
Total		4162446	5160871	6072086	7245079	6353952	5900837	100.00	8376780	100.00	7592258	100.00	7978117	100.00	18754937	100.00	8120586	100.00

supplimented by the silt laden tidal water where the silt settles and thus raises the river bed indicating higher shape factor (Kulti, Hooghly estuary) stated elsewhere. Very likely this happens in case of other estuaries.

Deltaic areas thus increases and simultaneously fishing effort increases in comparatively shallower zone where fishing effort with traditional gear increases. However the fishery of the Hooghly-Matla estuarine system are discussed separately.

Table 10. Fishing efforts and their corresponding catch during 1960-61 to 1970-71.

Year	*Fishing effort.*	*% increase (+) decrease (-)*	*Actual catch**	*% (increase (+)/ decrease (-))*	*Weather*
1960-61	189752	-	3398457	-	Cyclone
1961-62	129804	(+13%)	3104090	(-8%)	Cyclone
1962-63	264397	(+13%)	3656611	(+17%)	Storm
1963-64	299797	(+13%)	4658852	(+27%)	
1964-65	323040	(+7%)	5769503	(+23%)	
1965-66	389330	(+20%)	5466188	(-4%)	
1966-67	302409	(-22%)	5101646	(-6%)	Flood
1967-68	360988	(+19%)	7039263	(+37%)	
1968-69	302459	(-37%)	6914220	(-1%)	Severe cyclone
1969-70	372093	(+23%)	6225116	(-9%)	
1970-71	422858	(+13%)	9023798	(+44%)	Severe storm
1978					Flood, Cyclone
1981		Large number of new fishes			
1984		Cyclone			

* Excluding hilsa.

Table 11. Number of commercial important species representing of seven centres during different years 1966-67 to 1970-71.

	Assembly centre	*Year*				
		1966-67	*1967-68*	*1968-69*	*1969-70*	*1970-71*
1.	Port Canning	18	18	17	18	18
2.	Hasnabad	22	18	21	19	18
3.	Diamond Harbour	20	21	21	20	23
4.	Kakdwip	24	23	22	22	22
5.	Kalinagar	22	23	23	24	23
6.	Namkhana	23	24	24	23	22
7.	Raidighi	22	20	20	22	20

The table shown above states that till 1970-71, the existing biodiversity of Sundarban delta even could not bring any significant changes in the species composition except possibly in the total catch (e.f. Table 11). [Out of 7 assembly centres port Canning was showing comparatively lower number of species (17 & 18), followed by Hasnabad (18 & 19) with 2 exceptions (21 & 22). At other 5 centres, Diamond Harbour, Kakdwip, Kalinagar, Namkhana and Raidighi were showing a little higher values (20-24). But 30 years later Chaudhury (2000) in his recent observation has noted that 86 species of fish and 46 species of crustaceous had either become extinct or endangered in the Sagar island which is undoubtedly alarming. Recently Sen (2005) has also presented a long list (Addendum 2) of endangered species of freshwater area of Bhagirathi-Hooghly.

Important species caught in the Sundarban estuarine system (Hooghly-Matla) between 1966-67 and 1970-71 are 26 in number. This excludes the miscellaneous group. Zonewise total species composition in different years are presented below:

Table 12. Number of Species recorded in different zones during 1966-67 and 1970-71.

Zone	*1966-67*	*1967-68*	*1968-69*	*1969-70*	*1970-71*
I Nabadwip-Khusigoli.	13	12	11	13	13
I Khusigoli-Diamond Harbour	17	19	19	19	15
III Entire estuarine area of Sundarbans.	24	24	24	24	24
IV Small Zone in the Matla estuary around Port Canning.	23	22	20	20	22

It is interesting to note that zone III was showing maximum number of species (24) followed by zone V (20-23) and lowest number was recorded at zone I (11-13).

6
The Hilsa Fishery

(I) THE HILSA FISHERY

The lower zone which includes the entire estuarine area of the Sundarbans and the tract below Diamond Harbour on the main channel where salinity varied from 10% to nearly neritic value and a smaller zone of the Matla estuary comprising the local area around Port Canning (Dutta *et al.* 1973). These are from Thakuran river and joining Canals, Matla river.

According to Pillay (1958) the lower estuaries and the insore areas of the sea from their natural home from where they migrate up to the river for the purpose of spawning. Hence the inclusion of the Catch at insore areas close to the confluence of the estuary with the sea is meaningful for Hilsa as we are including the same fish when it is caught up the river while ascending.

Hilsa being a very economic species, its fishery in the seasons of availability greatly replaces the other fisheries. There occurs a fall in intensity of operation of other gears. When Hilsa gears increasingly appear in the field particularly in upper areas. Hilsa being a highly economic fish, the Hilsa fishery continues almost as long as ascending or decending Hilsa is available in the estuary in the seasons (Winter or monsoon). Hence, the abundance of winter or monsoon runs can to a fair extent be judged from the catches of Hilsa during these seasons. The upstream migration of Hilsa particularly during monsoon appears to be determined to a large measure by the volume of fresh water discharge and generally monsoon runs are stronger. although the behavioral cues to migration of the winter run is not exactly known, physico-chemical conditions like temperature and salinity have been suggested (Jones and Sujan Singhani - 1951) by some workers as possible governing factors. However the point of relevance for the abundance studies that emerges shown, while at other centres like Canning, Kalinagar and Hasnabad showed much lower landing or negligible landing, as these areas are not influenced by sufficient flood water like Bhagirathi-Hooghly and hence were deprived of hilsa migration. A high landing of hilsa was also recorded during 1997 and 2001 at the confluence of the river Hooghly which was more than usual, to spawn in the freshwater. (The statesman 20.08.1997 : Bartaman 18.08.2001). It is unfortunate that the sizes of hilsa has never been recorded so far.

(II) HILSA FISHERY OF 2003

Interestingly this year the catch of hilsa is found to be different than the previous year. In July juveniles of hilsa of sizes 200-250-300 gr. are being caught. According to the District Fishery officer, West Bengal the catch this year is about 30 percent more than last year. Such low size catch is mainly due to the use of nets of 45 mm. mesh in place of 65 mm. used by the fishermen ignoring the specied law imposed by the Government of West Bengal (Kundu, 2003). The adverse affect of such large scale catch of young hilsa may have some impact in the comming year's breeding in the absence of inadequate matured hilsa.

Further, in the absence of inadequate freshwater flow in the Hooghly due to heavy silt precipitation the possibility of hilsa ascending upstream has become less and hilsa and other fishes prefer moving towards the adjoining Padma river (Bangladesh) where the Bangladesh Government has taken adequate steps for not allowing such small size mesh for catching hilsa and have declared illegal (Kundu, 2003).

But things are found to be different this year even in Bangladesh. The imported sizes of hilsa in West Bengal from Bangladesh which was 3.0 Kg. (average weight) in the past has gradually lost its size and has come down below 1 kg. This year according to hilsa exporters of Bangladesh, the catch of hilsa in Bangladesh was mainly from the rivers-Padma, Meghna and Jamuna. But this year the fishermen have engaged themselves for catching hilsa from the confluence of all these rivers in Bhola district near Barisal which is further down. as a result shoals of hilsa fishery has changed their direction towards Mayanmar. This is how young hilsa of sizes between 500-700 gms. only are available (Addendum - 1) this year against the large size 2-3 kgs. of the past (Bhattacharya, 2003).

It is thus clear that this year both in West Bengal and Bangladesh the catch of hilsa is mainly of juvenile size which is a warning for both the countries about the future production and size of the hilsa fishery.

Addendum -1

"Hilsa hunt faces five months ban".

(Statesman News Service ; The Statesman (Kolkata) South Bengal Plus 04.06.2004) :

1. The government as stated by the Fishery minister, West Bengal, was planning the move as a step towards preventing extinction of the variety.
2. The natural conditions that fecilitate the migration of the fish from salt water to rivers are almost non-existent as of now.
3. The variety of hilsa which is netted by fishermen between October and February is essentially the fry, coloquilly called "Khoka ilish" of "chandana". These are offsprings that are left behind by the shoals of the full grown variety that migrate to freshwaters to lay eggs during the month of September.
4. Catching of fry during that period would hasten the extinction of hilsa.
5. Last year, the National Export of hilsa fell : from 6000 metric tonnes to 5000 metric tonnes.

6. The minister further added, he would renew efforts to have the coastal laws on banning coastal fishing extended to April-August period from existing April-May duration. The ban till August already exists in the county's Western Coast. This has to be implimented in the Eastern Coast as well.

7
Winter Migratory Bagnet Fishery of the Hooghly Estuary

A special feature of the Hooghly estuary is the fishery exploitation during the winter season by bagnets. This fishery is commonly known as Winter Migratory Bagnet fishery and are mainly concentrated on Sagar island at the mouth of the Hooghly estuary and the others are around Freserganj, Bokkhali, Kalistan and Jamboo dwip (upper and lower, Fig. 3) during the four months November-January (Dutta *et al.*, 1973).

The factors which induce the migratory fishery are:

1. Calm weather prevailing in the extreme lower reaches favourable for operation of fixed bagnets till south wind starts towards the middle of February.
2. Catch per unit of effort is higher than the upper and middle zones, and
3. Plankton production reaches a peak in this region (Shetty, Saha and Ghosh, 1961 ; Saha, Ghosh and Gopala Krishnan, 1970).

The capture by the migratory bag net fishery during the four months accounts for 30-35% of the total production from zone III and 22-38% of the total production of the whole estuary (Table 14.). Bombay Duck (*H. nehareus*), anchovies (*S. phasa* and *S. taty*). Ribbon fishes (*L. Sabala* and *T. pantului*) and prawn dominate the catch (Table 13). Their percentage during.1966-67 to 1970-71 varies between 37 and 55%, 7 and 16%; 5 and 12%; and 0.01 and 10%. November and December are the peak months when the harvest varies between 70 and 80% Dutta *et al.* (1973).

Similar study was also made by Mitra *et al.* (1997) and subsequently by Paul *et al.* Three and a half months seasonal fishery for an average, estimated catch of 256. 36 t consisting about 67.2 % of the total annual fish yield from the entire estuary during the period 1984-85 to 1993-94 compared to only 32.7% about one and half decades ago during the period 1970-71 to 1974-75 (Mitra *et al.* 1997), while Paul *et al.* (1997) has shown catch by winter migratory bag net fishery during 1993-94 to 1996-97 which is six times more than that of the period 1970-71 to 1974-75. Comparative statement of total landings by Migratory Bag nets and their catch per unit of effort during different periods are shown in Table 15. Important gears used in different zones of the estuarine system during different seasons along with their principal catch are also shown in Table 16.

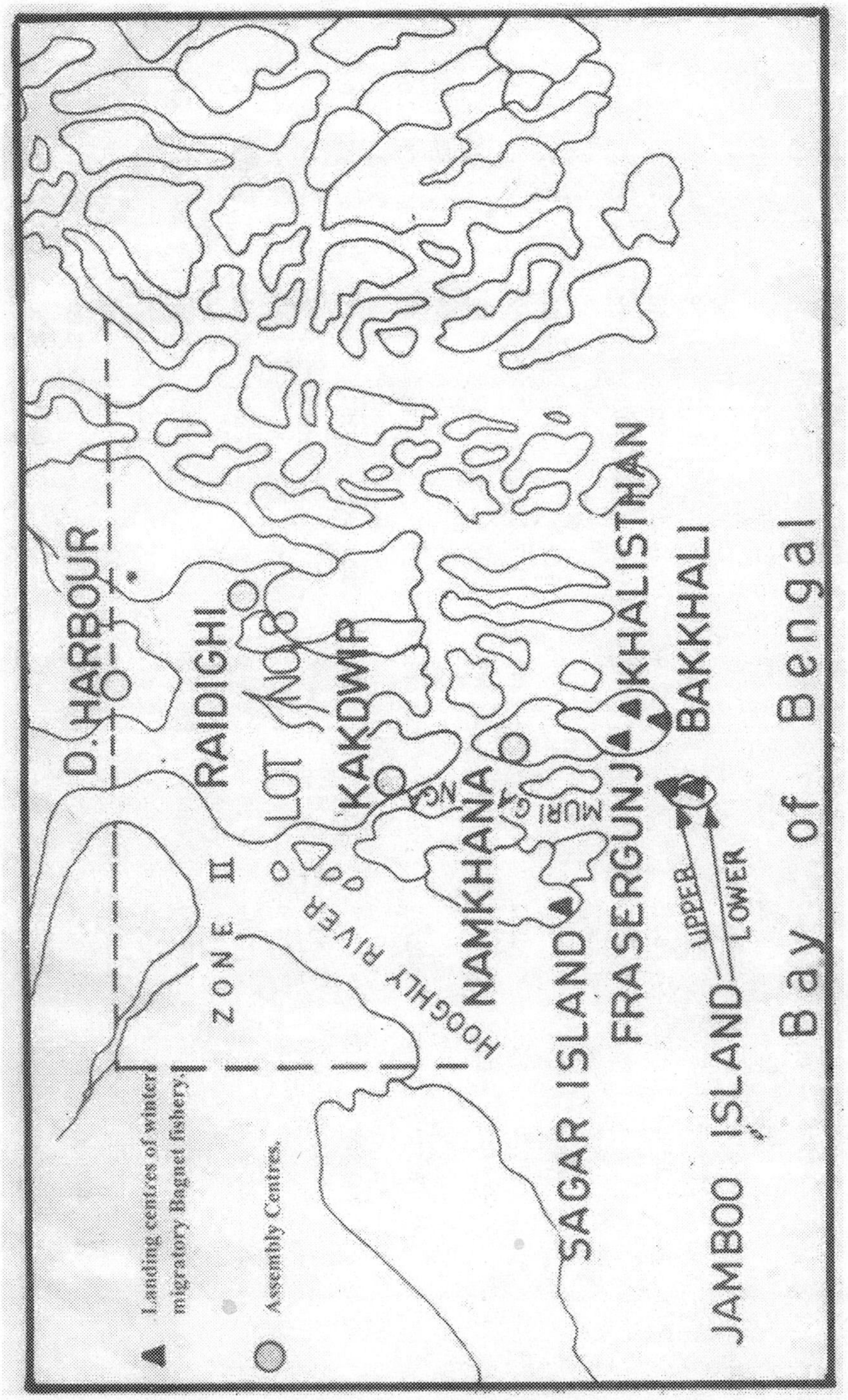

Fig. 3 : Mouth of the Hooghly Estuary showing the landing centres of winter migratory bagnet fishery

Table 13. Specieswise composition of catches from the Hooghly-Matlah estuary by migratory bagnet fishery in winter

Species	*1970-71*		*1969-70*		*1968-69*		*1967-68*		*1966-67*	
	Kg	*%*	*Kg*	*%*	*Kg*	*%*	*Kg*	*%*	*Kg*	*%*
M. parsia	634	0.02	239	0.01	56	*	-	-	2	*
S. panijus	364	0.01	706	0.04	216	0.01	1352	0.04	2136	0.12
P. paradiseus	7108	0.24	3247	0.18	4140	0.15	3194	0.10	4765	0.27
P. indicus	95	*	222	0.01	521	0.02	113	*	410	0.02
E. tetradactylum	105	*	-	-	-	-	-	-	245	0.01
B. biauritus	219685	7.37	76900	4.34	92531	3.24	55397	1.78	131391	7.41
S. miles	8566	0.29	8305	0.47	10703	0.37	4886	0.16	4942	0.28
P. pama	-	-	8076	0.46	10104	0.35	5976	0.19	2019	0.11
H. ilisha	21	*	613	0.03	2	*	39	*	60	*
H. toli	3268	0.11	1275	0.07	755	0.03	353	0.01	355	0.02
I. elongate	41030	1.38	20502	1.16	33417	1.17	16316	0.52	18793	1.06
C. ramcarati	1743	0.06	3215	0.18	575	0.02	1837	0.06	1703	0.10
C. borneensis	42	*	-	-	-	-	-	-	193	0.01
S. phasa *S. taty*	420113	14.09	248722	14.05	457183	16.02	23086	7.43	117951	6.65
P. pangasius	325	0.01	175	0.01	1260	0.04	5350	0.17	252	0.01
T. jella	21477	0.72	3947	0.22	4908	0.17	3140	0.10	-	-
O. militaris	9651	0.32	7579	0.43	5907	0.21	-	-	9110	0.51
T. savala *T. pantului*	331002	11.10	217429	12.28	353843	12.40	312554	10.05	84959	4.79
H. nehereus	1251728	41.97	660413	37.30	1263390	44.26	1725563	55.51	844164	47.61
S. cinereus	3600	0.12	-	-	173	0.01	1953	0.06	152	8.01
Prawns	173120	5.81	126979	7.17	112324	3.93	317068	10.20	142064	0.01
Miscellaneous	488530	16.38	382015	21.58	502481	17.60	422736	13.60	407419	22.98
Total	2982207	100.00	1770559	99.99	2854485	100.00	3108690	99.98	1773085	99.98

M. tade, *L. calcarifer* and *P. canius* were absent in the bagnet is catches from the Hooghly-Matlah estuary.

* less than 0.005%

Table 14. Monthly total landings and landings of major species by migratory bagnet fishery of winter during 1966-67 to 1970-71 (in Kg)

Months	*H. nehereus.*	*I. pantului & T. savala*	*S. phasa & S. taty*	*Prawns*	*Total*	*% contribution to the total catch of zone III*
1970-71						
November	506055	75183	174729	102382	1170505	39.25
December	283075	66077	106857	38910	723949	24.28
January	462598	189742	138527	31828	1087753	36.47
February	-	-	-	-	-	-
NOVEMBER-FEBRUARY					2982207	34.98 (27.73)
1969-70						
November	258201	48285	43100	64889	531689	30.03
December	296793	114901	134217	35042	844010	47.67
January	105419	54243	71405	27048	394860	22.30
February	-	-	-	-	-	-
NOVEMBER-FEBRUARY					1770559	30.78 (22.19)
1968-69						
November	478468	135611	122936	68779	1011517	35.44
December	561040	137410	223444	24681	1241497	43.49
January	185905	68832	80370	15135	485871	17.02
February	37977	11990	30433	3729	115600	4.05
NOVEMBER-FEBRUARY					2854485	51.10 (37.60)
1967-68						
November	639779	68311	57724	173625	1079784	34.73
December	617516	118636	58703	91193	1069573	34.41
January	461779	112538	90746	49039	892051	28.79
February	6489	13069	23690	3211	67282	2.16
NOVEMBER-FEBRUARY					3108690	51.95 (37.11)
1966-67						
November	582291	33303	52979	80047	1081365	60.99
December	191577	26002	28990	49198	480215	27.08
January	51707	23211	31369	10576	173610	9.79
February	18589	2443	4613	2243	37895	2.14
NOVEMBER-FEBRUARY					1773085	39.91 (30.05)

The percentages within brackets are percentage contributions to the total landings of the whole estuary.

Table 15. Showing comparative statemet of total landings by winter Migratory Bagnet and their corresponding - CPUE during different periods.

Year	*Total catch (t)*	*CPUE*	*Period of Operation*	*Remarks*	*References*
1966-67	1773.1	-	November to February	*Dominating species* *H. nehereus* *T. pantului* & *T. savala* *S. phasa* & *S. taty* Prawns	-
1984-85 to 1993-94	256.36 (average)	-	-	-	-
1993-94	1769.2	78.19	End of October to early February		1. Winter bagnet catches mainly consists of small size fishes. The entire landings are dried and sold. 2. Fishing intensity in recent years has been on the increase as is revealed by the rising number of fishing units from 953 in 1994-95 to 1247 in 1995-96 and further to 1629 in 1996-97. 3. Nets are operated in shallow coastal water and *net in deep water which results low yield.* 4. *Dominant species : H. nehereus.* Misc (19.16%) Setipinna spp (16.74%) Trichiurus Spp. (15.67%) *P. pama.* (11.34%) - Paul *et al.* (1997)
1994-95	20820.6	93.72			
1995-96	28185.9	89.46			
1996-97	35844.6	75.75			

8
The Gear

A variety of gears are operated in the estuary. Some of them are selective for a particular spacies but mostly a gear Catches a number of species consisting what is known as 'mixed fishery' or a multiple species of fisheries. A list of important gears with their local names, types, area of usual operation and main species caught are given in Table 13. The more important gears of the estuary are bagnets, seines, gile nets, long lines, trawls and barrier nets. The gearwise partison of the annual catch is shown in Table 4. Bagnets lead the total catch, others so far as the catches are concerned followed by seines, drift nets, hooks and lines, trawls and set barriers in the order mentiond. The leagnets are used in all part of the estuary where tidal force is sufficient for their operation. The main gears that are operated in the lower estuary (zone-III) are :

1. Been Jal.
2. Atone Jal (Bagnet).
3. Sele Jal.
4. Chhandi Jal.
5. Bhasa Jal (Drift gill net).
6. Naugar Jal.
7. Baranj Jal (Set gill net)
8. Kochal Jal.
9. Jangla Jal.
10. Ber Jal (Seines).
11. Pata Jal (Set barrier) and
12. Long Lines.

A. Gear Used and the Species Caught

1. *Been Jal and Ber Jal* catch varied types of fish. Atone and Barang have Bhetki (*L. Calcarifer*) as their main Catch.
2. *Chhandi Jal*
Bhasa Jal
Nagar Jal — These are mainly used for Catching hilsa.
Kochal Jal and
Jangla Jal.

3. *Sela Jal* – is used to catch *P. indicus* chiefly.
4. *Long Line* – Catches consists of cat fishes, Perches, Polynemids, Sciaenids and anchovies.

The availability of the species composition during March-February can be judged from Table 6 while their availability in different zones for different years (1967-71) can be seen from Table 9 (I-V).

B. Gear Used for Hilsa Fishery

This is the most important fishery of the estuary which is seasonal in nature. The two main seasons of this fishery are monsoon (July/August-October) and Winter (November-February) when fish ascends up the river.

i. *During monsoon* (July/August-October)

Catches are made Through grill nets	Chhandi Jal ; Nangar Jal etc.

ii. *During Winter* (November-February)

Catches in inshore areas of lower regions are made mostly through seines.	Kochal Jal, Jangla Jal.
iii. Also by different gill nets in the upper zone (I) besides gill nets small purse nets are also operated	Chhandi Jal Sanglo Jal.

Table 16. Important gears used in different zones of the estuarine system during different seasons alongwith their principal catch.

		Type of gear			*Principal Catch*	*Main seasons of operation*	*Zones where operated*
A.	1.	*Sine nets* (medium and small)	-	Berjal	Various types of fish	All seasons.	Zone-III.
	2.	*Bag net* (Stationary)	-	Been Jal.		All seasons	All zones.
B.	1.	*Bagnet* (Stationary)	-	Atone Jal	*L. Calcarifer* (Principal catch)	Nearly all seasons	III
	2.	Set gill net	-	Barang Jal sele jal	*P. indicees* (chiefly)	November-May	III
C.	1.	*Drift net*	-	Chandi Jal	-	Monsoon and winter (greater in monsoon)	III
	2.	*Drift gill net*	-	Bhasa Jal			
	3.	*Set gill net*	-	Kangar Jal	Mainly Hilsa	November-May	I, III

contd...

Table 16 - *contd...*

Type of gear		*Principal Catch*	*Main seasons of operation*	*Zones where operated*
4. *Scines* -	Kochal Jal.	-	Monsoon and winter	III
5. *Bag scines* -	Jangla Jal.		March to July November-February occasionally in Monsoon	Confluence of zone III (coastal region) -Do-
D. Set Barriers -	Pota Jal	All species particularly prawn, *M. perisa*, setipinna spp., *P. pama, E. tetradactylum*, *L. Calcarier* and *I. elongata*.	All seasons	All zones
E. Hook and Line		Cat fish, Parches, Polynemids, Scianids and setipinna spp.	All seasons	-Do-

* Based on Dutta *et al.* (1973).

References

Anonymous (1978) *Collected Reports from the 3rd workshop on "All India Co-ordinated Research Project on Brackish water-Prawn and Fish Farming."* ICAR, New Delhi.

Bhattacharyya Rasanaya Khara, Hai Boro ilish Tomar Din Giyeche.

Arunakkya (Bangaon) (2003) (The taste is lost, Oh. : Large size hilsa, your days are no more) Ananda Bazar Patrika, 17.8.2003.

Blasco, C.J. (1975) The Mangroves in India (Translated by Mrs. Thanikaimoni from LES MANGROVES DE LINDE), Institute Francais de Pondicherry, Inde, Shri Aurobinda Ashram.

Chakraborty, K. (1984) Dynamics of Flora and Fauna Diversity in the Mongroves of Sundarbans and Laterite Tracts of West Bengal. A Bio-ecological study. *Ind. J. For.*, 7(3) : 220-32.

Chakraborty, N.M., Karmakar, G.C. and Roy, A.K. (1981) Observations on the effect of supplementary Feed on Growth and survival of a Grey Mullet *Liza parsia* (Ham) Fry in Brackish water Nursery Pond at Kakdwip (abstr) *Proc. symp. Coastal Aquaculture,* Cochin.

Daugherty, H.E. (1975) Human Impact on the mangrove Forests of EI Salvador, ed. G.E. Walsh, S.C,. Snedaker and M.H. Teas, *Proc. Inst. Symp. on Biology and Management* of Mangrove (1974), University of Florida, U.S.A., pp. 816-24.

Dutta *et al.* (1973) Fishery Resources of the Hooghly-Matla Estuarine System, *Bull. No. 19*, December, CIFRI (ICAR) Barrackpore, West Bengal.

Ghosh, A.N. (1980) Prespects and Possibilities of Brackish water Aquafarming in Saline soils of West Bengal with special Reference to Sundarbans. *Proc. Sominar 24-Parganas Fish Prod. Association*, West Bengal.

Hall, D.N.P. (1962) Observations of Taxonomy and Biology of some Indo-West Pacific Penaedae (Crustacea, Decaponds), Colonial Office Fishery, Publicaton No. 17, HMSO, p. 229, London.

Jena, M.M. & S. Chatterjee (1974) Some aspects of Fishing in Sundarbans. Its problems and future development. *Geo. Res. 36* : 76-82.

Jhingran,V.G. (1982) Fish and Fisheries, of India (2nd edition), Hindustan Publishing Company, New Delhi.

Jones, S and K. Sujan Singhani (1951) The Hilsa Fishery of the Chilka Lake, *J. Bombay Nat. Hist. Soc.* 50(2) : 264-280.

Kundu, Debasis (2003) "BANGALEER PATE AR KATODIN ILISH, AASONKAYA MATSAJIBIRAI" How long the Bengalees will see the hilsa in their Plates - are teh fishermens' anxiety - Bartaman, 2.7.2003.

Luther, G. (1973) They Grey Mulled Fishery Resources in India, *Proc. Symp. on living Resources of Seas Around India*, CMFRI, Cochin, Sp. Publ., pp. 455-60.

Mac Nae, W. (1968) A General Account of the Fauna and Flora of Mangrove of Swamps and Prospects in the Indo-west Pacific Region. *Adv. Mar. Biol.*, 6 : 73-720.

Macnae, W. (1974) Mangrove Forests and Fishes, India Ocean Programme, Pub. No. 34, pp. 1-34.

Mitra, P.M., *et al.* (1997) Fisheries of the Hooghly - Metla Estuarine system - An appraisal. *Bull. Cent. Int. Cap. Fish. Res. Inst.* No. 67 : 49.

Naskar, K.R. (1981) Flovistic studies of the District 24-Parganas in West Bengal, India with Special Reference to the Mangrove Vegetation of the Sundarbans (Ph.D.) Thesis University of Calcutta (Unpublished).

Naskar, K.R. and Chakraborty, N.M. (1984) Studies on the Economic Fauna from the Sundarbans Delta in West Bengal. *J. Indian soc. coastal Agric. Res.* 2(1) : 56-62.

Naskar, K.R. (1987) Mangrove Swamps of the Sundarbans - An Ecological perspective. Naya Prakash, 206 Bidhan Sarani, Calcutta, pp. 1-264.

Naskar, K.R. and S.C. Santra (1986) Studies on Enteromorpha tubulosa in Brackish mixed sewage fed fisheries from the Sundarbans *Sci & Cult.*, 52 : 210.

Paul, S., *et al.* (1997) Winter Migratory Bagnet Fishery of the Hooghly Estuary Bull No. 76, Nov., 1997. Central Inland Capture Fisheries Research Institute (ICAR) Barrackpore 743101, West Bengal.

Pilley, T.V.R. (1954) The Ecology of Freshwater Bherij with special Reference to the Fish culture Practices and Biotic Interaction, *Proc. Nat. Inst. Sci. India.* 20 : 899-927.

Pilley, T.V.R. (1958) Biology of the *Hilsa ilisha* (Ham) in the Ganga River system. *Proc. Indi-Pacific Fish Coun.* 13(11) : 132-139.

Pantulu (Unpublished) In : Dutta *et al.* (1973). Fishery Resources of the Hooghly - Metla Estuarine System. Bull. 19 Dec. 1973. CIFRI, (IARI), Barrackpore, West Bengal.

Sewell, R.B. (1934) A study of the Fauna of the Salt Lakes, Calcutta *Recd. Ind. Mus.* 36 : 45-121.

Schuster, W.H. (1952) Fish Culture in salt water pond in Java. *Indo. Pacific Fisheries Counsil*, Bankok.

IV
Past and Present of Sundarban Delta

1
Kolkata Wetland – A Nightmare of the Past

There was a time when Kolkata was well known for its vast wetland located in the eastern part of the city. The Hooghly river flowing in the West at a higher level (average height of river bank was 6 m) than Kulti Gong flowing in the east at a much lower level (3 m or even less - Rudra (2001), had the advantage of disposing the west water through the wetlands that acted as an excellent natural purifier, since long, it was for this reason Kolkata had no sewage treatment plant like other big cities.

The Conservation of wetlands signed by Ramser, Iran, in 1971 was an intergovernmental treaty which provides the framework for natural action and International Co-operations for the conservation and better use of wetlands and their resources. The Ramser Bureau had thus declared the East Kolkata wet land as Ramsar site a status that placed it among 1229 other such sites all over the World (Goswami, Kolkata, 2002).

Although considerable time has passed, the State Government took interest, as stated by Goswami (1998) for the first time, set to prepare an action plan to help preserve the ecological balance in the East Kolkata wetlands which include sewage-fed fisheries covering an area of 12,500 ha. The wetlands consisted of 286 fisheries in the East Kolkata located at Bhangar, Bidhanagar, Sonarpur and Tiljala covering an area of 3832.27 ha. area out of which twenty two became imperative. A team comprising well known environmentalist, ecologists, engineers and management experts from all over the country was planned to prepare the action plan to workout a scheme to preserve the sewage-fed fisheries. These fisheries (locally known as bheries) clean the city's sewer water under natural condition. In case bheries are closed, a sewage treatment plant would have to be installed in the city.

It is unfortunate that due to sudden population influx te huge wetland areas could not maintain its natural ecosystem. The population influx was mainly from East Pakistan and subsequently during 1971 when East Pakistan emerged as independent Bangladesh. In addition to these major influx, the sporadic infiltration which is still continuing are the factors which completely changed the character of Kolkata wetlands and the adjoining Sundarbans. Urban development by hectic construction of houses and housing complex by filling up water bodies for settlement, Construction of Shanties, Creation of bheries and different industrial complexes, piling up of huge garbage and clinical wastes and more recently proposal of shifting of taunery complex to Bantola, wetlands in prime locations have gradually disappeared. Further, land developers have thrived in Salt Lake, Dum Dum and other areas in North and South 24-Parganas and more recently the State Government's new township project at Rajarhat, the entire wetland of the past has now become an ecological nightmere (*The Statesman* – a, 2003).

In the interest of saving the wetland a series of activities were initiated which are discussed below :

When the situation has grown up like this, the Government of West Bengal appointed one creative Research Group which however has prepared Management Action Plan of East Kolkata wetland in accord with Ramear Conservation guide lines (Goswami, 2002) the results of which are yet to be seen.

In a recent workshop organised by the Institute of Wetland Management for Ecological Design (IWMED) at Paribesh Bhaban. The minister for environment and information Technology announced that a menu driven wetland information system has been developed in Arc view Geographical Information System for West Bengal.

IWMED had participated in the Nationwide mapping and carried out the inventory and mapping of wetlands of West Bengal on a 1:50000 scale jointly with National Atlas and Thematic Mapping Organisation, Kolkata and Space Application Centre, Ahmedabad. It provides a complete detail of all the wetlands in the state and talks about the possible benefits of wetlands (*The Statesman* News Service - b, 2003).

Thus it is clear that no substantial work has been done so far except organising seminars and workshops by different organisations until finally the Govt. of West Bengal set up a twelve member committee (*The Statesman* News Service - C, 2003) and with the help of a Creative research group decided to make up a precise demarcation of East Kolkata Wetland and prepare a map and land schedule when the vast wetland areas have already been used indiscreminately to make room for housing complexes over the past few years.

Surprisingly the Government is now redying a blue print with this wetland potential to develop vegetable farming and floriculture also (*The Statesman*, South Bengal Plus, 2003). Thus the wetland of the past which is now considerably reduced over the past few years by indiscriminate use to make room for housing complexes etc. is redying a blueprint for floriculture along with the development of wetland (*The Statesman* News Service, 2003).

It is unfortunate that so far no definite steps could be taken when again the transport minister Mr. Subhas Chakraborty, West Bengal (2003) in his recent statement has justified filling up of water bodies in the city and the Subarbs. According to him except for ancient Kolkata which comprise a small fraction of the present city area, the rest have come to existance by filling up water bodies. This includes water areas such as Ballygung, Alipore and almost the whole of South Kolkata. The whole Salt Lake falls under this category and so does the upcoming Rajarhat township which covers a filled up area three times that of Salt Lake.

Switching over to the scientific justification Mr. Chakraborty reiterated "Relative humidity 95 and 99 percent) is Kolkata is caused by existing marshy land and cause severe health hazard, as maximum number of epidemics took place in Eastern part of the Country in the past 2000 years which can be attributed to this humidity.

Once again after a considerable laps of time a new committee comprising the Secretaries of the Department, environment, irrigation, fishery and forest was planned to protect the environmental balance of the place. Despite the State Government's endeavour to preserve the wetland of the East Kolkata, South 24-Parganas, the promoters and the real estate owners are on a spree to build up houses as usual in the area (*Statesman* News Service - a, 2004), until the most modern township covering a large wetland area has came into existance. After Rajarhat Complex, as usual, once again the management Committee for conservation of east Kolkata wetland has decided to evict farmer from around 30 ha. of Ramsar area only of the project which will be carried out on 12,500 ha of land in the entire Kolkata wetland area. (Ghosh, 2004), also Addendum 1.

Thus it is evident that when the huge wetland area (Ramsar site) of Sundarbans are gradually being reduced, various expert Committees formed one after another periodically, only to become silent witness of their deterioration.

Are all these not ecological nightmare :

"Microbial Cleaners"

Maureen Nandini Mitra in her recent article has shown how scientists have located easy to culture strains that could help remove Toxicity in East Calcutta Wetlands. Perhaps after Ramser Bureau (1971), the importance of wetlands is shown but with scientific background (Addendum 2).

In view of the above findings Mitra options that the East Kolkata West land (ECW) which comprises several water bodies spread over 5000 ha. in the absence of any sewage treatment plant, can treat the sewage as the wetlands ecosystem uses the refuge in fisheries (which meet on third of the city's fish requirement) and agriculture and in the process purifies it to a standard equal to many urban secondary – level treatment plans.

Addendum I
Save wetland, target farms

(Kaushik Ghosh (Kolkata). *The Statesman* 24-09-2004).

And finally when the modern Rajarhat township which has emerged by occupying a vast wetland areas of Sundarbans, the Govt. of West Bengal feels that the wetland now left has to be saved through a management Committee comprising representatives of forest, environment, fisheries, urban development and Municipal affair department who has decided to evict farmers from around 300 ha of the Ramsar area to turn the land into water bodies. This project will be carried out on 12,500 ha of land in the east Kolkata wetland area. The decision of digging water bodies out of agricultural land has been taken to restore the ecological balance of the wet land area in the East Kolkata.

Parts of the east Kolkata wetland have been destroyed by farming, water bodies were filled up and used for farming. Constructions in this place and adjoining areas also pose a threat to the ecological balance. Illegal constructions in the area have been pulled down and has been notified that no more buildings should be constructed.

Quite a large part of the wetland was lost as opioned by a senior Scientist of wetland and ecological design (Research wing of State environment department).

Since the ground in the area has already been dug up for Construction at Rajarhat township a lake of 300 ha area will be created here.

The importance of conservation of Ramser was decided in 1986. Later a status report of this wetland was prepared in 1992 and 2002 and the Conservation of the wetland began. Subsequently a status report of Ramser site was also prepared in 2003.

Addendum II
Microbial Cleaners

(Maureen Nandini Mitra)
8th Day, the Sunday, *Statesman*,
Magazine, Kolkata, October 1, 2006)

The Research team of West Bengal University of Technology's Microbiology Department while studying the Microbial population in the Wetlands soil and water found strains from 12 bacterial phyla. According to Dr. Shaon Ray Chaudhuri, Co-writer of the stdy, the team found 20 strains of bacteria that could grow in the presence of heavy concentrations of metals like lead, iron, cobalt, chromium, nickel and silver. These bacteria can remove metal toxicity and could help treat waste from battery factories and tanneries. The findings were published in the 25 July issue of current Science (Vol. 91, No. 2).

9 of the 22 strains can degrade oils while another 9 secrete extra-cellular protease, an enzyme that can be used in pharma and soap industries. Of these one is a secondary protese that, when added to detergents, helps remove tough grease and oil stains. Yet another strain was found to cause "de-hairing" of hides and might be useful in the leather tanning industry. (CSE/Down to Earth Feature service).

2

Canals of Kolkata – Their Impact on the Sundarbans Eco-System

In the past, river Hooghly used to contribute considerable volume of fresh water to Ichhamati and Bidyadhari through its tributary, R. Jamuna and subtributaries – Nowii and Sunti. (Rudra, 2001) has stated that average height of Hooghly river bank around Kolkata was about 6m, while that of Kultigong flowing in the east of Kolkata was about 3 m or less. The slope between Hooghly and Kultigong the 28 km. stretch was utilized for disposing the Kolkata sewage waste water by constructing net work of Canals. Initially, in the north Kolkata, the net work of Canal System carrying waste water were arranged in such a way that the wastes were carried through Circular Canal – new cut canal and Kispopeer Canal before it was finally discharged into the Kultigong. The entire waste water was pushed by additional water it received from the Hooghly through Baghbazar canal and thus a continuous flow was maintained. Simultaneously in the South Kolkata, such operation was done through Tolly's Nullah and discharged into the Bidyadhari river. In subsequent years more canals were excavated for disposal of various complex water (Sewage, tannery and other complex waste water).

With the diversion of the main Ganga flow to Bangladesh and shifting of Damodar's Course further down, the tributary Jamuna and subtributaries Nowii and Sunti fast deteriorated and their impact was felt very badly in the Ichhamati, Kultigong, Bidyadhari and other connecting rivers due to absence of any head water supply. Secondly, with the population influx urbanization and construction of bheries started growing up very fast in the wetland area of the eastern Kolkata from Salt Lake/EM By-Pass to Kultigong (20 km. stretch) and even canals of Kolkata, river banks and their beds gradually occupied by encroachers. Even for their habitation, agricultural purposes in some areas the shallow river leeds and banks of Ichhamati and Bidydhari were occupied by encroachers for construction of bheries, brick kilns and used for dwelling and agricultural purposes. As a result both the Kolkata canals and rivers had to suffer a set back as they could not maintain their normal flow. Except the seasonal rainwater there was no other source of fresh water supply. Salinity in Sundarbans rivers and cracks thus gradually went on increasing affecting the growth of bio-diversity. River beds gradually get silted up. Even in the Hooghly some ferry services (Ramkrishna Palli Ferry Ghat) were suspended for want of adequate depth. Wetlands are being lost, if it rains even tidal bore can incindate Kolkata. Flood has thus become a common phenomenon in the city even in low rainfall. High tidal bores erode banks causing further damage by raising the beds of rivers and creeks lying criss-cross the Sundarbans causing continuous silt precipitation.

The piting up of city garbage which includes dangerous clinical wastes and wastes of tannery are quite considerable and are neither treated nor utilized and thus cause additional problems.

The shifting of Tannery Complex to Bantola may cause further problems.

Note : The Supreme Court today fixed October 31, 2002 as the dead line for completion of all infrastructural work including that of effluent treatment plant of the Calcutta Leather Complex at Bantala. The Calcutta Leather Complex will house 500 tanneries and shoe factories that were housed in the residencial colonies in the east Kolkata. (Supreme Court sets date for leather complex. *The Statesman,* 28 September 2002). In view of the lack of infrastructure the tanneries are reluctant to shift their tannery to Bantala (*Statesman* News Service, 19.11.2002).

3

Metamorphosis of Sundarban Delta – Its Impact on Aquatic Life

(Impounded Water, Estuaries, Biodiversity and Other Related Problem)

The Sundarban delta after its formation has been facing problems mainly from natural calamities (Storm, Cyclone, flood, draught and sometimes associated with earthquake), and human interference (Table 1 and Fig. 1). These two factors have affected a series of changes both in Culture as well as in the capture fishery ecosystem mainly in 2 ways. First, with the growing population the large sized impoundments gradually reduced to smaller size with the passing of time and were being regularly used as fresh and brackish water impoundments of different categories (perennial, seasonal, fish-cum paddy culture and paddy-cum fish culture etc.) for fresh and brackish water fish culture. In the later years the shallow river beds were being used for construction of bheries and brick fields thereby affecting the flow. The estuarine ecosystem is however controlled by high saline water (during non-monsoon months) in the absence of fresh water supply. This has affected the growth of bio-diversity of forest. The size of forest is also affected by increased population due to its reclamation (increase of forest was only 6.7% in 40 years) and finally high silting which led the estuarine ecosystem in a moribund condition one after another thereby causing the delta to increase by 33.6% in 40 years. Both loss of forest and heavy silting of estuaries have affected the fishery to a great extent. The forest was and is being used as a

Table 1. Natural calamities experienced in the Sundarban Delta.

Year	*Storm*	*Cyclone*	*Floods*	*Earthquake*
1582-1852 (270 years)	16	-	-	-
1856-1904 (48 years)	8	17	5	1
1907-1956 (49 years)	11	8	9	1
1960-1981 (21 years)	2	3	1	-
1982-2000 (18 years)	4	1	1	-

natural habitat for brackish water and marine fishes including the prized Crustacean *P. monodan,* where they breed, grow and attain maturity. Again because of shallow depth of estuaries fishing intensity increased by simple gear and finally with a few feet ruse if water level, becomes inevitable and under such condition hilsa another prized fish cannot migrate upstream for breeding purpose. All these aspects are highlighted in Part-IV under different headings.

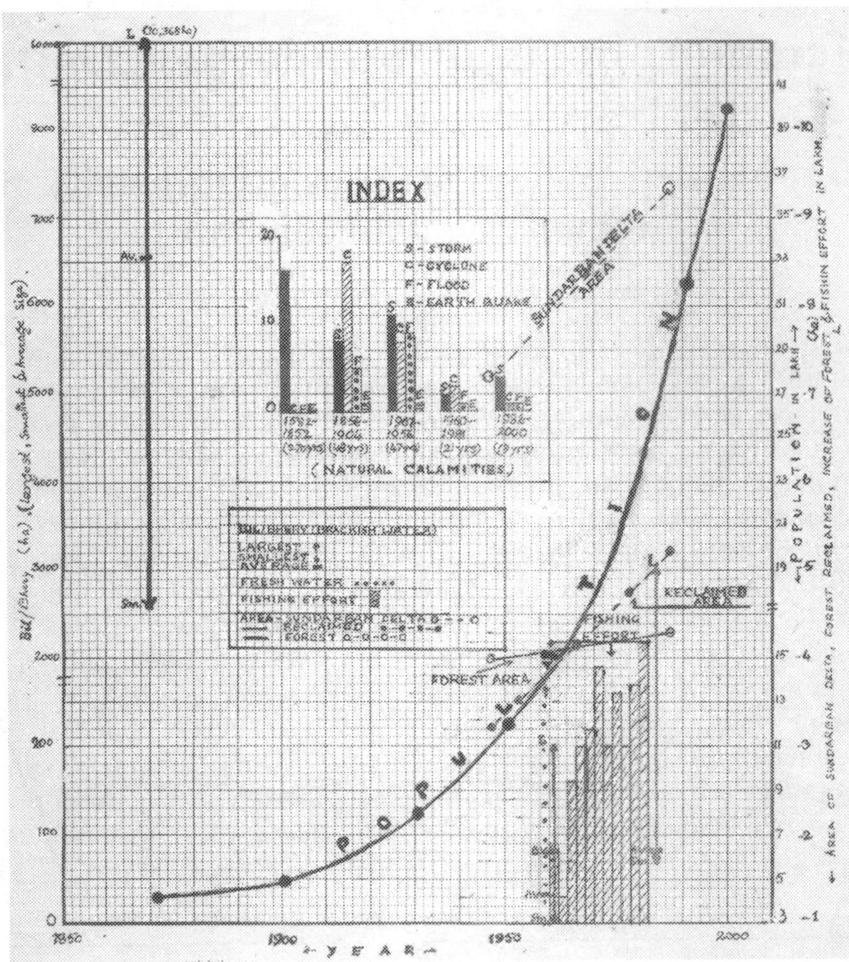

Fig. 1 : Damaged Ecosystem of Sundarbans

From 1856-1981, Cyclone, flood and storm were common phenomenon although instances of earth-quake on two occasions were recorded. Mukherjee (1959) in his valuable document "Drainage in greater Calcutta area" has mentioned that within the last 12 years, inundation has occurred more than 10 times and flooded hundreds of square miles of area with saline water resulting in almost complete destruction of crops. Such natural calamities are capable in altering the normal ecosystem which is quite different from the previous years. Such things are reflected with the passing of time through the gradual change in the morpho-ecology and their resultant effect in various fields.

4

Progressive Environmental Changes of the Sundarban Delta and the Estuarine Ecosystem vis-a-vis Fish Abundance and their Composition

Alongwith the formation of Sundarban delta a number of natural shallow inundated swampy water areas came into existance (Stated elsewhere) which were regularly inundated and dewatered by the tidal brackish water streams and creeks lying criss cross in the delta during tidal phases. As a result both fresh and brackish water species became available both in natural shallow and inundated water spread areas and estuarine fish in brackish water rivers and creeks. With the growing population, the forest areas were reclaimed by people who started using land and water for their habitation, agriculture and aquaculture. Natural inundated shallow areas (large and small) are modified by putting earth embankments and fed by both rain and river water during their tidal impulses. By this process fish population (seed, young and adult) of estuarine water carried by tidal water are stocked and cultured to a suitable size. In later years brackish-water fish culture was changed into fresh water and culture of Indian major corps were initiated. Soil being saline due to high salinity of brackishwater, agriculture could not come up like fish culture which proved to be more profitable than agriculture, grew up with a rapid pace. This system of fish culture subsequently termed as bhery fishery (shallow impounded water area used for fish culture). Such culture was done both in fresh and brackish water; the later was further classified as parennial (Recorded) system of fishery, Paddy-cum-fish culture (where paddy is the primary produce and fish-cum-paddy culture (where fish is the primary produce) system.

On the whole the entire fishery system of Sundarban delta comprises both culture and capture in character. Culture fishery is done in fresh water, low, medium and high saline water; while capture fishery represents low, medium and high saline zones of the estuarine rivers. The fishery complex of the entire Sundarban delta is shown below in Table 2.

Abundance of fish stock in rivers, estuaries and creeks are very much related to their environmental conditions. Added to it climatic conditions such as heavy rainfall/drought, cyclonic weather, tidal impulse, pollution forest growth etc. are equally responsible for increase or decrease of fish stock.

The deltaic Sundarban which was 720000 ha during 1947 increased by 33.75% in 1987 due to precipitation of silt during which period reclaimed area increased by 67.6% used for habitation, agriculture, aquaculture etc. by the population, increasing at a reapid pace (Population in 1981 was 2455370 which increased to 3700000 during 1989 i.e. 33.6% increase in 8 years). This shows that the tidal rivers of deeper zone and cracks which once used for fish capture, either has gradually silted up or became very shallow and unsuitable for specific type of net operation used earlier. Such progressive silting must have created new field for net operation until again such new areas formed after a few years for such operation and the old area gets very shallow or exposed. By this porcess both catch and effort simultaneously went on increasing. But if the forest area decreases instead of increasing (due to absence of adequate fresh water supply and human interference) the fishery may be affected as their habitats and breeding grounds will be affected.

Table 2. Fishery of Sundarban delta under complex ecosystem

		Culture fishery			*Capture fishery*
Fresh water Bhery		*Low Saline water Bhery*	*Medium saline water Bhery*	*High saline water Bhery*	
Sewage fed Bhery	Sewage free Bhery	1. Perennial Brackish water (Recorded) 2. Fish-cum paddy (Sewage fed) 3. Paddy-cum fish *Locality* Barasat Swarupnagar De Ganga Rajarhat Baduria Gaighata.	*Locality :* Haroa (Sewage fed) Minakhan Basirhat Hingleganj.	*Locality :* Canning Barasat Hasnabad Kultuli Gosaba Sandeshkhali Mathurapur Jaynagar. Parthapratima Nazat Kalinagar	

Fresh water - Low saline water:	*Low saline- Medium saline:*	*Medium saline - High saline:*	*Medium saline : Kulti estuary (sewage fed)*	*High saline : Lower zone of entire estuary of Sundarbans and tract below Diamond Harbour.*
Bhagirathi-Hooghly	*Hooghly*	*Port Canning*		
Area : Nabadwip to Khusiguli	*Area :* Khusigali to Diamond Habour.	*Area :* Small zones of Matta estuary comprasing local area around port Canning.	Bidyadhari *Area :* Chotokolagachi Ghatihar Barokologachi * Kalinadi from Hingle-ganj to Saheb khali to Raimangal Hasnabad to Barunhat. * R. Dasa Chaktang-ramari to Kumarmari * R. Dasa Chaktang-ramari to Kumarmari	*Area :* Fresherganj, Bokkhali Jamboo Island (Upper and lower) Khalisthan, Sagar island.

* Prawn seed collection centre.

Fishing gears used in such areas are very simple and are traditionally old and fishing areas being shallow, efforts also increases as observed by Ray (1998) in case of river Ganga. Perhaps in the absence of any modern gear technology neither the fish of the deeper zone could be trapped nor any Commercial fish could be segreated from the miscellaneous group. Simultaneously, the shallower zones or water spill areas or submerged banks during high tide are used as breeding grounds. Crustaceans, especially, prawns or crabs and adult fishes use such areas for their habitation and maturation. The matured prawns migrate to the sea for the purpose of breeding. The female prawns breed in millions every year. Young larvae of different species and prawns are drifted upstream with the high tidal water which are largely trapped by fisherman. From six month's catch (cf Table 3) based on 2 days/month observation both during high and low tide and total catch of 15487200 in six months by operation of 900 nets by 300 men (3 nets/man) from a stretch of 3 km. has shown that prawn maintained 25% of the total catch consisting of 12 species. The catch further indicates that for 1 prawn 4 other species are destroyed and this is a common phenomenon (Mukherjee, 1996).

The entire Sundarban delta (bheries and estuaries) producing various species of fish under different ecosystem are highlighted at the appropriate places. A more detailed picture of the Hooghly and the Kulti estuaries flowing east and wet of Sundarbans alongwith some bheries (fresh and saline water) producing fish fauna including crustaceous regularly are also highlighted.

Table 3 : Showing destruction of their species while netting prized prawn *P. monodon*

Species	*Fry caught (No.) during 6 month based on 2 day's observation during low and high tide*	*Based on this total catch by 300 fishermen using 900 nets*
Penaeus monodan	4593	4133700
Penaeus Indicus	1822	1639800
Metapenaeus monoceros	3161	2844900
Lates calcarifer	1591	1431900
Glossogobius giuris (Ham)	3276	2948400
Scatophagus argus (Bloch)	500	486000
Eel	820	738000
Barbus (*Punitius sophore*) (Ham)	3431	3087900
Mystus tengra (Ham)	135	1215200
Ambasis ranga (Ham)	1214	1092600
Apocryptus bato	163	146700
Mastacembelus pancalus (Ham)	20	18000
Coilia dusomieri/neglectal	1035	931500

From the above table it is clear that from a stretch of 3 km. the catch of *P. monodan* was 4133700 (from 900 nets) when the total catch of other 12 species was 15487200. This means to catch 1 prawn, 4 other species were destroyed (Mukherjee, 1996).

5

Sequence of Morpho-Ecological Events Leading to Delta Formation – Its Impact on Aquaculture

The Sundarban delta came into existence through a series of morpho-ecological changes-mainly silt precipitation which resulted suspension of fresh water supply, climatological (cyclonic weather associated with heavy rain/drought) and tidal impulse. All these factors have resulted in formation of a large member of shallow impoundments (Large and small) which is later years converted into bhery fishery in the form of industry. Large forest areas have become a source of potential breeding zone which maintained high yield of fish production and other aquatic organisms. The following are some of the eventful incidents which has brought changes of the delta, forest area, area reclaimed for habitation and ultimately for fish culture. Besides agriculture, the greater area of Sundarban delta are being used both for culture and capture fishery.

A. Culture Fishery (Creation of Bheries)

1870-73 Gangopadhaya (1983) (In : Statistical Accounts of Bengal.	Formation of some natural impoundments (Large and small) like Boira Bil : (40 Sq. miles), Datbhanga Bil (22 Sq. miles), Balli Bil (10 Sq. miles), Kulgachi, Dholkera, Gajalmari, Nagarghata and Khaliskhali were reported. All these no more exist as they are converted into agricultural land.
1922 Thakur (1986)	Under a cyclonic influence the entire Sundarbans was flooded by a 22 ft. high flood water of R. Bidyadhari which again resulted formation of innumerable number of saline water bheries of which Goberia, Sarbaria, Ramchack and Ranjoy are worth mentioning. Each of these was a mile wide and 2-3 mile long. In all these bheries, prawn (*P. monodan*) was the major fishery followed by Bhetki (*L. Calcarifer*) and Persia (*L. persia*). Prawn used to be exported. Gobaria bhery alone used to earn 4 crores of rupees as foreign exchange by exporting prawn. This tempted people, especially lotdars (Lessesses) who encouraged fish/prawn culture by inundating agricultural lands by

	introducing brackish water during high tide to convert agricultural land into fishery as it was more paying.
1939 David (1959)	There were extensive water logged swamps many of which were converted into rich bheries. Ramjog and Gobariabad fisheries together extending to over 7500 acres and the largest near the outfall that open into the sewage mixed waters of the Kulti.
Sewell (1934), Hora and Nair (1944), Pillery (1954), Bose (1944) and Nair (1944).	Several species of mullets and a number of prawns were extensively collected in these brackishwater fish farm locally known as 'Bhasabadha Fisheries' or Perinnial bheries. The affected forms extended to an area of nearly 75 Sq. miles to the North and North East of Calcutta within a radius of about 50 miles. These bheries which used to supply appreciable quantity of fish and prawn to Calcutta fish market can no longer do it, since, this time Calcutta Municipal wastes is disposed into the Kulti estuary from 1939.
1958-78 Fish Producers Association (1959) Pakrasi (1965) Ghosh (1973) Master Plan for Fisheries Development West Bengal, Part I and III (1974) Report of Fach Finding Committee (1984) Report of Task Force West Bengal (1987).	It was during this long period when an interesting change was noticed in behries. There has been proliferation of integrated paddy-cum-fish culture system, in the same plot of law was noticed in the 24 Parganas District particularly in the Sundarbans area. The other system i.e. Fish-cum-paddy culture was also reported. In case of former, paddy is the primary produce, while in case of later, fish is the primary produce.
Formation of Salt Lake	
1904-19 Ray (1993)	It was during this period that due to heavy selling river Bidyadhari lost most of its characteristics. The high silt ladden sewage waste of Calcutta City sealed the fate of the Bidyadhari resulting formation of a large water logged swampy area and used to flood the vast shallow areas during the wet seasons since then the entire area was termed as Salt Lake.
1930 Sewell (1934)	It was this time the saline water of Salt Lake was changed into fresh water.
Use of Sewage in Bheries	
1900 Ray (1993)	(1) For the first time sewage was introduced in embanked spill brackish water where *Lates Calcarifer, Mugil persia, M. tade* and

prawns were introduced. No result was achieved as the area was flooded during the wet season.

(2) Subsequently in 1930 a local zaminder of Salt Lake finding his vast area of land not yielding any revenue made an attempt to Culture carps (possibly for the first time by letting in sewage in small quantity into the enclosed spilled water. The result obtained was very encouraging and within a short period practically the whole area was converted into bheries for culturing fresh water fish by mixing the water with sewage. This is how sewage water is being used since 1930, up to 1951. Metropolitan wastes were mainly organic in character. Complex industrial wastes appeared around 1963 which increased considerably during the subsequent years.

Fish Production

1972-81 Das (1981)	A good crop (*P. monodan*) in 1000 bigha could be 45 t whose cost was Rs. 23 lakhs.
1984 Report of Fact Finding Committee (1984)	A person having 10 bigha of land would earn about 8000/- on an average from fishery, where as income from paddy would be Rs. 3000/- only.

Salt Lake (Fresh Water)

1959-85	1959		1970	
Ghosh (1959)	1968		1971	
	1976	1.0-1.47t/ha.	1978	2.08-2.77 t/ha.
	1980		1983	
	1985			

Saline Water Zone

	Salinity		**Production (kg./ha)**
Saha et al. (1986)	Low saline	0.1 - 9.5%	876
	Medium saline	0.27 - 15.8%	749
	High saline	6.6 - 36.2%	694
	Perennial Fishery		
Ray (1993)	High saline		1200 kg/ha.
	Medium saline		1 t/ha.
	Low saline		100 kg/ha.
	Paddy-cum fish		
	High saline		800 kg/ha.
	Medium saline		800 1 t/ha.
	Low saline		300 kg/ha.

	Sewage-fed fishery		Sewage free fishery
1965			
Pakrasi (1965)	Medium saline	3.82 kg/ha.	0.947 kg/ha. (high saline)

B. Capture Fishery

Besides Kulti, Hooghly-Matla estuarine system of the Sundarbans are the important source of fish supply. Such potentiality has significantly gone down due to considerable changes in the morpho-ecological conditions of the estuarine system and pollution.

1. The Kulti estuary

Sundarban delta is normally subjected to suffer frequently from natural calamities like floods and cyclones. From the available information it is stated that high flood of 1922 brought prawn (*P. monodan*) as the principal fishery followed by Bhitki (*Lates calcariper*) and Persia (*Liza parsia*). In a subsequent study by David (1959) al estuarine water within Bengal delta yield a rich fish and prawn harvest and Kulti was no exception at least until 1939. Subsequently, since the time of introduction of Calcutta Municipal waste into the Kulti estuary, both the fish seed as well as adult fish have almost disappear.

1966. Paramanik and Roy (1972)	The high shape factor value of Kulti indicated considerable degradation due to silting.
Ray *et al.* (1981), Report of Fact finding Committee West Bengal (1984), Mukherjee (1996)	Further down, some improvement was noticed. Prawn seed (*P. monodan*) always dominated the catch.

2. Hooghly-Matla estuarine system

(Marpho-ecological change and their effect).

1853-1937 Report, Govt. of West Bengal (1945)	Discharge Flow above Bhagirathi	1,64,500 cub/sec.
1869	At Berhampur	1,40,762 cub/sec.
Report, Govt. of West Bengal (1945)	At Jalangi	1,40,160 cub/sec.

	High water	Low water
1873-1932 (60 years)	Maximum 58.3 - 63.5	39.9 - 35.6
	Minimum 53.8 - 58.1	32.0 - 33.9

1919 (Oag, 1919) **Approximate width and Depth of R. Hooghly.**

Centres	Width (miles)	Maximum depth (ft.) at H.T. during monsoon)	Maximum depth (ft.) at L.T. during winter
Barrackpore	0.34	60	40
Calcutta	0.47	51	30
Uluberia	0.47	71	50

Falta	0.76	68	48
Diamond Harbour	2.91	91	69
Kulpi	2.87	58	34

Bansberia

Water level

	High water	**Low water**
1945 Report of Govt. of West Bengal	Maximum 15.2 - 23.4	9.7 - 18.4
	Minimum 8.0 - 18.8	5.7 - 16.8

1975 Bera (1984) — **Formation of an island** at the source of Bhagirathi-Hooghly resulted rise of river bed 7.5 m. from R. Ganga.

Prior to 1972 : **Soil brought with** water - 90 - 100 mf^3/year

1984 : **Alluvial soil** brought during monsoon (4 months) - 20 crore f^3

Bera (1984) — According to Irrigation and Power - **2.8 mm^3 silt brought annually; 2/3rd deposited at Bansberia, slope between Farakka and Nabadwip.**

Nabadwip - 1 : 23000
Slope - 3″ per mile.

1972 — 279 Brick fields flanked on both sides of R. Hooghly between Kalyani-Achipur.

1988 — Number of bore days = 180/year.
Prior to fiftees = 20 - 22.

Pollution Load disposed into R. Hooghly.

1973 (Ray, 1981)

Volume of waste water (m^3/d)	=	67337.19
BOD t/d	=	53.948
Total solids (t/d)	=	775.453
Suspended solids (t/d)	=	371.642
Dissolved solids (t/d)	=	405.80

Fisheries

A. Bhagirathi - Hooghly

1. Hamilton (1822) : Purely freshwater fish except Hilsa (Calcutta and Lower stretches).
2. Hora (1943) : Considerable changes in the fishery composition was noticed. Estuarine groups in and around Barrackpore.
3. Hora and Pilley (1953) : Mainly Brackish water fishes in and around Barrackpore.

4. David (1954) : Mainly Brackish water fishes (salinity range 1.204-1.156 ppt)

5. Ray (1981) : Mainly Brackish water forms, in and around Barrackpore. Highest salinity recorded - 4.0 ppt.

B. Hooghly - Matla estuarine system.

Number of important species caught in the entire Sundarban estuary during 1966-71 are 26 plus the miscellaneous group (cf. Tables 11 and 12 chapter, III Sundarban Delta). It is interesting to note that zone III (lower zone including entire estuarine area of the Sundarbans and tract below Diamond Harbour on the main channel) is showing the maximum number of species; i.e. 24 which is followed by zone V (small zone in the Matla estuary comprising the local area around port Canning) numbering 20-23, Lowest Number recorded at Zone I, (11-13 ; upper zone, - Salinity - trace round the year, Nabadwip - Khusigoli).

Again out of 7 assembly centres (Port Canning, Hasnabad, Diamond Harbour, Kakdwip, Kalinagar, Namkhana and Raidighi), Port Canning indicatd comparatively lower number of species (17 and 18), followed by Hasnabad (18 and 19) with two exceptions 21 and 22. At other 5 centres, Diamond Harbour, Kakdwip, Kalinagar, Namkhana and Raidighi, were showing a little higher value (20-24). From the above data it may be inferred that the existing biodiversity of Sundarban delta under the present condition could not change the important species composition appreciably during 1961-71. In a recent report Chaudhury (2000) states that the biodiversity of Sundarban delta which is affected very much, has resulted the 86 species of fish and 46 species of Crustaceans either extinct or endangered in Sagar island.

In case of winter fishery, since the same type of bagnet is used for fishing and although fishing units ranged from 953 (1994-95) to 1247 (1995-90) and further 1629 in (1996-97), Paul *et al.* (1997) further states that the catch by the winter migratory bagnet during 1993-94 to 1996-97 is about 6 times more than that of the period 1970-71 to 1974-75. The authors have attributed mainly due to deployment of mechanical boats and higher number of fishermen involved in the fishing operation. This means where depth is less obviously fishing effort increases as the operation of such net is more convenient in the shallower area than the deeper area where it results low yield as fishing effort is found to be much less

Paul *et al.*, 1997). This indirectly shows how steadily the ecosystem is loosing depth and the delta is heading towards deeper zone. This fact is revealed by the increase of delta by 33.6 per cent. in 40 years when forest area although increased by only 6.66 per cent, again showed signs of destruction of about 4000 Sq. km.

With the restricted supply of fresh water and consequent increase in salinity of brackish water in rivers and tributaries criss crossed the Sundarban delta, the forest area is very much affected (6.66 per cent in 40 years and further a sign of destruction of forest since 1998, the natural breeding ground may have been affected and this may lead to poor production in future.

6

Problems Associated with the Culture and Capture Fishery – Their Remedial Measures

A. Culture Fishery

1. It has been observed that fish culture is more paying than paddy culture. But integrated farming has double advantage as both paddy and fish can be cultured simultaneously irrespective of their salinity. Hence, integrated paddy-cum-fish culture may be encouraged.
2. Calcutta Municipal waste may be used for better growth, provided the waste is treated to make it free from any metallic contamination. After removing the contaminants the waste can be used in the bhery after determining the required quantity beneficial to enrich the water with fish-food organisms suitable for healthy growth and reproduction of fish and does not harm the physiological activity of the fishes. When waste-water is inadequate proper manure could be used (It is reported that fingulings of Nilolika, Rohu and Mrigal in the fisheries of east Calcutta (Borochailabi, uttar and Dakshin Gorumara, Raniful and Boro Bon-Gheri in Dhapa Menpur area, (P. S. - Bidhan Nagar and Boser and Ghoser bheries in Dharmatola Pachuria area (P.S. Bangore) were dying due to inadequate supply of sewer-water in which fingerlings of size 100-150 gms. were released. The death is due to inadequate supply of sewer water which resulted insufficient food. The loss was considerable (25000/- – 30,000/-) (Secy 24 Parganas, Fish Producers Association ; In : *The Statesman* News Service, 31.5.1999.)
3. Since prawn is highly paying (already explained elsewhere) some brackish water bheries may be kept seperate exclusively for prawn (*P. monodan*) culture as suggested elsewhere.
4. Other measures suggested for obtaining higher production in case of bheries (fresh and brackish water area) as suggested elsewhere may be tried.

B. Capture Fishery

1. Forest has to be preserved to maintain the habital for rich breeding grounds of different species of fish including prawn.

2. Some spill area and also reservoir construction (suggested elsewhere) on both sides of the main estuary can help storing of water during high tide which can be released to flush the precipitated silt. This will maintain the flow of the Creck/estuary and prevent their deterioration to a great extent.
3. Some arrangements for freshwater resources is essential to lower down the salinity of brackishwater which will maintain the growth of forest.
4. Suitable or improved mash of fishing gear may be tried after adequate experimentation to catch the proper sized fishes.
5. While catching prawn or other commercial important fish adequate care should be taken not to use fine mesh size net and thus destroy the other species of commercial important or miscellaneous group that remain associated in the catch as their percentage in the catch is appreciable.
6. Shallow river beds are largely used as brick fields for manufacture of brick kilns and also for construction of bheries by putting embankments across the river. These are leased by local goons (Ichhamati and Bidyadhari river beds). Such practices are extremely harmful in maintaining the life of the river. Local goons lease out stretches of Bidyadhari river to set up bheries and brick kilns. Irrigation department officials recorded 268 illigal bheries and a large number of brick kilns which function from September to May are set up in a stretch of 16 km. between Muktarpur and Talia by setting up bundhs made of bamboos. Further a survey conducted by land and Land reform Development said, around 48000 area of paddy field are affected. As a result the river bed lost its navigability and banks get flooded every monsoon when crops worth croves of rupees are destroyed (Tarun Goswami; Statesman News Service, 21.12.99).
7. In case of monoculture prawn (*Peneus monodan*) in the deltaic lands, the paddy fields and the mangrove-reclaimed areas require to be stopped in the interest of sustainable estuarine fishes in the Sundarbans areas. Present trend of *P. monodan* culture for its export potentialities are very much alarming and for the export values, the prawn and carb culture and prawn seed collection have damaged the mangrove ecosystem very rapidly. Nowadays several unidentified prawn disease have created heavy loss to the prawn farmers which needs proper scientific investigation.

7

Sundarban Forest – Its Impact on Fishery Resources

Observations recorded by various workers - Mac Nae (1968), Naskar (1996), Raha (1999), Chakraborty (1999), Ghosh Thakur (1996) etc. are presented here to clear picture of the importance of forest which is directly related to the variations in the catch and the corresponding efforts from centre to centre during different years. Mangrove vegetation which dominates the Sundarban forest plays a vital role in preserving the ecological niches where fish and other crustaceans safety take shelter, grow, mature and breed to maintain the fishery potentiality of the area.

The natural vegetation cover in water catchment helps in maintaining hydrobiological cycles, regulating and stabilising water run off and acts as a buffer against extreme events like flood and draught. It also helps to regulate underground water tables and prevent dry land stability. Thus a pointes to degradation of habitats is the depletion of groundwater levels.

Biodiversity also helps maintenance of soil structure and increases the moisture retention capacity as well as nutrient level of soil. Non existence of any sustained effort to preserve the ecology has resulted in denudation and soil erosion. Thus stabilisation of soils, leaching of nutrients, soil erosion etc. lead to overall decline in soil productivity. Trees also help in soil formation, their root system enables deep penetration. Nutrient cycling is an essential event concerned with maintenance of the ecosystem. Biological diversity is inevitable for this. Microorganisms in the soil by decomposing dead and decayed wastes replenish the soil nutrients. The formation of nitrate bacteria and that of nitrifying bacteria being different, diversity of micro-organisms is also essential. It provides natural habitat for large number of estuarine or marine fauna, fishes and other threatened endangers endemic fauna.

Forest provides grazing grounds and nursery beds for estuarine or off shore fish, prawn and other crustaceans. For having this unique potentialities Mac Nae (1968) stated that the mangroves are the nature's own aquaculture system and according to him "No mangrove - no prawn" (Hall, 1962 ; Mac Nae, 1974 ; Warman, 1977 and Naskar, 1996).

Due to population increase which resulted their large scale habitation by deforestation and large scale prawn (*Penium monodon*) culture/farming, excavation of prawn farms, renovation of forest land for prawn farming, large scale prawn seed collection (by more than 1 lakh people) and discharge of pollutants from their semi-intensive/integrated prawn farms have alarmingly degrading their precious coastal mangrove vegetation (Naskar, 1996).

Due to increase in siltation and also neo-tectonic till to Gangetic basin which occurred during early years, the freshwater inflow has gradually diminished and stopped at major places. As a result of which the fresh-water tidal saline water interface has been restricted, the brackish water impact dominates causing detrimental effects to the ecosystem especially Sundari and some other trees which always prefer low saline water for their survival and healthy growth, as they cannot withstand high saline soil resulted from brackish-water. Similar findings are also recorded by Jadav *et al.* (1981). According to them Sundarban saline soils are considered responsible for higher plant mortality. They have recorded high salinity which reaches as high as 40 m whose and some times more. While salt crusts are very often visible on the surface soil. The salinity reaches maximum in the middle of May and decreases with the onset of monsoon. As a result these trees show retarded growth until finally they die and reduces the area of the forest. Such phenomenon has already been detected through satelite picture since 1998 by Raha (1999) and the area detected is around 4000 Sq. km. on the western side of the Sundarbans and thus showing like a bald spherical area. Besides, area reclaimed is fast increasing (56500.75 ha in 63 years, Banerjee 1964 ; 536257.11 ha in 40 years, (1947-1987) i.e. 67.6 percent and thereby reducing the forest abemingly. Bank erosion of Sundarbans is on the increase. Trees thus get affected due to soil erosion. Such soil are brought down into the river bed as a result of which the depth of the river gets reduced and the water spreads which means higher value of shape factor. Such continuous operation may ultimately result raising the river bed and formation of new island.

Besides climatic factors, the above facts come into consideration while explaining the higher and lower catch and also their fishing effort. These facts are discussed earlier. In case of winter migratary fishery, since the same type of bagnet is used for fishing when fishing units ranged from 953 (1994-95) to 1247 (1995-96) and further 1629 in 1996-97. This means when depth is less, obviously the fishing effort increases as the operation of such net is more convenient in the shallower area than the deeper area where both the fishing effort and catch are found to be much less (Paul *et al.*, 1997). This indirectly shows how rapidly the ecosystem is loosing depth and the delta is gradually heading towards deeper zone. This fact is revealed when in 40 years the delta has increased by 33.6% while forest shows increase of only 6.66%. In case adverse effect on Sundari and other trees comes up due to high salinity in the absence of fresh water washings its impact will be serious, and in fact such impact has come to be true and is revealed by Raha (1999).

Thus to maintain the richness of fishery potentiality both freshwater and forest are equally important to maintain the healthy ecosystem, otherwise some day may come when the forest wealth will gradually loose its density (already there are enough indications) and likewise fishes alongwith other crustaceaus will also loose their attractive habitats and breeding grounds when the delta will simply encroach towards the sea due to rapid silt precipitation (Refer statements of Mukhopadhaya 2003, on Geologists concerned over Sundarbans erosion). Initially, the area being wide and shallow more catch is likely with the increased effort. But simultaneously it should be remembered that the natural habitat which provide not only shelter but also rich food to fish and other Crustaceaus due to presence of forest, will be lost in the long run and alongwith the deforestation, the fishes and other Crustaceaus will be heading fast towards depletion.

8

Sundarban – A Goldmine for Prawn Fisheries

Prawn breed in the sea producing millions of microscopic eggs. Each female releases about 3 of million eggs at a time. The eggs hatch into young larvae. The larvae migrate into Coastal estuaries by tidal currents. Nearly 80% of the prawn population is believed to be destroyed by predators and the rest manage to survive and find their way into estuaries and grow in brackishwater impoundments up to adult stage. After attaining maturity they migrate back to the sea for breeding and thus complete their life Cycle.

According to Saktival (1980) prawns – *Penaeus indicus* (Chapra) popularly known as 'whilt prawn' and *Penaeus monodon* (Bagda) known as 'tiger prawn' grow to more than 30 cm fetching the highest price. While prawn is very common, but the tiger prawn is a rare commodity for commercial exploitation. It is the king of prawns by virtue of its size (maximum recorded size is 32.5 cm.) colour, texture, weight and extra-ordinary tolerance capacity to salinity changes from fresh-water to sea-water. The Hooghly-Matla estuary with its unique environment is perhaps the only estuary in India harbouring this precious commodity for large scale commercial exploitation in brackish-water impoundments (Bheries). The ingression of tiger prawn post larvae into this estuary in millions along with high tide from sea is a rare gift for aquaculture. Hence, it becomes highly essential to utilise this natural gift in a rational way.

Hall (1962), Mac Nae (1974) and Namin (1977) have highlighted the importance of mangrove forest which is used as a shelter (habitat) and food for prawn and other estuarine fishes. Destruction of mangrove is one of the major constraints for declining trends of prawn catches (Daugharty, 1975).

The genera *Penaeus* and *Metapenaeus* are commercially important prawn of the Sundarbans. Adult penaeid prawns migrate towards the sea for maturation and breeding and on reaching their larval stage they return to these mangrove forests of Sundarban estuaries which serve as a nursery bed. Post larvae of prawns occur almost throughout the year in the mangrove estuaries though the abundance fluctuates, seasonally, Thus, it is clear that mangrove is essential for prawn as it provides natural habitats for large number of estuarine or marine fauna, fishes and other threatened, endangered and endemic fauna. For having their unique potentialities Mac Nae (1968) opined that "No Mangroves no prawns".

The following are some of the facts which speaks of the dominance of prawn in the Sundarbans.

1. Gobaria (P.S Haroa) a flood prone brackish water bhery which was infested with dense forest; during 1922 the entire Sundarban was flooded by high (22 ft.) flood water of River Bidyadhari. This resulted in creation of innumerable number of large saltwater bheries. In all these saltwater bheries prawn (*P. monodon*) was mainly cultured, followed by Bhetki (*L. calcarifer*) and Persia (*L. persia*). Gobaria bhery (7.5 thousand Bigha) used to earn foreign exchange of 4 Crores of rupees by exporting prawn alone. A good bagda crop (*P. monodon*) produced in 1000 bigha could be 45000 kg. whose costing was estimated Rs. 1625300.-
2. In the Kulti complex prawns were largely captured from the river Kalindi from Hingalganj to Sandesh Khali, Sandeshkhali to Raimangal, Kalindi from Hasnabad to Barunhat and lastly Dasa from Chaktangramari to Kumarmari of Dasa river.
3. Ray et al. (1981) recorded prawn seed collection at Nazat (Kulti estuary) during 1979-80 ranging between 1.906 million and 33.877 million of seed per 4 hrs. per month, the peak period being December (10.96 m) to June (17.310 m) attaining maximum in March.
4. The rapid survey made by Department of Fisheries, Government of West Bengal (1984) revealed that the Bidyadhari river was the major source of prawn seed collection. During the peak season about 48 lakh of prawn seed were collected with the help of 3900 nets.
5. The following are some of the facts which indicate the position of prawn compared to other commercial estuarine fishes of the Sundarbans of the Hooghly-Malta estuarine system.

In 1961 - 62	Prawn dominated the catch (1012592 Kg.)
1962 - 63	Prawn dominated the catch (797577 Kg.)
1964 - 65 to 1970 - 71	Prawn showed second position (5200 Kg.)

Fish normally assemble at 7 important centres

Port - Caning Hasnabad and Kalinagar	At all these centres prawn dominated between 1966-67 to 1970-71.
Kakdwip and Namkhana	Prawn showed second position in the total Catch.
Raidighi	Prawn was present during 1966-67 and 1967-68.
Diamond Harbour	Showed lower value in the landing.

6. During Winer fishery the production at 6 centres (Frazerganj, Bakkhali, upper and lower Jamboo, Kalisthan and Sagar island) further down the sea, although prawn (dried) showed fifth position in the total landing between 1993-94 to 1996-97 they showed their progressive increase i.e. from 139.6 to 558 t.
7. In 1975-76 the production of prawn was around 5000 m tonnes. In 1970 total prawn exported from West Bengal was 149 t worth of Rs. 21.44 lakh. Which in 1978 increased

to 2853 m tonnes, Coast of which was 16.67 crores of rupees. According to Saktival (1980) the export figure can easily reach a value worth Rs. 1000 m.

8. In case of integrated farming and the suggested technology is followed as suggested to Ray (1993) in his valuable document "Aquaculture in Sundarban delta-its perspective," and assuming an average yield of 5 t/ha/yr (in 3 crops) in place of 6 t is produced, the estimated production of prawn alone that can be achieved from perennial tanks (Bheries) (16195 t), paddy-cum-fishery (182120 t) and the converted monoculture paddy field (15176056 t) together will be 17159200 t which is highly encouraging.

The above information reveal that since past the entire Sundarban has been producing prawn in the ecosystem not only to a large extent but also dominating the catch both in large inundated swampy areas, as well as, in the main estuaries heading towards the sea and were being exported with the growing population, the need for prawn culture (*P. monodon*) has increased considerably. As a result people have converted the large natural swampy areas into Bheries of various categories and sizes for improved culture. Thus integrated farming, both paddy and prawn culture in the same area came into existance. Scientist experimented various ways and means and have suggested improved yield of prawn to the extent that the average yield of prawn can go up as high as 5t/ha/yr (in three crops) easily.

The above account is from bheries only. But various estuaries and creeks flowing criss-cross through the Sundarbans for example, Bidyadhari-Kulti complex and Hooghly-Matla estuarine system contributes substancially to the prawn fishery. As a matter of fact Sundarban is being used as a habitat for prawn culture. They breed and grow and move up and down along with the tidal flow. Hence Biodiversity conservation of Sundarban is essential.

With the population influx, their habitation, agriculture etc. the forest is gradually being reclaimed. It is reported that in course of 40 years from 1947 to 1987 the Sundarban delta has increased by 33.7%, reclaimed by 67.6% but size of the forest increased only by 6.66% which is quite alarming. The reason for such set back besides deforestation and reclamation, is the increased salinity of the estuarine waters flowing through Sundarban in the absence of any head water supply of freshwater flow during summer months which restricts the high salinity (Jadav et al., 1981). The impact of such high salinity has shown the detrimental effect on the biodiversity of Sundarban forest so much so, that, recently a large area of about 4000 sq.km. (400005 ha) has been detected to be affected since 1989 from a satellite picture (Raha, 1999).

Thus to protect the highly economic prawn fishery, the forest has to be protected, otherwise, the goldmine of prawn fishery will gradually head towards destruction along with other commercially important estuarine fishes. (86 species of fish and 46 species of crustaceans had either become extinct or endangered in Sagar Island (Choudhury, 2000).

The protection of forest may be done by adopting a few major steps indicated below.

1. Suspension of deforestation.
2. Removal of silt load.

3. Allow adequate flow downstream and
4. Supply of freshwater as much as possible from the Hooghly estuary especially during extreme summer.

Deforestation or reclamation especially on the banks of the river/estuaries which is alarmingly increasing has to be stopped, otherwise such attempt will erode considerable volume of siltload in the river/estuarine bed and thus deteriorate the water holding capacity of tributaries and subtributaries of the main river in the absence of adequate flow.

Local goons lease out river beds of two important rivers, Bidyadhari and Ichhamati to set up bheries and brickfields on the river bed itself. It is stated that there are 268 illegal bheries and large number of brickfields in a stretch of 16 km. by setting up bounds across the river Bidyadhari which has lost its navigability (Goswami, 1999; Mallick, 2000).

De (1929), anticipating silting problem that might result in future, suggested construction of two large reservoirs at the appropriate place about half a mile long each on either side of the estuary. The objective was, during flood tide the reservoirs would be full and while ebbing, such water would have scouring effect on the deposited silt of the estuary. In case this plan is effective it can be tried at other places also. Mazumdar (1953) offered another practical suggestion. This can be tried with some modification. According to him the only connection the Kultigong had with the Hooghly river was through Nowii and Ichhapur Khal. He suggested if 20,000 cusecs of water was allowed to pass down. It would have the effect of pushing down the flow tide. With the improvement of the Kultigong its tidal flow would be accelerated and it could reasonably be expected that the time of arrival of the flow tide up to the Kultigong would be advanced. As a result of these improvement leading to the retardation of the flow tides in the Hooghly and its acceleration in the Kultigong, the meeting ground of tides could be shifted inside the Hooghly, the improved Kulti canal would maintain itself by natural action and the problem could be solved. But in the present case since the extra 20,000 cusecs of water may not be possible from the Hooghly river the reinforcement activities may be managed by releasing the water from the newly constructed reservoirs and thus silt precipitation may be almost cleared. Simultaneously, the excess of water which may be possible during high water of the Hooghly may be pushed through Ichhapur Khal to Kultigong through a lock gate constructed at the appropriate place to lower down the salinity of the forest area. Besides, the increase of soil erosion of the Ganga basin that has resulted poor water holding capacity of rivers like Kalinadi, Raymongal, Harinbhanga, Bortola, Saptamukhi. Thakuran, Matla and Gosaba lying criss-cross the Sundarban have to be linked with the Hooghly somewhere down stream after their proper desilting (additional reservoirs may be arranged as suggested if required). Thus the fresh water or low saline water may be pushed from both eastern and Western Sundarban which may control the high salinity of the estuaries appreciably during extreme summer period.

Lastly, the existing 3600 km long earth embankment erected around the Sundarban delta for its protection may not prove to be useful in view of the recent Tsunami effect of high waves, opioned by various scientists of "The Centre for advanced study in Marine Biology." Annamalai University (Tamil Nadu) (Press Trust of India, *The Statesman*, 12.1.2005) and Chairman, "Engineering Services International" (The Sunday, *Statesman*, Kolkata, 23.1.2005. Their suggestions in the strategy - "Protection of forest's mangrove plantation and not wall" which is less expensive. Their experience are annexed in The Addendum 3 and 4.

ADDENDUM 3

Mangrove, Not Wall - Scientists.

Source : Press Trust of India : *The Statesman,* 12.01.2005

1. It was a dense belt of mangrove that saved the village of Pitchavaram around 40 km. from here from extensive damage caused by tsumani.
2. Scientists of the Centre for Advanced study in Marine Biology Annamalai University (Tamil Nadu) are not contemplating raplicating the belt along the entire coast line, a project which will not only be more coast effective than the 1076 km. long sea will proposed by Chief Minister Ms. J. Jaylalita, but will also sustain natural bio-fauna.
3. "The mangrove have long since been called the green wall of the sea, as they provide protection from the waves" - Director of the centre Dr. T. Balasubramayan said.
4. The coasal village of pitchavaram was saved from Tsunami due to the mangrove it is surrounded by. If this can be replicated all over the coast line, in case of similar catastrophe in future, safety can be guaranted.
5. Plantation of a bed of mangrove which is not very expensive and is a breeding ground of shell fishes, small fishes and larvae will augment the fishing industry.
6. A similar project has been undertaken at Pondicherry at a cost of Rs. 2 lakh. Dr. Balasubramanyam said, we have received proposals for 3 more afforestation project in the district after the tsunami, district collector Mr. Gagandeep Singh Bedi said.
7. We have carried out mangrove afforestation on an experimental basis in about 20 hectare at our Institute and the results are excellent. In 3-5 years the tree grow to the full height and the cost incured would only be in conditioning the soil and keeping the cattle out who feed on this.

ADDENDUM 4

Tsunami Packed Power of 10,000 Hiroshima Bombs Says Experts.

In : *The Sunday Statesman* (Kolkata). 23.1.2005.

At a personal discussion organised by the American Society of Civil Engineer's Association section, Institute of National Disaster Reduction Management, Kolkata - 700036, Dr. Anil Kar, Chairman, Engineering Service International said.

1. A survey of the devastd areas shows, trees could withstand the effects of tsunami's much better than buildings.
2. There are obvious reasons behind the success of trees in resisting the wall of water a tsunami brings.
3. Before tsunami waves hit the Coast, tremendous amount of energy is generated. Much of the energy is transmitted to the ground or shore structure and part of it is changed into potential energy when wave gains height.
4. The pressure of spilling water acts against a wall on one of its faces. As such huts, corrugated structure and brick work collapses. But the pressure acts on all faces of

a tree and the tree survives. The trees that were swept away had most likely broken after being hit by floating objects.

5. Based on these Dr. Kar suggested that one strategy for protection against tsunami's could be to plant mangroves along the shore and coconut trees further island for a belt of about 2 km. The survey shows, he says coconut trees or even well-drives poles near every house can help people survive.
6. India had before been hit by tsunami on 31st December 1881 and 28th November 1945. - SNS.

9

Natural Calamities and Human Interference Responsible in Altering the Ecological Balance of Sundarban Delta – Their Resultant Effect on Aquaculture

Natural Calamities like, storm, cyclone, high flood and sometimes earthquakes as observed from the record (Table 1 and Fig. 1) are the factors responsible in altering the morpho-ecological conditions of Sundarban delta. The faster the cyclone proceeds the lesser is the effect. If however the Cyclonic weather persists for a longer duration the effects would be more pronounced. The surging effect is felt not only in the area of the tract of the cyclone, but also in the surrounding region. The upsurge of the water level in the tract area will cause the water at higher levels to move towards the surrounding regions with lower levels and will thus set up a long wave motion which would penetrate into the areas outside the storm region. When this wave penetrates into the shallower region of the estuary and the river system its height may increase importantly. As a result surging effects may be experienced in areas far beyond the limits of the storm affected regions. This is how estuarine ecosystem is maintained. The cyclone associated with flood cause changes in the morphology of the ecosystem to the great extent by flooding the estuarine banks, resulting a number of large and small swampy waters where fish culture was being practiced since long. Such flood waters were helpful in case of bheries (Swampy shallow inundated area) as huge quantity of fish and prawns were brought alongwith the water i.e. self stocked. In the later years with the population influx these are converted into smaller ones where only commercial important fishes are cultured as industry.

In the estuarine region however, duration of ebb-tide in a tidal river being longer than that of flow tide the capacity of water to transport silt is unable to push back fully the silt that has been carried in these tidal rivers by flow tide. Slight deposition of silt will go on accumulating as the tide function twice daily throughout the year and the channel will begin to deteriorate. This will impede the propagation of tidal wave which would cause further deterioration until the channel is completely dead. Added to it the Construction of bheries and brick kilns in shallower areas of the river bed or in the adjoining areas and putting bundhs across the river as is the practice in recent years has further accelerated the destruction of the estuaries and thus help extending the areas of Sundarban delta and deeper zones become shallower. The important rivers like Ichhamati and Bidyadhari are suffering from such problems. (Number of bheries and brick kilns on the bank of Ichhamati are approximately 1000 and 140 lying between Swarupnagar and Basirhat). It is now full of high growth of water hyacinth. Even shops are running on the bamboo bridge constructed across the river. At places the silting of about 40 feet high has taken place which results flood during monsoon. In addition, an island formed on the river bed hs been converted into a park. In case of Bidyadhari river, there are 268 illegal bheries and a large number of brick kilns covering a stretch of 16 km. by erecting a bundh across the river. River Hooghly is also suffering from similar problems. Thus two aspects are evident from the above facts and they are :

1. Formation of shallow swampy areas (presently known as bheries) large and small which in latery years are reduced into smaller ones due to demand by the growing population and are being used for fish culture as industry.
2. Capture fishery in the estuarine system where tidal water flows. Thus the Sundarban delta is controlling both by culture and capture fishery. A clear picture is evident from Fig. 1. which shows that during 1870 there were only 3 large bheries ranging between 2600 and 10,400 ha. In 1922 the larger ones were reduced to 420-780 ha when population increased to 5 lakh from 3 lakh recorded in 1872. Again during 1950's and 1960's the larger sized bheries bheries were further reduced to 200-330 ha with only one large size 660 ha when population in 1951 sharply increased to about 12 lakh from 8 lakh recorded in 1931, which further increased to 32 lakh in 1991.

It is interesting to note that fishing effort in the estuarine system rapidly increased from 1 to 4 lakh during the period 1960-61 to 1970-71 and in 1970-75 a large group of people diverted their attention towards fish culture and brick kiln operation on the bank/river beds of Ichhamati and Bidyadhari. The rapid increasing of fishing effort in the estuaries is possibly due to the fact that the existing fish culture in the bheries already attained their saturation point and the growing population were forced to explore possibility of fish culture on the river banks/beds of Ichhamati and Bidyadhari as this was how these bheries and brick kilns came into existance since 1970 onwards. River Hooghly is also suffering from similar problems in the North and South 24-Parganas and also in the Nadia district.

Extensive deforestation has resulted in silting problem in the Ganga basin which intern has affected the rivers like Kalinadi, Raimangal, Harinbhanga, Bortola, Saptamukhi, Thakuran, Matal and Gosaba of Sundarbans by reducing their water holding capacity.

These rivers now have no connections with the Ganga and as a result the source of fresh water which once used to control the high salinity of the entire estuarine system and encouraged the forest growth, at present are completely suspended and as a result the entire forest has been greatly affected on account of high salinity. The repeated observations from satellite imagery has shown that about 4000 sq. km. has been affected since 1986 (Raha, 1999). The 3600 km. long embankment around Sundarbans, meant for protecting the island from high salinity is in a much deteriorated condition. Reclamation of forest are gradually increasing by rapid population growth and has attained to about 66.7% in course of 40 years (1947-87). Such high rate of influx of human population from different sides has to be checked.

Forest acts as a natural nursery for aquatic life. From an account it has been found that some commercial fishes like *L. Calcarifer, Liza parsia, Mystus gulio, Liza tade, Illutheronema tetradyctylum* etc. have appreciably reduced their population (Thakur, 2000). This is in addition to the prized crustaceans like *P. monodan* which once used to dominate in the catch is now gradually loosing their population.

In a most recent statement given by Aquatic resources wing of the Department of Forest Services, Sundarbans (Bangladesh) and Sundarban Development Affairs Department and some NGO (West Bengal) presented by Pinaki Roy and Kaushik Ghosh both in the Sunday Statesman, 13.4.2003 (ef. Addendum 4) and The Statesman, South Bengal plus, 22.4.2003) (ef. Addendm 5) under the headings "Bangla may ban fishing in Sundarbans" and "Deforestation plays havoc with Sundarbans mangrove" are quite alarming and are indicated below to appease the readers of the threatening situation Sundarbans is heading in both the countries.

Additional information of Jamboo dwip presented by the *Statesman* News Service, Kolkata are also add seperately under Addendum 6.

Addendum 5

Bangla may ban Fishing in Sundarban.

Pïnaki Roy (The Daily STAR/ANN), *The Sunday Statesman*, 13.04.2003.

1. As many as 80,000 to 150,000 people catch fish every day in some 450 riers and canals in the Sundarbans. According to Aquatic Resources wing of the Department of forest sources (Bangladesh), approximately 15000 to 20,000 tons of white fish, 3600 tons of shrimp and crab, 750 tons of hilsa and 15 crore shrimp fry are procured every years from the Sundarbans river system.
2. Some 210 species of white fish, 26 species of shrimps and prawn, 13 species of crabs, 3 species of turfle and 41 species of molluscs are available in the region. Besides 9 species of fish which includes Pangas (*Pangasius pangasius*), Koi (*Anabus testudinus*), Magur (*Clarius batrachus*), Rup chanda (*Chanda range*) and Vetki (*Lates Calcarifer*) have become endangered.
3. Over fishing, especially shrimp fry collection has been depleting fish population.
4. As fishermen kill 300-500 other fish fry to collect one strimp fry the Government is planning to impose the ban.
5. Since the water salinity is increasing, some species of sea fish and dol fishes are approaching towards the Sundarbans area these days.

Addendum - 6

"Deforestation Plays Havoc with Sundarbans Mangrove"

Kaushik Ghosh, (*The Statesman*, South Bengal Plus - 22.4.2003)

1. In a field of survey conducted by the Sundarbans Development affairs Department and some NGO's it is revealed that about 60 species of mangrove of Sundari grow in the forest.
2. But deforestation of the region including that of mangrove belt is causing the forest to loose the ecologial balance.
3. The vast populace of rurals living in the Sundarbans have to engage in deforestation in course of farming and fishing and increase of habitat.
4. Research show that during the last two centuries more than 5000 Sq. Km. of the mangroove forest area were cleared to make way for human settlement. Several hectores of mangrove zones at Jharkhali, Bakkhali, Jamboo dwip and other areas are on the verge of decline.
5. These mangroves are the major antidote against population around the area.
6. Research further shows that though pollution load discharged by the KMC area is about 4 lakh kg/day and half of it is discharged along the Hooghly river passing directly into the Sundarbans estuaries the mangrove plants containing a high percentage of tanin in their bank, absorb some of the pollutants, to reduce their detrimental effects.
7. Besides, siltation of river bed causing flood and overflow of the estuary river is another cause of decline of mangrove.
8. Further, as a result of negligence of the Coastal guard mangroves are sometimes destroyed by the people.

Addendum 7

Additional Information on Jamboo Dwip.

(In : *Statesman* News Service, Kolkata 16.11.2002)

1. Fish traders illegally have occupied 300 hactores and destroyed rare mangrove trees spread over 150 hactores in Jamboo dwip-reported by an official.
2. Jamboo island is one of the most ideal hatching ground of the endangered olive Ridley turtles in the state.
3. If the fishermen continue to damage the eggs during the process of drying their catch and moving around the island, the olive Ridley population will be seriously depleted.
4. Central intelligence reports have pointed out that a large number of 'Bangladeshis' have illegally settled in the island posing a security threat.

10
Problems Responsible for Low Production

Problems which appear to be directly or indirectly responsible for low production are discussed below :

1. No scientific or systematic study for fish culture has been made either in bheries or in the estuarine system.
2. Organically rich Calcutta Municipal wastes were being used for fish culture in bheries and in some estuaries since 1930 but without any scientific basis. Hence the proper growth could not be estimated. Again from around 1963 onward with the improved industrial technology, the organically rich waste gradually changed into complex industrial waste when by using such wastes fish growth was likely to be affected. In subsequent studies high concentration of heavy metals although was estimated in different parts of fish (both in bhery and estuarine fish) but no interpretation regarding the growth or maturity of fish was made. Sometime insufficient supply of waste water might be affecting the growth of fish which was never revealed.
3. Relay stocking system came into practice around 1970. Whether it was due to stunted growth (affected by heavy metals etc.) or for greater demand for fish due to increased population is not clear. Such relay stocking certainly do not permit fish to attain better growth in such a short period.
4. In the estuarine zone fishing are being done by using simple gear without considering the importance of improved scientific gear technology developed. As a result both large and small size fish (both Commercial and non-Commercial importance) are simultaneously trapped. By such practice the younger ones of commercial value cannot grow up to mature size. Such practices affects the growth and production of fishery as a whole. For example, hilsa the most prized fish suffer from such practice. Such practices are also very common when trapping prawn seeds of *P. monodan* another prized crustacean (ef. Table 3). There is no record of the size of some commercial important species, only catch is indicated.
5. Silt load has appreciably affected the morpho-ecological condition of Sundarban ecosystem. Supply of fresh water resources from up stream is completely suspended. The Bhagirathi-Hooghly carries about 80 crore tons of silt (Report from Irrigation Department) annually up to the confluence. The river bed of Ichhamati has come up to 40 ft. at some places. People wade waist deep water in Matla estuary for going

to Gorbeta, Canning and Basanti. Further, every spring tide erodes the banks of different estuaries especially of Garbeta, Basanti, Kulpi and parts of the vast area of Mollakhali, thereby affect the mangrove trees. Even Bhagirathi-Hooghly has considerably raised its bed. The use of shallow river beds/adjoining areas of rivers for construction of bheries in recent years is a further addition of silting of river bed (width of Ichhamati has come down to only 33 ft. from 333 ft.). All these have affected the ecosystem by raising the river bed, loss of fish population (during dry and monsoon floods) and loss of forest and thus has created ecological disbalance, the results of which are presented in the following way :

(a) Gradual approach towards destruction of estuaries and thereby increasing the land area.

(b) Aquatic life is greatly affected both in low (dry season) and flood water (wet season) associated with tidal bores.

In low water of estuaries, fisheries are greatly affected. Though there is not much of change in the commercial important species in the forest area their population is greatly affected. But Chaudhury (2000) from his study states that 86 species of fish and 46 crustaceans have either become extinct or endangered around sagar is land. Further, in another record it is stated that some commercial important species like *L. Calcarifer, L. persia, Mystus Julio, L. tade, Illutheronema, tetra dyctylum*, etc. have reduced their production (Ghosh Thakur, 2000). The raised river bed on the other hand can cause flood during monsoon in the low lying adjoining areas. According to the report of Irrigation Department 80 crore tonnes of silt are brought down in the Bhagirathi-Hooghly annually. Silt load deposited, has raised the river bed at Nabadwip in such a way that during dry season Bhagirathi flows about two feet higher than Nabadwip city. During monsoon a slight rise in the water level, the city gets inundated (*Ananda Bazar Patrika*, 5.11.2000).

Even siltation of Hooghly has reached to such an extent that it could force ferry services Closure (Maity, 2001, ef. Addendum B. Chapter 6 Bhagirathi-Hooghly). Further, lately it is stated that large area of the stretch between Budge to Birlapur of the Hooghly river is likely to be affected by causing flood from breaching the low banks, possibly due to high surging effect of tidal water over the silt precipitated river bed (*Bartaman*, 4.2.2003).

In case of bheries of commercial important species cultured, are likely to be lost and intern some miscellaneous group of fishes are likely to get into the bhery. In case of estuary such floods which are not unusual, with a few feet rise of water can overflow the adjoining areas and there is every possibility of loosing some fish.

(c) Affect the bio-diversity of Sundarban forest due to increase of salinity (during dry season in the absence of any dilution by fresh water supply). Natural nursery of Sundarbans an important source of estuarine fish breeding and rearing is thus heading towards destruction. It is unfortunate that Sundarbans which was once a gold mine for *P. monodan* is gradually loosing its wealth due to considerable loss of bio-diversity on account of high saline water and human interference (Ray, 2000).

11
Overall Suggestions to Improve the Sundarbans Ecosystem

It is an uphill task to suggest means for conservation and management of the river-line and estuarine system whose problems are enormous. However some suggestions are put forward keeping in view of the present economic situation of the country.

On analysis of various factors involved for the ecological imbalance affecting the biotic and abiotic status of Sundarbans, it is evident that conditions are well set and is fast deteriorating and unless efforts are made to improve over it, the situation at the later stage may go beyond control. Natural calamities, like flood, storm, cycline, earthquake etc. cannot be controlled. It will occur from time to time. Keeping in view of these facts some suggestions on some important aspects to improve the ecosystem are indicated below. These are :

1. Control of population influx.
2. Conservation of bio-diversity reserve.
3. High silt load and resulted problems - their control.
4. Pollution arising mainly from Calcutta Municipal wastes, Tannery waste, solid waste i.e. Garbage, clinical waste and complex Industrial wastes - their control.
5. Experimentation of new fishing technique.
6. Extensive scientific experimentation in bheries of different categories etc.

Subsequent to the partition of Bengal in 1947 as East Pakistan and later on its emergence as free Bangladesh in 1971 are the two major incidents when large population migratd to West Bengal, a certain percentage of which were agriculturist and fishermen community who rehabilitated themselves in the Sundarbans (*Statesman News Service,* 2.11.2002). Added to it the sporadic cross border infiltration is a regular affair and is still continuing. It is stated that fish traders have occupied 300 ha and destroyed rare mangrove trees spread over 150 ha in jamboo dwip to use the land for the purpose of drying fish which takes place between November and March (Winter fishery). The empowered committee on forest and environment appointed by Supreme Court has asked the State Government to stop the drying of fish by unauthorised fish traders in the island and discuss the measures that may be taken to step up for the island security. (*The Statesman News Service*, Bengal Plus, 16.11.2002).

From around 1950-2000 a significant increase in population was noticed (Fig. 1). The major influx of 1947 and subsequently in 1971 could not be controlled but in the absence of effective checking of this, sporadic infiltration through different routes are still continuing.

The rapid population explosion and their interference in various ways especially by denudation of forests resulting silt erosion, construction of bheries, eruction of bundhs, brick kilns and encroachments etc. are the major factors that has led the Sundarbans ecosystem to such a desastrous condition. Population has exploited the situation and is still continuing to its maximum and so far no effective measures has been taken to control it. However, though it is difficult, if some measures are taken by stopping further ingress of population step by step the situation may improve to a great extent.

Conservation of bio-diversity of Sundarbans forest is essential especially from the banks of estuaries, crecks etc. as it not only helps maintain biodiversity but also acts as a buffer against natures fury. Further, it acts as a natural habitat for large number of estuarine or marine fauna, fishes and other threatened endangered and endamic fauna. Mac Nae (1968) opioned that "No mangrove no prawns", Also, the very existance of mankind is dependent on the existance of adequate forest cover and man alone is the potential enemy of forests. The existing 3600 km. long expensive earth embankment which was made to protect the Sundarbans may not be possible to maintain properly as from time to time it may pose problems and in its place the protection of forests by mangrove plantation which is less expensive but more effective is suggested by various scientists from the back ground of heavy damage caused by the recent tsunami effect of high waves.

Silting which results from denudation of forests is a major problem in rivers, estuaries and cracks, create problems for easy flow of water. This is mainly due to deterioration of head water supply and denudation of forests especially from the catchment area. The resultant effect is the increase of salinity of water which is maintained by tidal waters. Such high saline water is not suitable for growth and reproduction of most of the biotic life.

The flushing effect resulted from tidal bore cause severe damage to the banks of the shallow estuaries and creeks due to erosion resulting unconsolidated changed silt in suspension, over flow the banks and add further silting on the river bed while ebbing. Mangrove roots although acts as an effective tide brakers and creek, land run off are helpless. Areas of delta thus increases, while the forest growths are affected. All these affect the migration of some commercial important fish, especially hilsa which are largely caught only at the confluence area of deep waters only during monsoon recorded during August of 1997 and 2001 (*The Statesman*, 20.08.1997 and *Bartaman*, 18.8.2001) as they fail to migrate further up in the fresh water region for breeding purposes, as they did in the past.

Besides, local people explore the possibility of trapping tiger prawn (adult the seed) in the spilled water. By doing so large number of other fishes are also caught and destroyed. Hence, dredging operation is essential to allow some freshwater flow during non-monsoon months. The silt being fertile may be used elsewhere for agricultural purposes.

R. Ichhamati

Silt precipitation at places of Ichhamati has gone up to 40 feet. The Dhippi park established in the mid river upstream, Mirjapurer char (Basirhat) and about 1000 bheries and embankments erected across the river and 140 brick kilns lying between Barirhat and Swarupnagar of Ichhamati (*Ananda Bazar Patrika*, 3.11.2000) has to be cleared. Hence, a link between Mathabhanga and Ichhamati has to be established after proper dredging.

The project* undertaken for dredging Ichhamati bed has run into rough weather, as owners of illigal houses, slums and fisheries refusing to vacate the river bank at Lalpur, Tentulia and Charghat. Dwellings have come up on the silted river bed as well. As a result the Ichhamati had been regularly flooding its banks at Bangaon and parts of Basirhat sub-division over the past two years (*Statesman News Service*, 10.6.2000). It is unfortunate that a stretch of 142 km. only, between Majdia nad Rajmahal being so close to the sea does not have any tidal influence.

1. Analysis and evaluation of the extent, effect and causes of floods of 2000 which had affected Bangladesh and South-West Bangladesh.
2. Necessary field surveys in the flood affected areas of West Bengal and South-West Bangladesh to assess the present drainage condition and identification of revent spots or channels to improve drainage condition. These are : Taking necessary measures like re-excavation, dredging of the settled channels to facilitate smooth drainage in the area and identify which are now lying used for other purposes unlawfully.

R. Bidyadhari

Further, there are 268 illigal bheries and large number of brick kilns in a stretch of 16 km. by setting up bundhs across the river Bidyadari which has lost its navigability (Goswami, 1999). All these are to be cleared to maintain the flow round the year. Added to it some extra flow is essential to push back the silt brought by tidal waters. For this constrution of reservoirs, half a mile long, one on each side of the bank of the estuaries at the appropriate place is necessary (Ray, 1993).

R. Hooghly

R. Hooghly is no exception from such problems. Due to low flow, a number of islands have come up in the bed itself despite the existance of a number of brick kilns along the bank of the estuary. Subsequent to the pact of sharing the Ganga water between India and Bangladesh in December 1996 such scarcity of water flow was never felt earlier. To save the

---

* Following the flood on 2000 a joint task force was formed both by India and Bangladesh to review the condition of the river after the flood along the embankment of Ichhamati flowing to Bangladesh along North 24-Parganas in 2000. After a preliminary survey in 2001 it was decided to make a thorough survey on the following points and submit a report to the Joint Commissioner.

The programme is nice and welcome, but unfortunately the task force which visited Ichhamati in West Bengal has not yet submitted any report to the Joint Commission (*The Statesman*, South Bengal Plus, 12.6.2003). Further, the Government of West Bengal ordered the North 24-Parganas District Administration to remove all the Fisheries from the bank of Ichhamati after high floods (*The Statesman News Service*, 26.10.2001). But this has not been complied so far. Subsequently, in May 2002, a seven member joint task force comprising officials from Centre and State has been formed by the Central water resources ministry to control floods in Ichhamati river Basin in North 24-Parganas district, only the future can tell the outcome of the task force.

navigation of the port, the requirement of water flow which should have been 40 thousand cusecs has come down to 15 thousand only.

It is alarming that due to scarcity of water flow, the NTPC is compelled to stop one 500 mega watt power unit, while in case of other four units, the power generation is restricted to 600 mega watt from 1600 mega watt (Gupta, 2001). Besides it has been noticed that the water level of the river in the Eastern Channel of Hooghly point-about 40 km. South of Kolkata has gone down to only 1.6 m. and the steamers cannot move unless at least the depth of 2.5 m. is allowed. Similar situation was also experienced in 2000 (Ananda Bazar Patrika, 6.5.2002). In view of this, Lieutenant Governor, N.N. Jha (Andam and Nicobar Island) reportedly expressed the view that passangers board for the island should begin their voyage from Haldia, as vessels find it difficult to navigate the Kolkata-Haldia stretch of the waterway due to inadequate depth of Hooghly (The Statesman 15.8.2002). Even in the main stream Hooghly, the ferry ghat authorities has started warning public through notice board that unless adequate water level is reached vessels cannot move. (Addendum 2.A and B, Chapter 6 Bhagirathi-Hooghly, its past and present.)

Further, ritual performances which occur every year mostly in September, October and January e.g. 'Tarpan' - prayer to ancestors on the Mahalaya day, immersion of Durga and Kali idols, Paus Sangkranti etc. When people assemble in a large scale along The Ganga bank (mainly at The littoral zone) and at the confluence of Ganga and Sagar (Ganga-Sagar) for holy dip, cause immence displacement of precipitated silt from large area along with the tidal water which again settle elsewhere in the river bed to cause fresh problem.

In view of the presence of high silt load in the water, the brick industry may be encouraged for their operation along the banks of the river away from the city without damaging the river banks. The water can be drawn a little interior, allow the sediment to precipitate (even by adding some chemicals to enhance the precipitation) and send back the clear harmsess water to the river by pumps.

Besides, there are a number of private Brick kilns on both sides of the estuary (stated earlier) and are being operated regularly. The river beds may be based out stretch by stretch to a number of private enterprising organisations to remove the earth free of cost. This will not only improve the condition of the river but also save crores of rupees that are being spent every year towards flood affected victims as flood relief. Such deposited alluvial soil are supposed to be rich in nutrients (rich organic matter and also metallic ions) and if used by agriculturists at low cost in their way in the river during rains. Whether such soil have any adverse effect on the crop may be tested and then used. Webber (1972) studied effects of toxic metals in sewage on crops, opioned that, so long as the soil retained a high organic matter content and the lime status was kept high no problem may arise. But if the soil was allowed to become acidic and if the organic matter content fell following a change in the cropping system, trouble could arise. Further, the contaminated soil can be diluted with uncontaminated soil from below by deeper cultivation. It is worthwhile to mention that solid garbeg of Calcutta Metropolitan City were largely used as fertilizer for raising good crop (Ray, 1993) which were being supplied regularly to the city and its suberbs. All these efforts may solve the problem to a great extent arising from silt precipitation in the river bed and also city garbage, especially when crores of rupees are spent every year towards flood relief measures.

Besides, some ancillary reservoirs constructed at a number of suitable points of the rivers or tributaries as suggested by the eminent scientist M.N. Saha (Chattopadhaya, 1993) for holding maximum water during monsoon and releasing them during off seasons according to the requirement to maintain a constant flow would have helped the situation.

In this context a similar proposal from brick kiln owners of Kolkata was placed before the Government to build a sand bank on payment of a royalty. This was an excellent and encouraging suggestion and could be taken up easily. This would have reduced the incidence of flood in the rural Begnal as excavation would have increased their capacity to carry more water. By allowing excavation and dredging the navigability of the deretict rivers and canals would increase, besides helping in prevention of floods which occur due to excessive accomulation of silt on river beds. Although in the proposal several rivers identified for excavation which include, Nonagong, Suti, Noaikhal, Jamuna, Ichhamati and Chaiti, the ultimate effort could not materialise and the proposals are now collecting dust. (Gupta, 2002). Such practices of removing silt are already in practice privately from several places on both sides of the Hooghly estuary by some brick kiln owners.

It is unfortunate that not much of interest is paid in any of the proposals. Even earlier a number of proposals were undertaken during different periods since 1955-56 by different experts, Commissions, Agency, Paribesh Unayan Samity etc. for dredging, desilting, demolishing illegal settlement along the river bed etc. of Ichhamati and Bidyadhari rivers and Canels, Nowi, Jamuna, Paskhali, Karotia - Nagerta, Ganga-Padma, Nonaganj, Sarorat Khali and Haroganj etc. as indicated under the caption "Silted rivers pose danger to North 24-Parganas" (*Statesman News Service,* October 21, 1998) remained in proposal only.

In view of construction of Farakka Barrage and subsequently the pact between India and Bangladesh on the Ganga water, it is difficult to allow even 20,000 cusecs of water as suggested by Mazumder (1953) to flow through Hooghly-Matla estuarine system. As such the tidal water of the Hooghly may be allowed to pass through Ichhapur Khal after necessary dredging. The reinforcement activities may be managed by releasing the water from newly constructed reservoirs about half a mile long each on either side of the estuary as stated by Ray (1993). Simultaneously, the excess of water which may be possible during high water of the Hooghly may be allowed to pass through Ichhapur Khal to Kultigong through a lock gate constructed at the appropriate place to lower down the salinity of the forest area. Along with this the brick kiln owners' suggestion if implemented will be dieal. Thus the density of forest can be protected to an appreciable extent. Besides, increase of soil erosion of the Ganga basin that has resulted poor waterholding capacity of rivers like Kalindi, Thakuran, Matla and Gosaba lying Criss-cross the Sundarbans have to be linked with the Hooghly some where down stream through Tolly's Nullah (Adi Ganga) after proper dredging. This fresh water or atleast low saline water whichever is possible may be pushed from both eastern and western Sundarbans which is likely to improve in Controlling the high salinity of estuaries to an appreciable extent during non-monsoon period.

Calcutta Canal system due to their connections with the Hooghly should be made free from encroachers, desilted and properly maintained as was in the past, as all these have some contribution in maintaining a flow by providing additional water.

The recent proposal of the Centre to transfer water on the Ganga for its partly diversion to Sundarbans through the river system Jamuna and Ichhamati during non-monsoon months for navigation, irrigation and other purposes and some water to the Hooghly for Kolkata port is welcome.

According to the Centre's (Govt.) recent proposal : Manas-Sankosh-Tista-Ganga link for which existing Farakka Barrage is proposed to be utilised for diversion of water. The water is proposed to be carried through the existing Canal (38.3 km.) from Farakka Barrage to Jangipur barrage on Bhagirathi river for 323 km. A barrage on Bhagirathi river is proposed to be constructed near Mukta Tribeni. This barrage will divert 450 MCM of water to the Jamuna river and then to Ichhamati river. A diversion structure near Hasnabad on Ichhamati is also proposed for diversion of water to Bidyadhari river. Balance quantity of water will be carried along the Hooghly river till the Chitpur lock gate and to the point into the river Bidyadhari through Kestopur Canal. From Chitpur lock gate 7000 MCM of water will also be released in the Hooghly river for Kolkata Port (Statesman News Service, (b) 26.1.2004).*

Further, it is interesting to note that simultaneously the Government of West Bengal has initiated a programme to maintain a link from Kolkata to Sundarbans a stretch of 430 km. through a canal system from Chitpur to Kultigong mainly for the purpose of navigation (Bartaman 16.1.2004).

In case the centre's recent proposal is effected, Sundarbans can get enough freshwater to lower down the highly salinity when the lost flora and fauna of Sundarbans can be revived to a great extent. The above two proposals are undoubtedly very encouraging. But it is an uphill taks to be compiled, especially the earlier one.

Culture Fishery

In the Culture fishery system in bheries all types of culture, viz. sewage fed, sewage free, perennial, fish-cum-paddy, pady-cum-fish, low saline, medium saline and high saline zone (including Meen dwip) are to be exploited scientifically as suggested elsewhere. Simultaneously, integrated fish farming should also be encouraged. Proper size of fingerlings and density of commercial important fishes with proper ratio of species to be cultured should be stocked. Restocking can be done only after removal of mature size. Equal care should be given to prawn culture. Standardisation of growth rate including crustaceans is essential. The fishes including prawn should be kept free from disease. Special effort should be made in case of prawn culture (especially stocking density) as they are more susceptalde to disease like bacterial or viral infection. Suitability of use and abuse of complex metropolitan wastes for fish culture needs to be studied.

---

* (The demand by the working president "Paschim Banga Ichhamati Sanskar Committee", is, such diversion is likely to cause the Ichhamati river to dry up near its mouth. In its place if the diversion is planned from Farakka to Jalangi and then to Mathabhanga river, the Ichhamati could also be revived fulfilling the navigation, irrigation, desilting and other needs of Sundarbans. In case the Union Government does not comply with their request they are set to launch agitation against the decision taken by the Union Government with the consent of the State Government).

Pollution Problems – Their Control

Pollution, another important aspect has to be controlled particularly Calcutta Municipal wastes, taunery wastes (these are largely used in bheries for fish culture), hazardous clinical wastes and city garbages that piles up at Dhappa and finally the complex industrial wastes released all along the Bhagirathi-Hooghly including industrial complex waste released from Falta.* Besides, the normal pollution arises from various industries, municipal wastes and particularly pollution problem caused by shipwreck around sand heads which occur during May to July almost every year and release toxic material should be properly studied and remedial measures suggested for the safety of aquatic life. Formerly hilsa used to ascent in river in search of fresh water of adequate depth for the purpose of breeding and now they have gone down towards sand-heads for breeding. In future fish may have to move to Padma river (Bangladesh) for breeding purposes.

Although non-polluting energy resources are being deployed and technologies being diverted to minimise impact of pollutants on environment they are not enough to cope with the situation. Hence, besides the existing traditional system of treatment of such complex waste water there is a tree "*strychnos postatorin*" locally known as "*chilla Ginjalu*" which largely grow in the forests of Andhra Pradesh (India) has excellent purifying capacity for clearing the effluents discharged from various factories as well as from Atomic Power plant, may be tried. A detailed information of the "Wanter tree from Andhra Pradesh controls pollution" is enclosed in Addendum 7. A large scale plantation of this tree at appropriate places may be encouraged both by the Government and other welfare agencies. A sincere effort on this direction along with excavation and dredging of estuaries etc. may help the ecosystem free.

Addundum 8
Wonder Tree from Andhra Controls Pollution

A substance isolated from the seeds of "Chilla Ginjalu", a tree growing in the forests of Andhra Pradesh, may provide the answer for controlling global pollution, scientists say. The substance also binds 18 metals including uranium and thorium. Besides, it helps in cleaning effluents from factories as well atomic power plans.

For centuries, tribes in Andhra Pradesh have used the powdered seeds of the "Chilla Ginjalu" (which also grows in Burma and Thailand) to purify drinking water. The powder had the ability to bind itself with impurities and remove them - phenomenon called "flocculation".

By adding modern biology to this tribal secret, scientists of the Andhra Pradesh Girijan (Tribal) Cooperative Corporation, have now isolated from the seed a "bio-flocculant".

The tree (Strychnos Posta-torm) that was hitherto unknown outside the region, is now up for international exposure with industries world wide and the International Atomic Energy Agency in Vienna showing interest in the product.

---

* Effluents of Haldia Industrial Complex contains high mercury level besides other carsinogenic heavy metals like lead and Chromiam above permissible level and may affect Meen dweep fishery.

"It is non-toxic and bio-degradable", says Mr. Durgaprasad, one of the scientists behind the operation. Apart from uranium and thorium, the substance easily binds with toxic mercury and cadmium that are usually present in factory effluents.

It also tags on to metals such as silver, gold, copper and nickel, thereby providing mining companies an entirely new method of ore recovery, according to Mr. Prasad. He said the metal binding nature was because of proteins present in the bio-flocculant.

According to the State Chief Minister, Mr. Vijaya Bhaskar Reddy, the product will be soon commercialized and profits ploughed back for tribal development. Andhra Pradesh has a tribal population of 4.2 million people and over 44 percent of them live in forests.

According to an American scientist, moves were on for collaboration aimed at further purifying the proteins in the bio-flocculant and improving its metal-binding ability. Dr. C.E. Furlong of the American University at Seattle says that the protein isolated from the Indian tree was unique because some of the substances developed in the United States could at best bind with one or two metals, and not 18.

Dr. Furlong added his university plans to use genetic engineering methods to synthesize the proteins found in the seeds of the Indian tree. Dr. K. Mahadev Rao of the IAEA says substances isolated by GCC scientists had great promise in removing traces of uranium and other metals in reactor effluents.

He said its ability to remove cobalt was particularly useful to nuclear engineers as the steel used in nuclear industry must be cobalt-free. Dr. Rao said the substance can also bind with "actinides" a group of metals in nuclear dumps which emit radiation for a long time. These can be selectively removed using the GCC's bio-flocculant.

According to the GCC managing director; Mr. T. Vijay Kumar, the research on "nut-cleaning" was started three years ago by himself and four of his colleagues in a car shed. There was now a stock of 2,000 tons of seed which can produce the substance on a semi-commercial scale.

The local scientists have also developed the technology for immobilizing the bio-flocculant on a suitable polymer which means that it can be re-cycled. DPA from major pollution problems and allow not only the aquatic life to have a healthy and reproduction but also conserve the biodiversity of Sundarbans and other forests. It can purify the drinking water and the tribes of Andhra Pradesh have been using the powder of the seed for centuries.

Fishing Technique

Experiments on new fishing techniques are to be experimented with a view to improve the existing fishing methods as done in case of reservoirs (Gulbadmov, 1961, 1962). The object is to evolve suitable techniques for a rational exploitation of the fishery. In the present case the effectiveness or the efficiency of a particular net may be guaged not by relative abundance of fish it captures in the estuarine system, but the size it captures. This will enable only the larger fish to be caught by allowing the younger commercial and non-commercial fish to attain larger size.

In the existing system, in general efficiency is possibly considered in terms of relative abundance of fish a net can trap irrespective of their sizes. For example, hilsa, the most prized fish are trapped in smaller sizes along with other fishes in large quantities by operating simple nets. The large size nets mentioned by Dutta (1973) and De (1980) which used to be operated in the Hooghly-Matta estuarine system prior to 1970 for catching large sized fish are not being operated these days due to shallow depth of the river on account of heavy silt deposition. Hilsa are now largely caught only at Diamond Harbour area during high floods. Likewise, in case of prawn fishery (*P. monodon*) a prized crustacean, while operating traps for seed collection a large number of important and unimportant seeds of other fishes are also caught and destroyed. Mukherjee (1996) in his recent observation has shown the size of destruction of other species while trapping seeds of *P. monodan* in a stretch of 3 km. based on 6 months observation (Table 3). He has shown that when the catch of *P. monodan* seed was 4133700 the total catch of 12 other species was 15487200. This means for one piece of *P. monodan* seed four other species were trapped which could have satisfied the demand in case they were allowed to grow to a bigger size. Similar problems has been experienced in Bangladesh (ef. Addendum 4). In case of capture fishery system there can be restricted fishing effort after some improvement of the ecosystem. More prawn seed (*P. monodan*) collection centres may be exploited for culture in bheries keeping in view that seeds of other species are not destroyed.

Finally, instead of making effort to protect the 3600 km. long/earth embankment around the Sundarbans cannot be a permanent solution, rather a mangrove wall by extensive plantation, which is less expensive and takes 3-5 years only to grow up, can be done more effectively as experienced by Dr. T. Balasubramanyan, Director of the centre for advanced studies in Marine Biology, Annamalai - University for a permanent solution (ef. Addendum 2). Similar opinion is also given by Anil Kumar, Chairman, Engineering services international at a panel discussion organised by the Amercian society of civil engineers, Indian Section, Institute of Natural Disaster Reduction and environment, Kolkata - 36, that coconut trees or even poles and not concrete walls can save people from tsunamis. Dr. Kar suggested that strategy for protection could be to plant mangroves, along the shore and coconut trees further island for a belt of about 2 km (Addendum 3).

12
Rapid Limnological Changes of Sundarbans Concerns its Existance

Sundarban delta, because of its high potential value invited a population of 2,96,045 by 1872 which rapidly went on increasing especially during the period 1950 and 1990 (175%). For their habitation and living, they exploited the land, water and forest of Sundarbans to its maximum. The Sundarban delta which was 72000 ha in area during 1947 was reclaimed by 320000 ha and the actual forest area came down to 400000 ha. Forty years hence, the reclamation of forest area increased to 67.6 per cent to accommodate such high population. As a result the increase of forest was only 6.66%. But interestingly it is noticed that the delta area increased by 33.75 percent. The extensive deforestation and other undesirable use or river and estuarine beds (Construction of bheries, occupation of river banks by encroachers, brick kilns etc.) has resulted in silting problems in the Ganga basin which in term has affected rivers, tributaries, sub-tributaries like Jamuna, Nowii and Sunti connecting the river Bidyadhari and estuaries like Ichhamati, Kalindi, Raimongal, Haringhata, Bortola, Saptamukhi, Thakuran, Matla and Gosaba of Sundarbans by reducing their water holding capacity (some are already dead). It is unfortunate that since 1955-56 a number of proposals were submitted by different experts for dredging, desilting, demolishing illegal settlement along the river bed of Ichhamati and Bidyadhari and other canals, but all are collecting dust due to strong opposition by the occupants. Elevated river bed has become a common phenomenon. These rivers now have no connections with the Bidyadhari-Hooghly. Even situation of Hooghly has reached to such an extent that siltation could force ferry services closure in some case and as a result the source of fresh water which once used to control the high salinity of the estuarine system and encouraged the forest growth are now greatly affected. Soil salinity reaches maximum in May (electrical conductivity of the surface soil reaches up to 40 m. mhos and sometimes more). Such high salinity has further indicated signs of destruction of bio-diversity of forest by another 4000 Sq. miles noticed since 1989 is also recorded through satellite imagery.

Similar findings is also supported by Rakhi Chaudhury (Kolkata, 2004) in his recent article "Siltation cloud on Gangasagar" in *The Statesman* 2004, where it is stated that uncertainly boons large over whether the famous 'Gangasagar mela' will be held in January 2005. The reason is heavy siltation in the Muriganga river, particularly on the eastern bank. It has made it difficult for vessels to ply between Hardwood point on the main land and Kachubaria on

Sagar island. During low tide, boats get stuck mid-stream. The people have to wait for eight hours for the hightide to move from main land to the island.

Sundarbans was once a gold mine for prawn fishery especially for highly Commercial important crustacean like *P. monodan* and hilsa. But little did the people dream when they found themselves happy for their extending habitation in the delta by their progressive resettlement, colonisation, agricultural development, construction of roads etc. The affected forest area which resulted extensive silting in the estuaries and crecks lying criss-cross the delta and further down, thereby causing their deterioration in the absence of upland freshwater resources may sometime cause problem to the inhabitants when they are destined to explain many matters they are going to face in the near future ever since the creation of Sundarbans delta is now revealed recently by a group of Scientists, Department of Geological Science, Jadavpur University shown in Addendum 8.

Addendum 9

Based on the research in physical dimension of coast line ecosystem of the Sundarbans over a period of 10-20 years, the Conclusions drawn by a group of Scientists Drs. Sugata Hazra, Tuhin Ghosh, Rajashree Das Gupta and Goutam Sen - from the School of Oceanographic studies under the Department of Geological Sciences, Jadavpur university are quite alarming (Mukhopadhaya, 2003). The scientists developed a mathematical model which was then applied to predict the future shore line position of the Sagar island in different sea level rise situation. The findings are as follows :

1. An analysis of the erosion-accretion pattern in Sundarbans that emerge out to remote sensing date and filled mapping confirmed that a significant reduction of land area (around 86 Sq. Km.) has occurred in the estuarine island system over the past three decades. The total erosion during this time span is around 163 Sq. km.
2. The cause of this has mainly been attributed to the relative sea level rise in the region which during the study period 1986-98 has registered a minimum 4 cm. rise. Precise calculations reveal that the annual mean sea level rise is 3.24 mm. which may go upto 3.5 mm. over the next few decades.
3. This will pose serious problems during the pre and post phases during which most of the cyclonic storms occur.
4. The effect of the storm surge is likely to move devastating in areas which were not vulnerable earlier. Even Kolkata is expected to be affected by inundation during monsoon instability coupled with high rainfall and high tide in the river Ganga.
5. The mangrove forest cover would diminish further alongwith degradation of existing species - is a further prediction.
6. Among other observations made in the report are that the Sagar island has suffered the bulk of erosion with an areal loss of 30 Sq. Km. and that islands the Lohachora and Bed ford have totally vanished from the map. Around 30,000 perople are feared to be homeless from Sagar island alone.

7. In addition alarming growth of population in this ecologically sensitive and fragile niche has posed a major threat for its very existance.
8. Frequent embankment failures, submergence and flooding, beach erosion and siltation of jetties and navigational channels cyclone and strom surges are all making this area increasingly vulnerable.
9. The scientists have also expressed their concern over the decline in fisheries resources in the Sundarbans over the past 15 years including a significant decline in the region's major catch hilsa.
10. By the year 2020 nearly half a million people living in the Sundarbans may be rendered homeless on account of several coastal erosion caused by a rise in the sea level in the region.
11. Its far reaching impact would not even spare the state capital which till now is considered to be not vulnerable to such whims of nature.
12. The study also revealed that in keeping with the existing trend of global warming, the area experiences a temperature rise of 0.019°C per year.
13. In view of high temperature during summer followed by fall in water level with rise in water temperature, it is quite likely that aquatic life especially fish and crustacean animals may not feel comfortable to live in the existing environment which may show fall in the cath. These fishes, except some which may acclimalises are likely to move towards a deeper area for their safety where catch may temporarily increase. Hence a separate study to reveal the actual facts may be taken.

    Incidentally, it may be mentioned that even in the sea such cases are observed in case of North sea fishes. A study covering 25 years of data has found that nearly two thirds of North sea's fish species including commercially important fish (cod and haddock) have because of global worming shifted either further North or to colder depths - reports steve Connor. (The Independant, London, in the Statesman 16.05.2005). (Also addendum 9 A.C.) (On Global Warming and its effects).

Addendum 10 (A)

Dese Dese Garom Ato Barche Keno

(Why there is increase of such warming from country to country)

Chir Dutta, Bartaman 20.6.2005.

American Scientist of Colorado University clearly stated that the opinion made by IPPC group prior to 2001 on the rising of sea water level due to global warming will exceed their estimation several times more by the end of this Century.

This is due to global warming which is responsible for increase of melting of global ice cap.

The melting of global ice cap causes increase of sea water level by 0.14 mm every year. 10 years before the sea water level also increased in a similar manner (0.14 mm). This means in a period of 10 years the rise of sea water level will exceed 1 meter. If such condition continues, half of the Bangladesh will come under water.

Dr. Swadesh Misra an environmentalist of West Bengal has stated that Gangotri the origin of the Ganga has been shifted 1 km. away due to reduction of Himalayan glacies.

Chir Dutta has discussed in details the effect of warming up in West Bengal.

(B) Danger Signals - II

(Global warming could cause serious damage in India)
R.B. Brahmachari (Professor of Applied Physics, Calcutta University)

(*The Statesman*, 17.1.2006), Dr. R. Brahmachari in his article Danger signal II has cautioned saying "Global warming could cause serious damage in India in which he has indicated that due to influence of warming, winter is comming late and spring is arriving nearly two weeks early in Canada. Fishes and birds, adapted to cold weather have started migrating to the north and 1998 has been recorded as the hottest year in the history of mankind, followed closely by 2003 and 2004.

There is no doubt that temperature of the artic region is rising fast and many apprehend that, if things are allowed to go as it is, the polar ice-cap would disappear completely within 2080 leading to considerable rise in the sea water level and hence the valuable coastal land all over the world would be inundated. Experts also predict that cities like Mumbai, Hongkong, Singapore and New York would be flooded, considerable land mass of low lying countries like Bangladesh would be submerged under the sea water and small island nations like Maldives would go into extinction. For us Indians, the phenomenon of global warming and the rapid melting of artic ice is an extremely serious matter. Many apprehend that it would be catastrophic, since such developments would weaken the monsoon causing serious damage to Indian agriculture.

(C) Baro Bipader Mukhe Kolkata, Sundarban, Bangladesh, Tamil Nadu, Andhra Pradesh

Anup Kar. In : Weekly Bartaman 25.02.2006.

Anup Kar in his article has revealed alarming information on the effect of Global warming on Himalayan Glaciers.

In a report - "World wide fund for nature" it is stated that due to Global warming, The Himalayan Glaciers are melting. This will result the rise in the water level of rivers both in India and Nepal. The principal rivers that will be affected are Ganga, Brahmaputra and also Mekong and Yiengsew.

Some of the Himalayan Glaciers are 70 Km. long. Due to Global warming to what an extent the Glaciers are melting are shown in a report which states, that the Glaciers are reduced annually by 10-15 metres. Gangotri Glaciers are reducing by 23 metres annually.

The impact of this melting will increase of sea water level which may inundate Sundarban and many other cities of India including Kolkata.

Is it possible to conserve such potentiality rich biodiversity of the delta and estuarine network of Sundarbans or leave it to her own destiny.

Quo-vades - Naturalist, Scientists and people of Sundarbans.

References

Ananda Bazar Patrika, 2000, *Apon Bege Cholte N Diya Pali Jome Ichhamati Aaj Sarbanasha.* (By not allowing her normal flow, the deposition of silt has invited the calamity of Ichhamati) Ananda Bazar Patrika 3.11.2000.

Banerjee, A.K. 1964, Forests of Sundarbans. Centenary Commemoration volume. Writer's Buildings, Calcutta.

Bartaman 2004, *Poli tolar Kaj Sesh Holai Khal Pothe Lonch Jabe Uttar Kolkata Theke Sundarban.* (After completion of dredging, Launch will ply from N. Kolkata to Sundarbans through the Canal.) In : Bartaman. 23.01.20004.

Bera, S. 1984, *Nadibigyaner Katha.* (The Science of Rivers) 10, Ramkrishna Street, Calcutta - 700006.

Bhattacharya, S.K. and S.M. Chatterjee, 1972, Brick fields and Pollution. All India Conference Abatement of environmental Pollution Proceedings, Calcutta, December 14-16. Organised by Calcutta Metropolitan water and Sanitation Authority, Jadavpur University, in Collaboration with CPHERI (Calcutta) : ISI : AIIH : Geological Survey of India, IPHE (India) and Indian Standard Institute).

Bose, P.C. 1944, Calcutta Sewage and Fish Culture. *Proc. Inst. Sci.* india, 10 : 443-454.

Chakraborty, Rikhi (Kolkata), 2004, Siltation cloud on Gangasagar, In : *The Statesman,* South Bengal Plus 3.11.2004.

Chakraborty, Sanjoy. 1999, *Jal Bhaga Bhagir Janye Sundarbone Misti Jaler Samasya.* (Division of water resources resulted problem of sweet water), *Bartaman,* 6.12.1999.

Chakraborty, N.M., G.C. Karmakar and A.K. Ray, 1961, Observations on the effect of supplementary feed on growth and survival of grey mullel, *liza parsia* (Ham) fry in brackish water nursery pond at Kakdwip (Abstract) *Proe symp coastal aquculture,* Cochin, p. 64.

Chakraborty, Subhas, 2003, "Swamps and epidemic" In : The Statesman, South Bengal Plus, 9.9.2003.

Chattopadhaya, Enakshi, 1993, *Meghnad Sahar nadi-Chinta* (The thoughts of rivers of Meghnath Saha) In : Desh Magazine, October, 9, 1993.

Chaudhury, Amalesh, 2000, Threat to Marine life in Sundarbans In : *The Statesman News Service,* 28.3.2000.

Das, Ganesh, 1981, *Haroate Dhan O Macher Chash Samridhi Ene Diechhey.* (Paddy and Fish culture has brought affluence in Haroa) In : "Bhumilakshmi" - *Ananda Bazar Patrika,* 22.7.1981.

Dangharty, H.E. 1975, Human Impact on the Mangrove Forest of E1 Salvador ed. G.E. Walsh, S.C. Snedaker and H.S. Teas. *Proc. Inst. Sympon on Biology and Management of Mangrove* (1974). University of Florida, U.S.A. pp. 816-24.

David, A. 1954, A preliminary survey of the fish and Fisheries of a 5 mile stretch of the Hooghly River near Barrackpore *Indian J. Fish.* 6 (2) : 327 - 34.

David Augustine 1959, Effects of Calcutta Sewage upon the fisheries of Kulti estuary and the connected cultivable Fisheries. *Journal of the Asiatic Society.* Vol. 1, No. 4, 1959.

De, B.N. 1929, Kulti outfall scheme, A Report, Calcutta Corporation.

De, D.K. 1980, Fishing gear and Fishing method of Hooghly-Matla estuarine system. *J. Inland Fish Soc.* India, *12* (1) : 54-63.

Datta, S.N. 1973, On the Gill nets of the Hooghly-Matla Estuarine system (West Bengal) with special reference to their specifications and operation. *J. Inland fish Soc.* india, Vol. 5, December.

Dutta, P. *et al.* 1973, Fishery Resources of the Hooghly-Matla estuarine system. Bull No. 19, December. "Central Inland Fisheries Research Institute (ICAR), Barrackpore, West bengal.

Gangopadhaya, Naresh Chandra, 1983, *Paschim Bongyer Matsa Samasya.* (Fishery problems in West Bengal Deficiencies in Development Planning in Agriculture of West Bengal, Souvenir 1983. Organised by Institute of Applied Humanities and Socio-economics Development Consortium. Calcutta.

Ghosh, H. 1973, *The Fisheries and wet lands in the District of 24-Parganas.* Settlement office. Calcutta.

Ghosh Kaushik 2000, Enchroachers stall River Dredging In : *The Statesman News Service,* 10.6.2000.

2003, Ichhamati may flood its banks - say experts. In : *The Statesman, South Bengal,* Plus 12.6.2003.

2003, Green talk and lot of garbage. In : *The Statesman,*.10.6.2003.

Ghosh, Kaushik (Kolkata) 2004, Save Wetlands, target farms In : *The Statesman,* 24.9.2004.

Ghosh, Premtosh, 1959, Reconnaissance survey of fresh water impoundments in the North and South Salt Lake areas. Secretary, Fish producers Association 24-Parganas, West Bengal.

Ghosh Thakur, Deb Doot, 1999, *Sundarbone Misti jal Komaya Sankote Gach, Bagh.* (Reduction of Sweet water in Sundarban has resulted problems to forest, Tiger. In : *Ananda Bazar Patrika,* November 17, 1999.

2000, *Nona Jale Dubeche Sundarban, Paribesh Bipanna, Bipanna Manush.* (Sundarban is sinking in saline water, Environment is affected and so the human beings) In : *Ananda Bazar Patrika,* 15.6.2000.

2000, *Briksha Nidhane Sundarbone Bhumikshya, Nadirjal Dharan Kshamata Kome Paribesh Bipanna.* (Deforestation encourages soil erosion, water holding capacity in rivers is reduced resulting problems of the environment) In : *Ananda Bazar Patrika,* 24.4.2000.

Goswami, Tarun. 1998, Conservation Plan for East Calcutta wet lands. In : *Statesman News Service,* The Statesman, 29.9.1998.

The Statesman News Service, 1999, "Goons choke Bidyadhari by leasing. December - 21, Calcutta, 1999.

The Statesman, Bengal Plus 2002, Kolkata wetland on World environment map. Kolkata, 28.11.2002.

Government of West Bengal, 1945, Report of the Committee to enquire into the Drainage condition of Calcutta and adjoining areas.

1987, Report of the task force on the Best land use options for the saline soil areas in coastal districts of West Bengal.

Gulbadamov, S.B. 1961, Improvement of Fishing Techniques in Inland Reservois. Report to the Govt. of India, No. 1342 F.A.O., EPTA : 1-13.

1962, Improvement of Fishing Techniques in Inland Reservoirs. Report to the Govt. of India, No. 1499, F.A.O. EPTA : 1-37.

Gupta, Gautam, 2001, *Gonga-Jal Kome Bondar Songkot, Bipade Farakka Bidyut Kendrao.* (Calcutta port and also Farakka Electricity generating station are in trouble due to fall in water level of Ganga). In : *Ananda Bazar Patrika*, 11.4.2001.

2002, *Sukha Morsume Chollis Hazar Kusek Jal Chaiche Bondar.* (Port demands 40 thousand cusecs of water during dry season). In : *Ananda Bazar Patrika*, 6.5.2002.

Gupta, Subhasis, 2002, Brick-field owners' proposal gathers dust. In : *The Statesman*, Bengal Plus. 30.7.2002.

Hall, D.N.P. 1962, Observations on Taxonomy and Biology of some Indo-West Pacific Penaedae (Crustacea, Decaplods) : Colonial office Fishery, Publication No. 17. HMSO. p. 229, London.

Hamilton, Francies, 1922, An account of the fishes found in the river Ganges and its branches. Edinburg, Archibald constable and Company, p. 405.

Hora, S.L. 1943, Evidence of distribution of fishes regarding rise in salinity fo the river Hooghly. *Curr. Sci.* 12 : 89-90.

Hora, S.L. and K.K. Nair, 1944, Suggestion for the development of salt water bheries of Bhasabadha Fisheries in Sundarbans. *Fish, Dev. Pamphlet.* Govt. of West Bengal. Calcutta.

Hora, S.L. and T.V.R. Pilay, 1953, Further evidence from Fish distribution of the rise in Salinity of the river Hooghly, *Current science.* 22 : 49-50.

Hunter, W.W., *Statistical Accounts of Bengal.*, In : *Paschim Bongyer Matsa Samasya.* Fish problem in West Bengal - Naresh Chandra Gangopadhaya, 1983 : Deficiencies in Development Planning in Agriculture of West Bengal. Souvenir, 1983. Organised by Institute of Applied Humanities and Socio-economic Development Consortium, Calcutta.

Jadav, *et al.* (1981), In : *Mangrove Swamps of the Sundarban.* - An ecological peropective. - K.R. Naskar & D.N. Guha Bakshi. (1987) Naya Prakash, 206, Bidhan Sarani, Calcutta.

Jones, S. and K.H. Sujansinghani, 1951, The Hilsa fishery of the Chilka Lake. *J. Bombay Nat. Hist. Soc.* 50 (2).

Mac Nae, W. 1968, A general account of the fauna and flora of Mangrove Swamps with special reference to the Indo-West Pacific region. *Adv. Mat. Biol.* 6 : 73-270.

1974, Mangrove forests and Fishes, Indian Ocean Programme. Pub. No. 34 : pp. 1-3.

Maity, Subhendra 2001, Siltation could force ferry services closure. In : *Statesman News Service*, 23.3.2001.

Mallick, T. 2000, *Ichamatir Macher Bherir Jonney Nadi Aro Choto Hoye Gache.* (Bhery Fishery in Ichhamati has further reduced the river) Bartaman.

Mazumder, S.C. 1953, Ganga Barrage and the Bhagirathi-Hooghly river-Problems. *Indian Jour Pond River valley Dev.* 3 (3).

Mukherjee, G.D. 1959, Drainage in Greater Calcutta area. *The Journal of Institution of Engineers* (India) Vol. XXXIX, July, No. 11 Pt. I, pp. 1061-1083.

Mukherjee, Sonjoy, 1996, *Sundarbon Elakaya Paribeser Bharsamya Biparjasta.* (Ecological Imbalance of Sundarban has resulted disaster in the ecosystem). In : *Bartaman,* 27.8.1996.

Mukhopadhyay, Sougata, 2003, Geologists concerned over Sundarbans erosion. (City may not be spared nature's wrath). In : Statesman News Service, *The Sunday Statesman,* Oct 5, 2003.

Nair, K.K. 1944, The effects of Calcutta Sewage on Fish culture. *Proc. Nat. Inst. Sci.* India, 10:455-458.

Namin 1977, Mangrove Swamps of the Sundarban (An Ecological Perspective). K.R. Naskar and D.N. Guha Bakshi, Naya Prakash, 206, Bidhan Sarani, Calcutta, India.

Naskar, K.R. 1996, Status of Coastal vegetation in West Bengal with special reference to the Sundarbans. Integrated coastal zone Management, Govt. of West Bengal with Assistance from Dept. of Ocean Dev. Govt. of India.

Oag. T. M. 1939, Report of the river Hooghly and its head waters. The Commissioners for the port of Calcutta. In : Observations on the Hydrology of the Hooghly estuary. - Benoy, B. Bose (1956) *Indian J. Fish.* 3 : 101-118.

Pakrasi, B. B. 1965, A report on the preliminary survey of Brackish water imfoundment in West Bengal. Survey Report No. 3, March. CFRI, Barrackpore, West Bengal, India.

Paramanick, H. R. and P. B. Roy, 1972, Conservation of River Kulta Gong. *Seminar on Estuarine Pollution.* March, 9-10, Sponsored by CMDA and CPHERI, Nagpur.

Paul, S., P.M. Mitra, D.K. De, H.K. Sen, N. C. Mondal and P. Singh 1997, Winter Migratory Bagnet Fishery of the Hooghly Estuary - An economic evaluation. Bull. 76, Nov. 1977, CICFRI (ICAR) Barrackpore, West Bengal.

Pilley, T.V.R. 1954, The ecology of Freshwater Bheri with special reference to the Fish culture practices and Biotic interaction. *Proc. Nal. Inst. Sci.* India.

Press Trust of India, 2005, Mangrove, not wall : Scientists. In : *The Statesman,* (Kolkata) 12.1.2005.

Raha, A. (1999), *Sundarboner Mistijal Komaya Sankate Gach, Bagh.* (Reduction of freshwater supply to Sundarban has resulted problem to forest and Tiger) - Debdut Ghosh Thakur. In *Ananda Bazar Patrika,* Nov. 17, 1999.

Ray, P. 1981, Impact of man on the Hooghly estuary. Water pollution and Management Reviews. South Asian Publication Pvt. Ltd., New Delhi.

1993, Aquaculture in Sundarban Delta - its perspective (An Assesment). International Books and Periodical Supply Service, Delhi 110034.

1998, Ecological imbalance of the Ganga River System - its impact on aquaculture. Daya Publishing House, Delhi.

2000, Sundarban - A gold mine for Prawn fisheries. Celebration of Fourth Sundarbans day. Calcutta Wild Life Society (An organisation for the cause of conservation of environment and ecology) at Birla Industrial and Technological Museum (BITM) on 3rd June, 2000.

Ray, P., S.B. Saha and R.K. Banerjee 1981, A case study of use of Calcutta Municipal Wastes for fish culture in Bidyadhari-Kulti Complex, West Bengal. International symposium on wastes resources conservation, Pollution and Abatment. December 11-13, Department of Civil Engineering University of Roorkee.

Report of Fact Finding Committee, Govt. of West Bengal, 1984, The problems of Paddy-cum-Brackish water Fish/Prawn cultivations and forcible conversion of Paddy lands into fisheries in parts of 24 Parganas District.

Rudra, Kalyan 2001, Wet lands are being lost, if it rains even high tidal surging can inundate Kolkata. In : *Ananda Bazar Patrika*, 17.9.2001.

Saha *et al.* 1986, Ecology and Fishery Management of Brackish Water Bheries in West Bengal. CIFRI, Bull. 46 : 1-23.

Sakthivel, M. 1980, Sundarban - A gold mine for Prawn fisheries. In : *Matsachasi Dibas Palan O Seminar.* Fish Culturish Day observation and Seminar West Bengal.

Sewell, R.B.S. 1934, A study of the Fauna of Salt Lakes, Calcutta *Rec. Indian Mus.* 36 : 45-121.

Statesman News Service Bengal Plus 2002, Panel order on fish drying on Jamboo dwip. In : *The Statesman*, 16.11.2002.

2003, Wetland Information system for State In : *The Statesman*, 15.2.2003.

Steve Connor, Driven to Despair. The Independent, London. In : *The Statesman*, Kolkata, 16.5.2005.

Statesman, News Service (c) 2003, Wetland panel to suggest remedy. In : *The Statesman*, South Bengal Plus, 27.8.2003.

2003, Chrimes boom along with infiltration. District Focus, North nd South 24-Parganas. In : *The Statesman*, 4.11.2003.

2004 (b), Drying Ichhamati fuels agitation. In : *The Statesman*, South Bengal Plus 26.1.2004.

2004 (a), Govt. takes steps to conserve wetland. In : *The Statesman*, 3.2.2004.

Thakur Kishalay 1986, *Gobarier Sangkot Govir Biparjayer Ingitbahi.* (The problem of Gobaria is an indication of great disaster). In : *Ananda Bazar Patrika*, June 27, 1986.

The Statesman (a) 2003, Operation Wetlands (editorial) 10.2.2003.

Wabber, J. 1972, Effects of toxic metals in Sewage on crops. Agricultural Development and Advisary Service, Leeds. 404, Water Pollution Control.

Yadov, T.S.P. *et al.* 1981, Management of Coastal Saline soils of Sundarbans. Bull. 7 (CSSRI - ICAR), Karnol, India, pp. 1-32.

Plate 1 : Birds eye view of bheris in the north salt lake area.

Plate 2 : Fishing operation in a bheri.

Plate 3 : Fish assembly centre at Bantala where catch from bheris are brought for disposal.

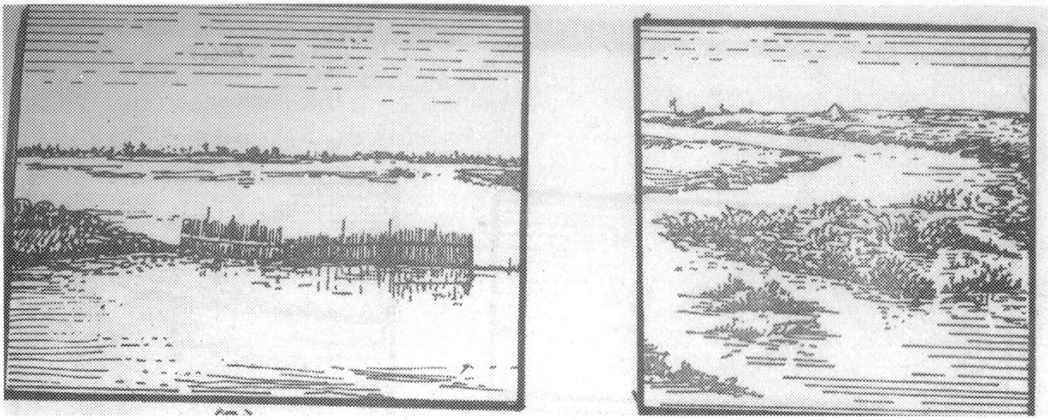

(a) (b)

Plate 4 : (a) A view of a brackishwater bheri. (b) Layout of a feeder canal for lething in tidal water.

Plate 5 : Sluice box used as water inlet, and gate for controlling the water.

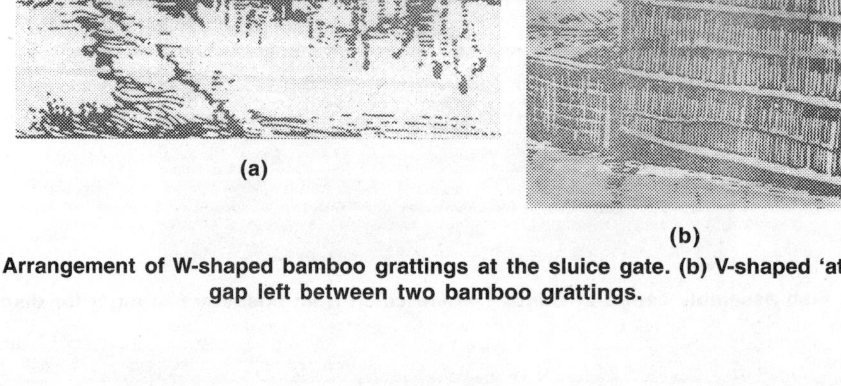

(a) (b)

Plate 6 : (a) Arrangement of W-shaped bamboo grattings at the sluice gate. (b) V-shaped 'atols' to fill the gap left between two bamboo grattings.

(a)

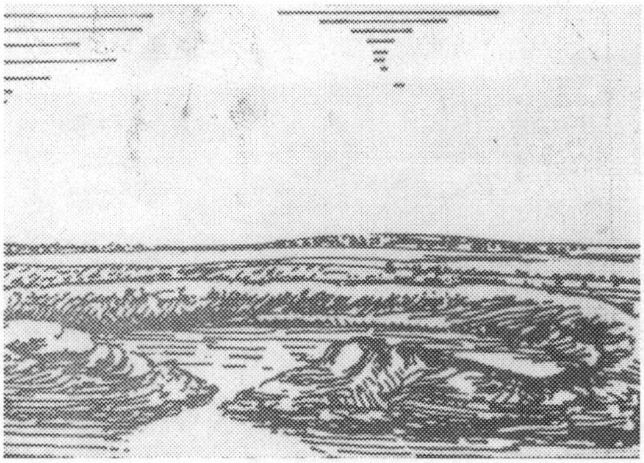

(b)

Plate 7 : (a) Device to prevent escape of fish by providing emergency gap in the embankment of a bheri (indicated by an arrow). (b) Subsidiary pond to stock immature stock of.

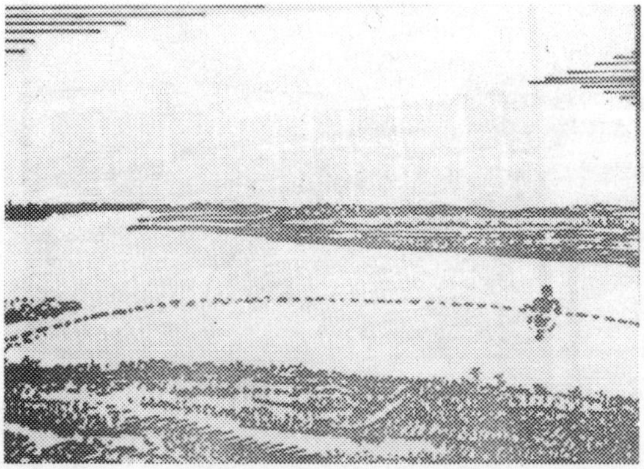

(a)

(b)

Plate 8 : (a) Drag netting operation in a bhery. (b) Cast net fishing in a bhery.

(a)

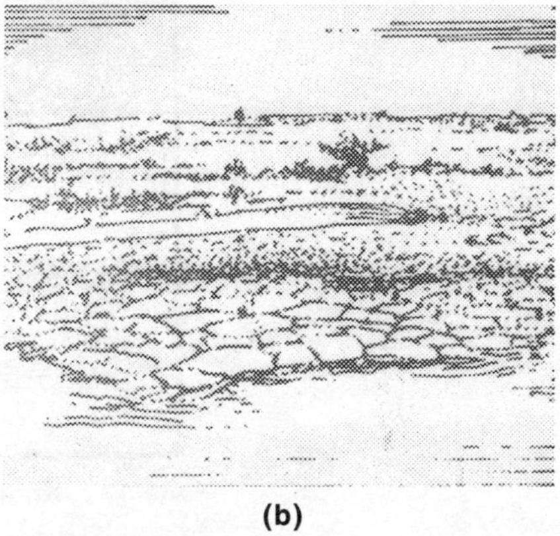

(b)

Plate 9 : (a) Dry hedges planked at the bottom of bhery to prevent poaching during summer months.

Plate 10 : A view of paddy-cum fishery.

Plate 11 : Shallow wells dug near the margin of bheri to keep fish alive till marketted.

Plate 12 : 'Hapar' used to keep fish alive till marketted.

Plate 13 : 'Khonga' used to trap crabs.

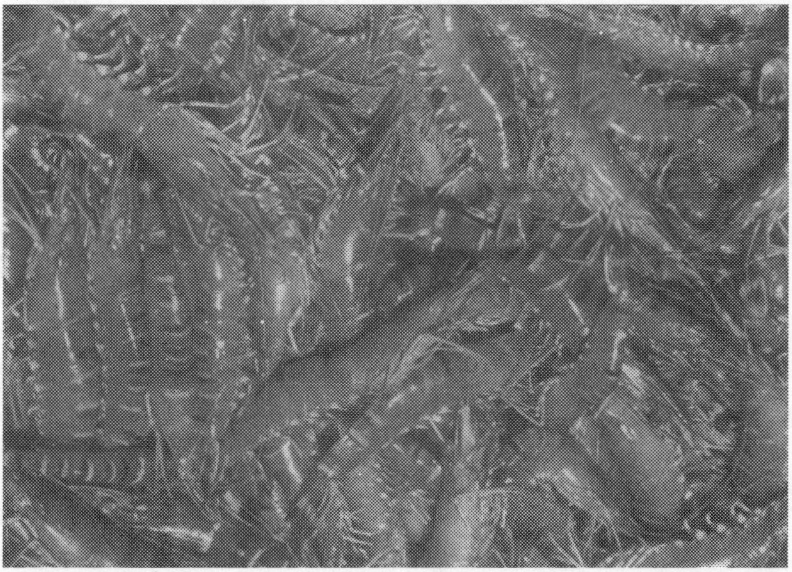

Plate 14 : Tiger prawn (*P. Monodon*) an important fishery of Sundarbans is a substancial foreign exchange earner1111111111111111111111111111111111111111111111111

Plate 15 : Bhetki (*Lates Calcarifer*) a commercially important fishery of Sundarbans.

Plate 16 : Indian major Carps (Rohu, Catla and Mrigal) can also show a high production in the integrated farming system.

Author Index

CULTURE FISERY (AQUACULTURE IN FRESHWATER IMPOUNDMENT)

AQUACULTURE IN BRACKISHWATER IMPOUNDMENT

CAPTURE FISHERY – THE KULTI ESTUARY

THE BHAGIRATHI – HOOGHLY ESTUARY

THE SUNDERBAN DELTA

PAST AND PRESENT OF SUNDERBAN DELTA